Computational Exome
and Genome Analysis

CHAPMAN & HALL/CRC
Mathematical and Computational Biology Series

Aims and scope:
This series aims to capture new developments and summarize what is known over the entire spectrum of mathematical and computational biology and medicine. It seeks to encourage the integration of mathematical, statistical, and computational methods into biology by publishing a broad range of textbooks, reference works, and handbooks. The titles included in the series are meant to appeal to students, researchers, and professionals in the mathematical, statistical and computational sciences, fundamental biology and bioengineering, as well as interdisciplinary researchers involved in the field. The inclusion of concrete examples and applications, and programming techniques and examples, is highly encouraged.

Series Editors

N. F. Britton
Department of Mathematical Sciences
University of Bath

Xihong Lin
Department of Biostatistics
Harvard University

Nicola Mulder
University of Cape Town
South Africa

Maria Victoria Schneider
European Bioinformatics Institute

Mona Singh
Department of Computer Science
Princeton University

Proposals for the series should be submitted to one of the series editors above or directly to:
CRC Press, Taylor & Francis Group
3 Park Square, Milton Park
Abingdon, Oxfordshire OX14 4RN
UK

Published Titles

An Introduction to Systems Biology: Design Principles of Biological Circuits
Uri Alon

Glycome Informatics: Methods and Applications
Kiyoko F. Aoki-Kinoshita

Computational Systems Biology of Cancer
Emmanuel Barillot, Laurence Calzone, Philippe Hupé, Jean-Philippe Vert, and Andrei Zinovyev

Python for Bioinformatics, Second Edition
Sebastian Bassi

Quantitative Biology: From Molecular to Cellular Systems
Sebastian Bassi

Methods in Medical Informatics: Fundamentals of Healthcare Programming in Perl, Python, and Ruby
Jules J. Berman

Chromatin: Structure, Dynamics, Regulation
Ralf Blossey

Computational Biology: A Statistical Mechanics Perspective
Ralf Blossey

Game-Theoretical Models in Biology
Mark Broom and Jan Rychtář

Computational and Visualization Techniques for Structural Bioinformatics Using Chimera
Forbes J. Burkowski

Structural Bioinformatics: An Algorithmic Approach
Forbes J. Burkowski

Spatial Ecology
Stephen Cantrell, Chris Cosner, and Shigui Ruan

Cell Mechanics: From Single Scale-Based Models to Multiscale Modeling
Arnaud Chauvière, Luigi Preziosi, and Claude Verdier

Bayesian Phylogenetics: Methods, Algorithms, and Applications
Ming-Hui Chen, Lynn Kuo, and Paul O. Lewis

Statistical Methods for QTL Mapping
Zehua Chen

An Introduction to Physical Oncology: How Mechanistic Mathematical Modeling Can Improve Cancer Therapy Outcomes
Vittorio Cristini, Eugene J. Koay, and Zhihui Wang

Normal Mode Analysis: Theory and Applications to Biological and Chemical Systems
Qiang Cui and Ivet Bahar

Kinetic Modelling in Systems Biology
Oleg Demin and Igor Goryanin

Data Analysis Tools for DNA Microarrays
Sorin Draghici

Statistics and Data Analysis for Microarrays Using R and Bioconductor, Second Edition
Sorin Drăghici

Computational Neuroscience: A Comprehensive Approach
Jianfeng Feng

Mathematical Models of Plant-Herbivore Interactions
Zhilan Feng and Donald L. DeAngelis

Biological Sequence Analysis Using the SeqAn C++ Library
Andreas Gogol-Döring and Knut Reinert

Gene Expression Studies Using Affymetrix Microarrays
Hinrich Göhlmann and Willem Talloen

Handbook of Hidden Markov Models in Bioinformatics
Martin Gollery

Meta-analysis and Combining Information in Genetics and Genomics
Rudy Guerra and Darlene R. Goldstein

Differential Equations and Mathematical Biology, Second Edition
D.S. Jones, M.J. Plank, and B.D. Sleeman

Knowledge Discovery in Proteomics
Igor Jurisica and Dennis Wigle

Introduction to Proteins: Structure, Function, and Motion
Amit Kessel and Nir Ben-Tal

Published Titles (continued)

RNA-seq Data Analysis: A Practical Approach
*Eija Korpelainen, Jarno Tuimala,
Panu Somervuo, Mikael Huss, and Garry Wong*

Introduction to Mathematical Oncology
*Yang Kuang, John D. Nagy, and
Steffen E. Eikenberry*

Biological Computation
Ehud Lamm and Ron Unger

**Optimal Control Applied to Biological
Models**
Suzanne Lenhart and John T. Workman

**Clustering in Bioinformatics and Drug
Discovery**
John D. MacCuish and Norah E. MacCuish

**Spatiotemporal Patterns in Ecology
and Epidemiology: Theory, Models,
and Simulation**
*Horst Malchow, Sergei V. Petrovskii, and
Ezio Venturino*

Stochastic Dynamics for Systems Biology
Christian Mazza and Michel Benaïm

**Statistical Modeling and Machine
Learning for Molecular Biology**
Alan M. Moses

Engineering Genetic Circuits
Chris J. Myers

**Pattern Discovery in Bioinformatics:
Theory & Algorithms**
Laxmi Parida

**Exactly Solvable Models of Biological
Invasion**
Sergei V. Petrovskii and Bai-Lian Li

**Computational Hydrodynamics of
Capsules and Biological Cells**
C. Pozrikidis

**Modeling and Simulation of Capsules
and Biological Cells**
C. Pozrikidis

Cancer Modelling and Simulation
Luigi Preziosi

**Computational Exome and Genome
Analysis**
*Peter N. Robinson, Rosario M. Piro,
and Marten Jäger*

Introduction to Bio-Ontologies
Peter N. Robinson and Sebastian Bauer

Dynamics of Biological Systems
Michael Small

Genome Annotation
*Jung Soh, Paul M.K. Gordon, and
Christoph W. Sensen*

**Niche Modeling: Predictions from
Statistical Distributions**
David Stockwell

**Algorithms for Next-Generation
Sequencing**
Wing-Kin Sung

**Algorithms in Bioinformatics: A Practical
Introduction**
Wing-Kin Sung

Introduction to Bioinformatics
Anna Tramontano

**The Ten Most Wanted Solutions in
Protein Bioinformatics**
Anna Tramontano

**Combinatorial Pattern Matching
Algorithms in Computational Biology
Using Perl and R**
Gabriel Valiente

**Managing Your Biological Data with
Python**
*Allegra Via, Kristian Rother, and
Anna Tramontano*

Cancer Systems Biology
Edwin Wang

**Stochastic Modelling for Systems
Biology, Second Edition**
Darren J. Wilkinson

**Big Data Analysis for Bioinformatics and
Biomedical Discoveries**
Shui Qing Ye

Bioinformatics: A Practical Approach
Shui Qing Ye

**Introduction to Computational
Proteomics**
Golan Yona

Chapman & Hall/CRC Mathematical and Computational Biology Series

Computational Exome and Genome Analysis

Peter N. Robinson • Rosario M. Piro
Marten Jäger

CRC Press
Taylor & Francis Group
Boca Raton London New York

CRC Press is an imprint of the
Taylor & Francis Group, an **Informa** business

A CHAPMAN & HALL BOOK

CRC Press
Taylor & Francis Group
6000 Broken Sound Parkway NW, Suite 300
Boca Raton, FL 33487-2742

First issued in paperback 2020

© 2018 by Taylor & Francis Group, LLC
CRC Press is an imprint of Taylor & Francis Group, an Informa business

No claim to original U.S. Government works

ISBN-13: 978-1-4987-7598-4 (hbk)
ISBN-13: 978-0-367-65774-1 (pbk)

Visit the Taylor & Francis Web site at
http://www.taylorandfrancis.com

and the CRC Press Web site at
http://www.crcpress.com

To Elisabeth, Sophie & Anna
-PNR

A Maurizia, für Patrizia
-RMP

Für Jule, Johanna & Julius
-MJ

Contents

Preface   xv

Contributors   xxi

PART I   Introduction

CHAPTER   1 ▪ Introduction   3

CHAPTER   2 ▪ NGS Technology   11

CHAPTER   3 ▪ Illumina Technology   21

CHAPTER   4 ▪ Data   47

PART II   Raw Data Processing

CHAPTER   5 ▪ FASTQ   57

CHAPTER   6 ▪ Q/C: Raw Data   67

CHAPTER   7 ▪ Q/C: Trimming   81

PART III   Alignment

CHAPTER   8 ▪ Alignment   95

CHAPTER   9 ▪ SAM/BAM   111

CHAPTER 10 ▪ Postprocessing the Alignment   129

CHAPTER 11 ▪ Alignment Data: Quality Control   149

PART IV **Variant Calling**

CHAPTER 12 ▪ Variant Calling & Quality-Based Filtering   163

CHAPTER 13 ▪ VCF   183

CHAPTER 14 ▪ Jannovar   203

CHAPTER 15 ▪ Variant Annotation   209

CHAPTER 16 ▪ Variant Calling QC   229

CHAPTER 17 ▪ Integrative Genomics Viewer   233

CHAPTER 18 ▪ De Novo Variants   247

CHAPTER 19 ▪ Structural Variation   259

PART V **Variant Filtering**

CHAPTER 20 ▪ Pedigree Analysis   297

CHAPTER 21 ▪ Intersection and RVAS Analysis   315

CHAPTER 22 ▪ Variant Frequency   321

CHAPTER 23 ▪ Variant Pathogenicity   329

PART VI **Prioritization**

CHAPTER 24 ▪ Prioritization   349

CHAPTER 25 ▪ Random Walk   355

CHAPTER 26 ▪ Phenotype Analysis   367

CHAPTER 27 ▪ Exomiser   387

CHAPTER 28 ▪ Medical Interpretation   407

PART VII Cancer
_____

CHAPTER 29 ▪ A (Very) Short Introduction to Cancer   419

CHAPTER 30 ▪ Somatic Variants   429

CHAPTER 31 ▪ Tumor Evolution and Sample Purity   467

CHAPTER 32 ▪ Driver Mutations and Mutational Signatures   477

APPENDIX A ▪ Hints and Answers   499

References   503

Index   547

Who is this book for?

This book is intended for bioinformaticians, computational biologists, and computer scientists who would like to learn about computational analysis of human whole-exome and whole-genome sequencing (WES/WGS) data. The field of computational exome and genome analysis has matured since the initial disease-gene discoveries using exome sequencing and has become an extremely heterogeneous and often challenging area of bioinformatics. One book cannot cover the entire field, nor can one single computational pipeline fulfill all possible needs. We therefore do not attempt to provide coverage of everything one will ever need to know about exome and genome analysis. Instead, our goal is to provide readers with an understanding of the technology, data, formats, and foundational algorithms that will allow them to perform exome and genome analysis in an informed and effective fashion.

The various chapters of the book cover all steps required to analyze exome or genome data in a medical context, from basic quality control to alignments, variant calling, and interpretation in medical genetics and cancer contexts. We assume that readers are comfortable with a Linux command line and are able to compile programs, and execute simple scripts. Numerous examples and exercises are provided so that readers without formal training in computer science or bioinformatics will have a place to start. For many of the topics, we present specific software tools, including some of our own, but we have attempted to present them in a way that clarifies the methods and approaches that many of the available tools use to address the analysis challenges.

Preface

The knowledge that we convey in this book draws from complementary sources and experiences. The Robinson Lab at the Institute for Medical Genetics and Human Genetics of the Charité University Hospital in Berlin, Germany, established computational pipelines for the analysis of whole-exome sequencing (WES) and (later) whole-genome sequencing (WGS) data starting in the early days of WES. Rosario Piro spent important years of his scientific career at the German Cancer Research Center (DKFZ), Heidelberg, which has a longstanding record of achievements in cancer genomics.

The Robinson lab published one of the first exome-based gene discovery papers in *Nature Genetics* in 2010, the identification of disease-causing mutations in *PIGV* by identity-by-descent filtering of exome sequence data [226]. We subsequently contributed to the characterization as disease genes of *PIGO* [224], *PGAP2* [225], *IL21R* [219], *TTC8* [140], and *PGAP3* [169]. We have additionally used WES analysis to address questions in hereditary cardiomyopathy [377], Vici syndrome [112], and other diseases [215].

The lab has developed the Human Phenotype Ontology [146, 209, 372] (HPO), which is being used by many of the major WES/WGS translational research projects, such as Genomics England's 100,000 Genomes Project, the Wellcome Trust Sanger Institute's DECIPHER and Deciphering Developmental Disorders (DDD) projects, the National Institutes of Health (NIH) Undiagnosed Diseases Network, and many others [214]. We have developed software that exploits HPO-based phenotypic similarity algorithms to prioritize genes and variants in WES and WGS analysis [373, 399, 403], as we shall examine in several chapters of this book. We have additionally developed the annotation tool Jannovar [184], and contributed to a range of other algorithms designed for the analysis of quality control of WES/WGS data [161, 162], as well as algorithms for ChIP-seq [159], ChIP-nexus [158], NGS-based T-cell receptor profiling [230], and RNA-seq [182].

Rosario Piro's research at the DKFZ contributed to the remarkable finding that all secretory meningioma harbor the same single nucleotide change in the pluripotency-related transcription factor *KLF4* [363], a finding that was independently and simultaneously confirmed by another research group [80]. A fusion gene (*NAB2-STAT6*) was identified in meningeal hemangiopericytoma and solitary fibrous tumors that allows these tumors to be distinguished from anaplastic meningiomas by simple STAT6 immunohistochemistry [391], having an immediate impact on cancer diagnostics [70, 391]. The integration of DNA sequencing and epigenomic data led to the finding that poor-prognosis hindbrain ependymomas harbor an extremely low mutation rate and are instead characterized by a CpG island methylator phenotype (CIMP), leading to a transcriptional silencing of differentiation genes [271].

Marten Jäger established and implemented the exome and genome pipelines in the Robinson lab at the Institute for Medical Genetics and Human Genetics of the Charité University Hospital. His research has concentrated on methods for exome and genome bioinformatics including approaches to VCF annotation and variant calling with non-linear genome assemblies, a method for composite transcriptome assembly of RNA-seq data, and contributions to many other research projects.

All three authors held courses on exome analysis in the Bioinformatics Department of the Free University of Berlin, out of which this book grew.

This book, therefore, reflects the experiences made over multiple years in establishing first an Illumina Genome Analyzer (GAIIX) and later a HiSeq 1500 and other devices in Berlin as well as working with HiSeq 2000 and MiSeq sequencing data in Heidelberg. It reflects the experiences of developing software for phenotype analysis that is being used internationally by groups involved in translational genomics. A major rule both in Heidelberg and Berlin was "eyeball the data", and we place an emphasis in this book on understanding how data is represented in the various file formats common in WES/WGS analysis, and how to recognize errors and artifacts (or indeed, how to recognize high-quality data). With several exceptions, this book does not attempt to explain in detail the algorithms or statistics involved in WES/WGS analysis, but instead presents a practical guide to the many steps of WES/WGS analysis with intuitive algorithmic explanations and pointers to the literature.

A NOTE ON HOW TO USE THIS BOOK

Bioinformatics is a fast-moving field. Websites and GitHub pages may change, and arguments required by programs may be altered in new versions. We will therefore keep track of this and also provide additional files, scripts, and what hopefully will remain a short errata page at the book website:

```
https://github.com/TheJacksonLaboratory/Computational-Genome-Analysis
```

Readers are encouraged to post questions or bugs on the book website.

We show commands that are to be typed into the shell with a dollar sign ($). For instance, the following command will print a list of files in the current directory.

```
$ ls
```

Some commands are too long to fit onto a single line. In these cases, we use a slash (\) to indicate that the command is continued on the next line.

```
$ command argument-1 \
    argument-2 ...
```

We have assumed that readers will adjust the paths of the commands shown in the book as needed. We have also assumed they are familiar with the basic Unix utilities such as `wc` and `cat` or are able to find information about them on the Web.

We will define the abbreviations used in each chapter — with the following exceptions that are used throughout the book.

- WES: whole-exome sequencing

- WGS: whole-genome sequencing

- PCR: polymerase chain reaction

- nt: nucleotides

- bp: base pairs

- b, kb, Mb, ...: bases, kilobases, megabases, etc.

Thanks

This book, and indeed much of the work done on genomics in Berlin, would not have been possible without the work, conversations, and advice of many people. Professor Stefan Mundlos, the Director of Medical Genetics and Human Genetics of the Charité University Hospital in Berlin, provided a challenging and rewarding environment for the Robinson lab, and well deserves our special thanks. Dr. Jochen Hecht set up the NGS lab in Berlin from the beginning, solving one problem after another by persistence, experience, and, yes, inspiration. Dr. Sebastian Köhler, who began as a Master student in the group, moving up through the ranks as a graduate student and a postdoc, was indispensable for multiple projects and is the co-founder and co-driving force behind the HPO. We have been extremely lucky with the people who became members of the lab, and would like to thank (in no special order) Peter Hansen, Peter Krawitz, Leon Kuchenbecker, Verena Heinrich, Na Zhu, Raghu Bhushan, Claus-Eric Ott, Johannes Grünhagen, Layal Abo Khayal, Valerie Johnston, Angelika Pletschacher, Robin Steinhaus, Max Schubach, Michal Schweiger, and Manuel Holtgrewe. These thanks also form a kind of farewell for Peter Robinson, who has relocated to the Jackson Laboratory (JAX) for Genomic Medicine in Farmington, CT, in August 2016 to pursue translational genomics and computational biology in a new setting. And finally, ... thanks to Daniel Danis, Hannah Blau, and Leigh Carmody and the entire Jackson Laboratory for being so awesome!

We would also like to extend thanks to Rosario Piro's colleagues at the DKFZ (in no special order) Rainer König, Natalie Jäger, Barbara Hutter, Roland Eils, Peter Lichter, Stefan Pfister, Andreas von Deimling, Susanne Gröbner, Sonja Hutter, Marc Zapatka, Volker Hovestadt, Bernhard Radlwimmer, David Reuss, David Capper, Felix Sahm, Leonille Schweizer, Yvonne Schweizer, David Jones, Marcel Kool, Paul Northcott, Benedikt Brors, Matthias Schlesner, Frank Westermann, Marcus Oswald, Hendrik Witt, Josephine Bageritz, Violaine Goidts, Martje Tönjes, Moritz Aschoff, and many others that were not directly involved in his research projects but provided invaluable backing, guidance and above all friendship.

Our book is based on over 10 years of experience in exome and genome analysis. This book grew out of course notes prepared for a course on exome analysis held in the Bioinformatics Department of the Free University of Berlin, and we would like to thank the students of that course for insightful questions that kept us on our toes. Many of the exercises in this book were adapted from exercises in our course.

Finally, we would like to extend special thanks to Knut Reinert, Alexander Bockmayr, and all the other colleagues at the Institutes of Bioinformatics, Computer Science and Mathematics of Free University of Berlin (FUB) for creating an amazing atmosphere for learning, teaching, and practicing bioinformatics.

—Peter Robinson, Farmington, CT

—Rosario M. Piro, Berlin, Germany

—Marten Jäger, Berlin, Germany

Contributors

Peter Hansen
Institute for Medical and Human
  Genetics,
  Charité-Universitätsmedizin
  Berlin
Berlin, Germany

Jochen Hecht
CRG-Centre for Genomic Regulation
Barcelona, Spain

Manuel Holtgrewe
Berlin Institute of Health
Berlin, Germany

Julius O.B. Jacobsen
Queen Mary University of London,
  London E1 4NS, UK; Genomics
  England Ltd.
London, UK

Sebastian Köhler
Institute for Medical and Human
  Genetics,
  Charité-Universitätsmedizin
  Berlin
Berlin, Germany

Peter Krawitz
Institute for Medical and Human
  Genetics,
  Charité-Universitätsmedizin
  Berlin
Berlin, Germany

Max Schubach
Institute for Medical and Human
  Genetics,
  Charité-Universitätsmedizin
  Berlin
Berlin, Germany

Dominik Seelow
NeuroCure Clinical Research Center,
  Charité-Universitätsmedizin
  Berlin
Berlin, Germany

Damian Smedley
Queen Mary University of London,
  London E1 4NS, UK; Genomics
  England Ltd.
London, UK

Tomasz Zemojtel
Institute for Medical and Human
  Genetics,
  Charité-Universitätsmedizin
  Berlin
Berlin, Germany

Johannes Zschocke
Division of Human Genetics, Medical
  University of Innsbruck
Innsbruck, Austria

Introduction

CHAPTER 1

Introduction: Whole Exome and Genome Sequencing

THE successful completion of the human genome project was announced in 2003, fifty years after the discovery of the double-stranded structure of DNA [454], by the National Human Genome Research Institute (NHGRI), the Department of Energy (DOE) and their partners in the International Human Genome Sequencing Consortium. This landmark followed the announcement of a "rough draft" of the human genome in 2000 and publications of the initial human genome sequence in 2001 by a public and a private consortium [234, 442]. The human genome project was widely hailed as a blueprint for genomic discoveries [85], analogous in many ways to the anatomy atlas *de corporis humani fabrica*, published in 1543 by Andreas Vesalius. Vesalius' book (*On the Fabric of the Human Body*) was the first modern atlas of gross anatomy based on observation, and enabled Harvey's identification of the physiology of circulation in 1628 as well as subsequent discoveries [286].

The gross anatomy of Vesalius provided a catalog of the parts of the body that enabled subsequent researchers to investigate their function (physiology) and medical relevance. One can say that we are in a post-Vesalian era with respect to the human genome — we have a relatively complete knowledge of the anatomy of the human genome, but are just beginning to understand the functions of human genes, regulatory elements such as enhancers and promoters, and how the genes and other elements of the genome work together within genetic regulatory networks and how anomalies of genes and gene regulation can lead to human disease.

The human genome project was completed using a DNA sequencing method published in 1977 by Fred Sanger [384], one of the few individuals to win more than one Nobel prize. The introduction of the first automated sequencing machine (ABI 370) by Applied Biosystems in 1987 and technical advances such as capillary sequencing and dye-terminator sequencing steadily increased the throughput of "Sanger sequencing", and in the heyday of the Sanger era, about 400,000 nucleotides (400 kb) could be sequenced per day on one automated sequencing machine. Thus, roughly 45,000 runs would be needed to sequence one human genome at 6x coverage, and correspondingly, the price tag for the first human genome was roughly $2.7 billion [302]. A number of technological advances collectively referred to as Next-Generation Sequencing (NGS), which will be addressed in detail in Chapters 2 and 3, have brought the cost of sequencing a human genome down to under $1,000, which is enabling the widespread use of genome sequencing for an ever-growing list of biomedical questions.

The history of NGS is replete with competing technologies, takeovers, and mergers, as well as steady progress in throughput, quality, and read length. NGS can be traced back to at least 2000 with the release of Massively Parallel Signature Sequencing by Lynx Therapeutics. This technology became obsolete with the advent of technologies such as sequencing-by-synthesis (SBS) and the marketing of the Genome Analyzer by Illumina in 2007, and other machines such as the SOLiD sequencing system (ABI), the Life Technologies Ion Torrent sequencer, the Pacific Biosciences Single Molecule Real-Time (SMRT) sequencer, and more recently nanopore sequencing devices by companies such as Oxford Nanopore Technologies [31]. The market for human genome sequencing is currently dominated by Illumina, and the lion's share of bioinformatics pipelines and algorithms have been developed with this technology in mind. The Illumina HiSeq X Ten, which was launched in March 2014, has brought the price of human whole genome sequencing to under $1,000 and has a throughput of roughly 1.8 Tb of sequence per instrument run.[1] As this book was going to press, Illumina had just released the NovaSeq series, which will achieve an even higher throughput (see Chapter 3).

---

[1] 1.8 terabases (Tb) refers to 1.8 trillion (1.8×10^{12}) nucleotides.

Bioinformatics, like genomics, is a relatively young field (in fact, the first use of the word "bioinformatics" only dates back to 1970 [166]). This book is about the bioinformatics of human genome and exome sequencing, which has gone through a similarly spectacular development over the last decade. This introductory chapter will set the stage for the rest of the book by introducing the main subjects of analysis by exome and genome sequencing: Mendelian disease, cancer, and precision medicine. Chapters 2 and 3 will explain NGS technologies, and the remaining chapters of the book will cover all relevant steps of the bioinformatic analysis of WES and WGS data from low-level data processing to bioinformatic clinical decision support algorithms.

1.1   INVESTIGATING THE HUMAN GENOME FOR SCIENCE AND MEDICINE

The human genome (Table 1.1) is still poorly understood. The best studied portion of the genome is represented by the ∼1.5% of the genome that encodes proteins, but even the functions of the proteins in the human genome are only partially understood. The functions of most of the ∼98.5% of the genome that does not code for protein are less well understood. It even remains unknown what part of the genome actually has any function, with estimates ranging from ∼8% based on evolutionary constraint [353] to ∼80% of the genome with a biochemical function [115]. Medical analysis of the human genome usually focuses on variants, i.e., differences from the reference sequence. However, we will see in Chapter 23 that the medical significance is unknown for the majority of variants that have been sequenced or could theoretically occur.

For these reasons, the computational analysis of exome or genome sequence data for biomedical research or clinical care can be divided conceptually into a discovery phase (that is likely to continue for the foreseeable future) that aims to identify the medical significance of all functional elements of the genome and to characterize the medical relevance of variants in these elements, as well as efforts to exploit this knowledge to improve diagnostics and clinical care of patients. Currently, genomic medicine has substantially contributed to the care of individuals with genetic (mainly Mendelian) disease and of cancer patients, but it is expected that genomics will contribute to almost all areas of medicine in coming decades.

Table 1.1. The Human Genome

Chrom.	Size (Mb)	Genes	Chrom.	Size (Mb)	Genes
chr1	249.0	5078	chr14	107.0	2055
chr2	242.2	3862	chr15	102.0	1814
chr3	198.3	2971	chr16	90.3	1920
chr4	190.2	2441	chr17	83.3	2432
chr5	181.5	2578	chr18	80.4	988
chr6	170.8	3000	chr19	58.6	2481
chr7	159.3	2774	chr20	64.4	1349
chr8	145.1	2152	chr21	46.7	756
chr9	138.4	2262	chr22	50.8	1172
chr10	133.8	2174	chrX	156.0	2158
chr11	135.1	2920	chrY	57.2	577
chr12	133.3	2521	chrMT	(16.6 kb)	37
chr13	114.4	1381			

Note: The statistics are derived from the NCBI Genome resource for genome build GRCh38.p9. The total length of the haploid genome is ∼3.088 billion base pairs, with 53,853 genes currently annotated (including protein-coding and non-coding RNA genes). The exact number of genes has yet to be determined.

1.2 MENDELIAN DISEASE

The two major areas in which WES/WGS have been used in clinical care are for the diagnosis of suspected Mendelian (genetic) syndromes and cancer. Almost all human diseases are influenced in one way or another by genetic variation, but Mendelian diseases are exceptional in that they are caused by mutation in one gene and may run in families. In Chapter 20, we will discuss the modes of inheritance and pedigree analysis. Leaving aside for now differences in the genetic basis of diseases with autosomal dominant, autosomal recessive, X chromosomal, and other modes of inheritance, it has been comparatively easy to understand the genetic basis of Mendelian disorders because of the fact that a mutation in only one gene causes a Mendelian disorder (which are also called monogenic, or single-gene diseases).[2]

In order to understand just how much of an advance WES and

---

[2]A Mendelian disease is called heterogeneous if a pathogenic mutation in one of a set of genes leads to a clinically identical or highly similar disease.

WGS represent for the study of Mendelian disease, it is useful to consider briefly how research and diagnostics were performed prior to the NGS era. The first identification of the cause of a Mendelian disease was Linus Pauling's finding in 1949 of differences in the hemoglobin of healthy individuals and those with sickle-cell anemia. Pauling inferred that the genetic variant responsible for sickling is heterozygous in those with sickle trait and homozygous in those with sickle-cell anemia [325]. It was shown seven years later that the observed difference was due to the substitution of the glutamic acid by a valine residue in position 6 of the hemoglobin β-chain [178]. Progress remained relatively slow until the introduction of positional cloning into research practice in the early 1980s. Positional cloning involves the collection of samples from one or more families affected by a given disease and genetic linkage analysis to define a critical DNA region of interest [345]. The first disease gene to be discovered by positional cloning, *CYBB*, was shown in 1986 to be associated with chronic granulomatous disease [380], and subsequently, positional cloning has led to the majority of the thousands of discoveries of disease genes in the last three decades. However, linkage analysis generally identifies a genomic interval of roughly 0.5–10 cM containing up to 300 genes [45], resulting in the need to sequence large numbers of candidate genes in order to find the actual disease gene, meaning that positional cloning projects could take years to complete (as one of the authors of this book can confirm [433]).

Although Mendelian disorders are individually rare, because there are so many different Mendelian disorders they are collectively common. It has been estimated that about 0.4% of all newborns are affected by a clinically recognizable Mendelian disease and that up to 8% of all individuals have some genetic disorder recognizable by early adulthood [26, 72]. The extent to which this represents a significant public health issue has been much overlooked — the overall prevalence of Mendelian disease is, for instance, much higher than the number of new cancer diagnoses per year at approximately 0.45%. Despite the medical need, the disease genes associated with only about half of the approximately 7,000 currently described Mendelian diseases have been discovered to date [72]. Many distinct genetic diseases tend to be pooled under functional or descriptive categories such as intellectual disability, hypothyroidism, congenital heart defect, or osteoporosis, and probably comprise a large number of etiologically distinct conditions. Thus, today the overall number of rare conditions is likely to be grossly underestimated and can be expected to grow with increasing knowledge about their molecular and genetic pathology. Therefore, one of the major applications of WES/WGS in translational research in medical genetics involves the characterization of all Mendelian disease genes.

In addition to research applications, WES/WGS is being increasingly used for genetic diagnostics whereby diagnosticians examine the WES/WGS data for mutations in known disease genes. Although this is conceptually simpler than using WES/WGS to search for novel disease genes, the task is far from trivial, and current studies suggest that an overall diagnosis rate of around 25% can be achieved in unselected patient cohorts [242, 342, 473, 474].

1.3 CANCER

Cancer is a disease of the genome, and NGS technologies have revolutionized our ability to characterize the somatic alterations present in cancer genomes [277]. It was only in 2008 that genome sequencing was performed on a tumor-normal pair (i.e., a tumor sample and a control sample from the same patient) in order to search for genes with acquired (somatic) mutations [249]. This kind of analysis allows researchers to survey genes that are likely to be important in tumorigenesis and the progression of cancer, and is increasingly being used diagnostically to search for therapeutically actionable mutations in individual tumor genomes. The evolution of a normal cell to a cancer cell can be thought of as a sequential accumulation of many independent lesions to the genome. These lesions may be somatic point mutations as well as chromosomal deletions or duplications, or other structural variants. As lesions accumulate, they do not generally affect all of the cells in a tumor; instead, clones give rise to subclones in an iterative fashion. During the evolution of tumors, new subclones can originate with novel mutations related to disease progression. The comparison of allele fractions may allow the clonal history of a tumor to be reconstructed. Subclones that derive from the same ancestor subclone should share all mutations of the original founder but can additionally harbor mutations that confer a proliferative advantage [106].

Cancer therapy options have traditionally included surgery if part or all of the tumor can be removed from the body, as well as radiation therapy and chemotherapy to shrink the tumor or eliminate migrating cancer cells. The latter two therapies preferentially damage the DNA of quickly dividing cells, which is effective against many types of tumors because cancer is characterized by increased cell division. This also explains many of the side effects of these treatments. Increasingly, other forms of therapy are being developed and introduced to the clinic that use drugs or immunotherapy to target cancer cells in a more or less specific manner, ideally with increased efficacy and reduced side effects. Whole genome sequencing followed by integrative bioinformatic analysis is being increasingly used to guide treatments in research settings [187], and smaller panels of known cancer genes are being used in diagnostic settings [301]. The next decade is likely to witness the increasing adoption of comprehensive genomic profiling and identification of actionable mutations as a part of standard clinical management of cancer [220].

1.4 PRECISION MEDICINE

The Precision Medicine Initiative (PMI) of the United States was launched in 2015. The basic concept of precision medicine is not new; its overall goal is the prevention and treatment strategies that take individual variability into account. However, the term precision medicine is now being applied to data-driven methods for characterizing patients such as proteomics, metabolomics, genomics, diverse cellular assays, and even mobile health technology. The PMI and similar programs in other countries will require substantial investments in computational tools for analyzing large sets of data. It is likely that exome and genome sequencing will be used at a broad range of clinical contexts as our ability to provide precision medical care advances [22, 86].

FURTHER READING

Ashley EA (2016) Towards precision medicine. *Nature Reviews Genetics* **17**:507–522.

McKusick VA (2001) The anatomy of the human genome: a neo-Vesalian basis for medicine in the 21st century. *JAMA* **286**:2289–2295.

CHAPTER 2

NGS Technology

ADVANCES in technology drive science by giving scientists the ability to observe things that were previously hidden. For instance, Anton van Leeuwenhoek (1632–1723), the inventor of the modern microscope, was also the first human to observe bacteria. Advances in DNA sequencing technologies over the past quarter-century have enabled substantial reductions in the cost of genome sequencing, leading to a veritable revolution in genomics. One of the authors of this book remembers his time as a graduate student performing radioactive Sanger sequencing. The process involved the generation of amplicons by polymerase chain reaction (PCR), pouring a sequencing gel (this was an especially daunting challenge for the author, who is not known for his manual dexterity!), performing the sequencing reaction with radioactive reagents, running the gel, exposing it to radiographic film, developing the film, and finally reading the sequence from the pattern of bands in the G, A, T, and C lanes. The entire process involved about a week of working and waiting and enabled a few hundred nucleotides to be deciphered. Today, it still seems hardly less than miraculous that a single run on a modern next-generation sequencing (NGS) machine can output up to 1.8×10^{12}, or 1.8 trillion nucleotides of sequence information in less time than this.

In this book, we will cover the entire gamut of computational analysis of whole-exome and whole-genome sequencing (WES/WGS). It is essential that bioinformaticians understand not only the salient biological and medical aspects of the data they analyze, but also that they know how genomic technologies have generated the data. We will therefore begin this book with an explanation of how NGS technologies work, and we will concentrate especially on the Illumina technology, which currently is by far the most commonly used platform for WES/WGS analysis.

2.1 SANGER SEQUENCING

The British biochemist Fred Sanger won a total of $1\frac{1}{4}$ Nobel Prizes. In 1958, he won the Nobel prize in chemistry "for work on structure of proteins, especially that of insulin", and in 1980, he shared half of the Nobel Prize in chemistry with Walter Gilbert "for determination of base sequences in nucleic acids". The method Sanger developed came to be known simply as *Sanger sequencing*, and it was the major method of sequencing nucleic acids until very recently. The key idea of Sanger sequencing, which is also known as the chain-termination method, involves the incorporation of dideoxynucleotides by DNA polymerase during *in vitro* DNA replication.

Each DNA nucleotide contains the five-carbon sugar deoxyribose in addition to a nitrogenous base and one, two, or three phosphate groups. The carbons of deoxyribose are numbered from 1' to 5' (pronounced "1-prime", etc.). The word dideoxynucleotide means that two ("di") carbon atoms do not have a hydroxyl group ("deoxy") but rather just a single hydrogen (Figure 2.1). DNA differs from RNA in that the 2'-hydroxyl group in RNA is absent. With dideoxynucleoside triphosphates[1] (ddNTPs) the 3'-hydroxyl group is also absent.

To perform Sanger sequencing, the DNA template is denatured (made single-stranded), and DNA sequencing primer complementary to a region in the template is added together with DNA polymerase, normal dNTPs, and a smaller amount of "chain-terminating" ddNTPs. DNA synthesis proceeds from the 5' end of the growing strand toward the 3' direction, whereby DNA polymerase adds a nucleoside triphosphate (dNTP) to the 3' hydroxy group of the last nucleotide of the growing DNA strand, forming a phosphodiester bridge between the 3'-hydroxyl group of the last nucleotide of the growing strand and the alpha-phosphate of the incoming dNTP and thereby releasing the beta and gamma phosphates as pyrophosphate. Since ddNTPs do not have a 3'-hydroxyl group, if a ddNTP is incorporated, it "terminates" further extension of the growing DNA strand. Therefore, ddNTPs stop DNA synthesis at specific nucleotides (Figure 2.2).

---

[1]Recall that the word "nucleoside" refers to a nucleotide without a phosphate group.

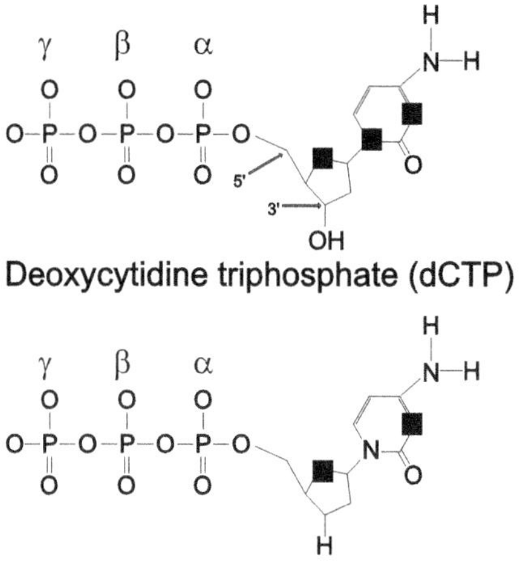

Deoxycytidine triphosphate (dCTP)

Dideoxycytidine triphosphate (ddCTP)

Figure 2.1. Dideoxynucleotides. A dideoxyribonucleotide does not have a 3' hydroxyl group (-OH), and thus no further chain elongation can occur once this dideoxynucleotide is on the chain. In the figure, 2'-deoxycytidine triphosphate (dCTP) and 2',3'-dideoxycytidine triphosphate (ddCTP) are shown.

In the original version of Sanger sequencing, the sequencing product was marked radioactively. For instance, dATP–[α-^{33}P] would be added to the sequencing reaction, so that the chain-terminated molecules would incorporate the radioactive marking at random 'A' sites in the sequence. Alternatively, the sequencing primer itself could be marked. Four separate reactions were carried out, one for each ddNTP. The reactions were subjected to gel electrophoresis and the location of the products became apparent after autoradiography as bands in the gel.

Applied Biosystems (ABI) introduced a series of automated sequencers that employed fluorescent dyes to mark each of the four ddNTPs with different labels (different wavelength), allowing one reaction with all four ddNTPs to be performed, rather than four separate reactions with each ddNTP as in the radioactive method. Electrophoresis was performed as before, or the samples were injected into an array of capillaries. A laser beam excited the fluorescent dyes as the DNA

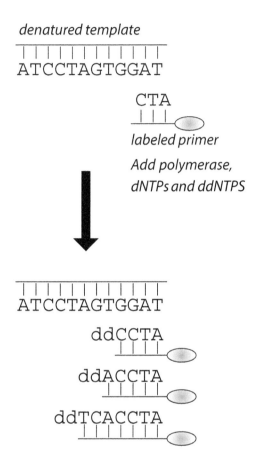

Figure 2.2. Chain termination in Sanger sequencing. Chain extension is terminated at defined locations corresponding to the sequence of the template. For instance, a ddATP can terminate the chain by pairing with one of the thymines (T) of the template. The chain terminated products possess well-defined lengths and are labeled with radioactivity or fluorescent dyes. Upon gel electrophoresis of the labeled products, the template sequence can be inferred based on the location of the bands or peaks representing the products.

fragments passed a detection window at a time that was proportional to the sequence length. The intensity of each wavelength was plotted against electrophoresis time to create a "chromatogram" that represented the DNA sequence of the product (usually) in the colors green

(A), red (T), blue (C), and black (G). Sequence analysis software inferred the DNA sequence from the chromatograms, often requiring only minimal manual revision. Sanger sequencing powered the characterization of the human genome, culminating in the human genome project. At the height of the Sanger era, the ABI 3700 DNA analyzer could sequence up to 400 kb per machine per day. Roughly speaking, about 45,000 runs were needed to cover the human genome at 6x coverage in the first iteration of the human genome project.

2.2 IT'S THE ECONOMY, STUPID

Bill Clinton won the 1992 United States presidential election thanks in part to his campaign slogan about the economy — apparently, his intuition was correct that the economy is the single most important issue to most voters. Next-generation sequencing does not do anything that wouldn't have been possible in principle using Sanger sequencing — it is the fact that NGS is enormously cheaper and faster than Sanger sequencing that has enabled its widespread use in diagnostics and translational research. Further reductions in price are likely to expand the utilization of WES/WGS in human medicine even further.

WGS can be used for *de novo* genome assembly, but this is not a topic we will cover in this book. With rare exceptions, WES/WGS in medical contexts are *resequencing* applications — that is, they require a reference genome in order to map sequence reads prior to downstream analysis. As such, WES/WGS would be impossible without a complete (or nearly complete) human genome sequence. The draft human genome sequence was announced by two groups in 2000 [234, 442].

Ten years later, the leaders of the public and private efforts, Francis Collins and Craig Venter, when asked what benefit medicine and human health had experienced from the human genome sequence, both answered essentially "not much" [303]. Perhaps the major reason for this answer was that the high cost of sequencing was preventing the full use of the human genome sequence for translational research and diagnostics. Starting in 2004, therefore, the National Human Genome Research Institute (NHGRI) of the US National Institutes of Health (NIH) committed more than $100 million to research with the goal of reducing the cost of human genome sequencing to US $1000 in ten years [387]. This was a major incentive toward the development of NGS technologies, which can be broadly defined on the basis of three major improvements over the automated Sanger sequencing methods that had been used for the Human Genome Project [439]:

1. Instead of requiring bacterial cloning of DNA fragments, NGS methods rely on the preparation of sequencing libraries in a cell-free system.

2. Thousands to many millions of sequencing reactions are produced in parallel (as opposed to hundreds of reactions with automated Sanger sequencing).

3. Sequencing output is directly detected without electrophoresis in a highly parallelized and cyclical fashion.

Between 2005, when the first NGS technology was brought to market, and today, there have been incremental and steady improvements in technology that have brought us very close to the goal of the $1000 genome (Figure 2.3).

2.3 OVERVIEW OF CURRENT NGS TECHNOLOGIES

In 2005, the first commercial NGS technology was released by 454 Life Sciences, which at the time generated about 20 Mb of sequence per run. The 454 family of platforms was based on a pyrosequencing approach with emulsion PCR and generated relatively long reads [278]. After subsequent development, read lengths of up to 700 nt or more were achieved. The 454 platform was purchased by Roche, which developed the platform as the GS FLX+System with the GS FLX Titanium Sequencing Kit XL+, which was able to produce about 1 million reads of up to 1000 bp in length per run. The platform, however, was not able to compete with the Illumina platform, and was taken from the market in 2014.

In 2006, the Solexa system was commercialized, and in 2007 Solexa was bought by Illumina [439], which currently is the leading provider of NGS technology. The Illumina technology is clearly the leader in the field at the time of this writing (2017), and will be discussed in detail in Chapter 3. Illumina has brought a number of different sequencers to market, that currently range from the MiniSeq systems designed for targeted gene panel sequencing to the HiSeq X series designed for human genome sequencing at a population scale, and very recently the NovaSeq device.

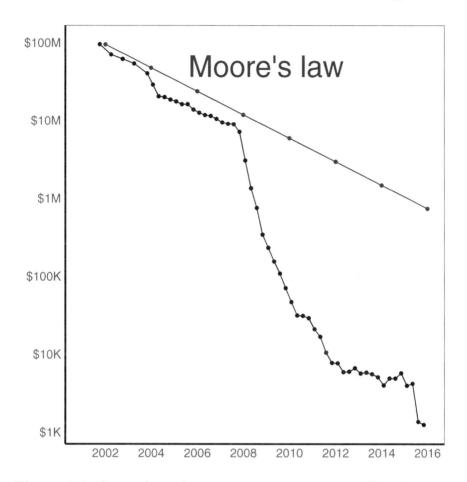

Figure 2.3. Cost of one human genome sequence. From 2008 to 2014, the rate of reduction in cost greatly exceeded that predicted by Moore's law, which states that there is a doubling in computing power per unit price every two years. Data were taken from the website of the National Human Genome Research Institute (NHGRI).

Applied Biosystems introduced the Sequencing by Oligo Ligation Detection (SOLiD) in 2007 [436]. SOLiD sequencing involves a multi-round, staggered ligation approach that uses a two-base encoding and fluorescently labeled 8-mer probes that "read" the two 3' bases, forming a kind of "color-space" [276]. The SOLiD system has relatively long run times and has not been able to move past a read length of 75 nt, which has limited its popularity.

The Ion Torrent system employs a semiconductor "chip" that de-

Table 2.1. Overview of NGS sequencing technologies.

Platform	Amplification	Chemistry	Reads
454	Emulsion PCR	Pyrosequencing	700–1000bp
Illumina	Bridge amplification	SBS	2×150bp to 2×300bp
SOLiD	Emulsion PCR	SBL	75bp
Ion Torrent	Emulsion PCR	SBS	up to 400bp

Note: Abbreviations: SBS, sequencing by synthesis; SBL, sequencing by ligation.

tects alterations in pH due to the release of protons upon addition of bases to a growing DNA strand. Ion Torrent offers both a small, "Personal Genome Machine" intended for gene panel investigations as well as a larger machine ("Proton") that can sequence exomes. The Ion Torrent platform has aimed since 2012 to introduce a new chip technology that would enable a human genome to be sequenced at a depth of 20x for less than $1,000. Table 2.1 provides an overview of the major short-read NGS platforms.

Single-molecule, real-time sequencing developed by Pacific BioSciences (PacBio) offers substantially longer read lengths than the previously mentioned next-generation sequencing technologies. PacBio sequencing captures sequence information during the replication process of the template DNA molecule, which is a closed, single-stranded circular DNA created by ligating hairpin adapters to both ends of a target double-stranded DNA molecule. Four different fluorescently labeled nucleotides are added to the PacBio sequencing cell, such that a light pulse that corresponds to each nucleotide is generated as the base is held by the polymerase [364]. PacBio's main advantage over previous NGS technologies is its much longer read length, which is extremely useful for phasing variants and for characterizing structural variants. PacBio still has higher error rates, lower throughput, and substantially higher costs than competing methods, but new technological developments may mitigate these shortcomings in the near future.

In 2014, the first commercial sequencer using a nanopore method was released by Oxford Nanopore Technologies [269]. As with PacBio, one of the major advantages of nanopore sequencing is the substantially longer read-length as compared with the Illumina platform, but at the time of this writing, it appears that the Oxford Nanopore technology is not yet as mature as PacBio. However, the entire field is moving so rapidly that it is hard to make a confident prediction about the future developments of NGS technologies, or of which platform(s) will dominate the market in several years time.

2.4 ENRICHMENT TECHNOLOGIES

The protein coding regions constitute ~1% of the human genome or ~30 Mb, split across ~180,000 exons. Currently, the great majority of known mutations responsible for Mendelian diseases in humans affect sequences within the coding regions of exons or are located within a few nucleotides of the exon boundaries. Exome sequencing is designed to capture the sequences in which mutations in Mendelian disease are most commonly located, and therefore is based on the enrichment of sequences corresponding to all (or nearly all) protein coding exons followed by next-generation sequencing (NGS).

Current methods for enriching exonic sequences all work in principle in the same fashion. Oligonucleotide probes are constructed to hybridize to ("capture") the target sequences from a library generated from fragmented total genomic DNA. After hybridization, the biotinylated probes with the captured targets are bound to streptavidin beads and the non-bound fragments are washed away. Enriched fragments are amplified by PCR using common adapter sequences added in the library preparation step before the hybridization (see Chapter 3).

A number of companies are offering target capture methods for whole-exome sequencing. The methods typically aim to capture all exonic and flanking sequences and may also include probes to target microRNA genes and other sequences of interest. Several reviews on the technical aspects of target capture methods are available [79, 275].

Currently, there are three major exome enrichment products, all of which use biotinylated DNA or RNA probes ("baits") that are complementary to the targeted exome:

1. Roche/NimbleGen's SeqCap EZ Human Exome Library

2. Illumina's Nextera Rapid Capture Exome

3. Agilent's SureSelect XT Human All Exon and SureSelect QXT

The main differences between these platforms comprise target region selection, bait length, bait density, the molecule used for capture, and the genomic fragmentation method. Published comparisons have not shown substantial differences in performance between these platforms, although Agilent XT and Illumina platforms may show slightly superior coverage for the targeted regions [68, 395]. It is important to be aware of the actual target regions for the capture kit when analyzing coverage, since each kit uses different and only partially overlapping definitions of the "exome".

FURTHER READING

Rhoads A, Au KF (2015) PacBio sequencing and its applications. *Genomics Proteomics Bioinformatics* **13**:278–289.

van Dijk EL, Auger H, Jaszczyszyn Y, Thermes C (2014) Ten years of next-generation sequencing technology. *Trends Genet.* **30**:418–426.

CHAPTER 3

Illumina Technology

Peter Robinson and Jochen Hecht

T HIS chapter will present a slightly simplified version of the Illumina technology for next-generation sequencing (NGS). There are three major steps in the sequencing protocol: library preparation, flow cell preparation, and sequencing by synthesis. Each step can introduce characteristic errors into the sequencing data, and it is thus essential for bioinformaticians to be familiar with the sequencing protocol and to be able to recognize characteristic error profiles.

3.1 LIBRARY PREPARATION

The library preparation ("library prep") step creates a collection ("library") of random DNA fragments that are ready to be sequenced. This is a major difference to Sanger sequencing, which requires some form of targeted DNA amplification: either by specific PCR primers or by cloning into sequences with universal primer binding sequences.

Most Illumina protocols require that the DNA to be sequenced is fragmented to sizes of less than 800 nucleotides (nt). A main reason is that the amplification efficiency for longer fragments is too low for proper cluster generation on the flow cell surface.

There are four major substeps for library prep (about 6 hours of lab work):

1. Fragment DNA

2. Repair ends and add A overhang

3. Ligate adapters

4. Select ligated DNA

3.1.1 Fragmentation

The main methodology for fragmenting DNA is sonication, which utilizes ultrasound waves in solution to shear DNA. The ultrasound waves pass through the DNA sample, rapidly expanding and contracting the liquid to create tiny bubbles. This process is called cavitation. The bubbles in turn generate focused shearing forces that fragment the DNA. Ideally, the DNA fragments should have a uniform size, and current sonicators, including the Bioruptor[1] and Covaris devices,[2] allow users to adjust the mean fragment size.

Nebulization is an alternate method for fragmentation by which genomic DNA is physical sheared by repeatedly forcing input DNA solution as a fine mist through a small hole in the nebulizer. The size of the fragments can be determined to some extent by the speed at which the DNA solution passes through the hole, the pressure of the gas blowing through the nebulizer, the viscosity of the solution, and the temperature. In general, however, sonication provides a better and more uniform fragmentation than nebulization.

3.1.2 End repair

Following fragmentation, the ends of the DNA fragments can have 5' or 3' overhangs, which would prevent some downstream enzymatic steps from working (Figure 3.1). The ends of the DNA must therefore be "polished" so that an A tail can be added. Phosphorylation of 5'-ends is strictly required for enzymatic ligation of nucleic acids. Ligases need a 5' phosphate group and a 3' hydroxyl (-OH) group to link two oligonucleotides by forming a phosphodiester bond. T4 DNA polymerase and the Klenow fragment of *E. coli* DNA polymerase I are used in this step. The Klenow fragment fills in the 5' overhangs by means of its 5' to 3' polymerase activity. Klenow also possesses 3' to 5' exonuclease activity, but this is relatively weak compared to the same activity in the T4 DNA polymerase, whose main job is thus to remove the 3' overhangs.

---

[1]https://www.diagenode.com/categories/bioruptor-shearing-device
[2]http://covarisinc.com/applications/dnarna-shearing-for-ngs/

3.1.3 3′ end adenylation

A single adenine (A) nucleotide is added to the 3' ends of the blunt fragments. This keeps them from ligating to one another during the subsequent adapter ligation reactions. Also, as we shall see shortly, the sequencing adapters are designed to have a thymine (T) base overhang that enables DNA fragments to ligate to the adapters. The adenylation step is performed by adding dAPT and Klenow exo⁻ fragment, whose 5' to 3' polymerase activity is responsible for adding the A overhang base that is not removed due to the lack of a 3' to 5' exonuclease activity in the mutant form (exo⁻) of this enzyme.

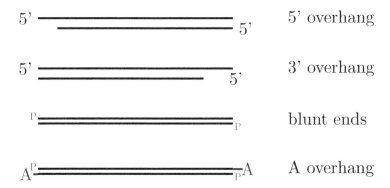

Figure 3.1. End repair. The end repair process removes the 3' overhangs, fills in the 5' overhangs, and adds an A overhang to the 3' end of both strands.

3.1.4 Adapter ligation

An adapter is a short, chemically synthesized double stranded DNA sequence that is attached to random DNA fragments. The adapters thereby provide a constant DNA sequence at the ends of the random DNA fragments that can be used to carry out some of the subsequent steps in the sequencing protocol. The design of the primer sequences is an example of truly elegant biotechnology.[3] The adapters serve several purposes in the Illumina protocol:

(i) The two adapters are designed to hybridize in a "Y-shape" to both

---

ends of the DNA fragments of the library in such a way that different specific sequences wind up at either end of the library fragment (c.f. Figure 3.5).

(ii) The adapters allow specific PCR enrichment of adapter-ligated DNA fragments.

(iii) They contain sequences that will later bind to complementary oligos on the flow cell.

(iv) The adapters can be designed to implement indexing or "barcoding" of samples.

There are many different Illumina adapter sequences, which differ depending on the type of experiment being performed (single end, paired-end, etc). Users should consult the Illumina documentation for details. Here, we will describe the TruSeq LT Adapter (P5) and the TruSeq Indexed Adapter (P7). We note that we present an idealized version of the Illumina sequencing procedure and sequences. There are many important differences for the various kits that have been used for Illumina sequencing. It is critical for bioinformaticians to be aware of these differences and they should consult the manuals and discuss the details with the sequencing facility. Nonetheless, the version of the sequencing protocol presented here will suffice to introduce the major ideas behind the process.

Figures 3.2 and 3.3 show the nucleotide sequences of the two primers and label the components. We will refer to them in the following text.

```
1. AATGATACGGCGACCACCGAGATCTACACTCTTTCCCTACACGACGCTCTTCCGATCT
2. AATGATACGGCGACCACCGAGA
3.                       TCTACACTCTTTCCCTACACGACGCTCTTCCGATCT
4.                                           GCTCTTCCGATCT
5. AATGATACGGCGACCACCGAGATCTACACTCTTTCCCTACACGA
```

Figure 3.2. Structure of the Illumina P5 universal adapter.
(1) The P5 adapter sequence; **(2)** the first 22 nucleotides are reverse complementary to the sequence of the oligonucleotides on the flow cell, and thus serve as the flow cell capture site; **(3)** the remaining 36 nucleotides of the P5 adapter correspond to the TruSeq Read 1 primer site. The sequencing primer for read 1 (the "forward" read) will anneal to this sequence; **(4)** the 12 nucleotides that are complementary to the 12 5' nucleotides of the P7 adapter plus the T overhang that is complementary to the A overhang of the library fragment; and **(5)** PCR primer 1.

```
1. GATCGGAAGAGCACACGTCTGAACTCCAGTCACNNNNNNATCTCGTATGCCGTCTTCTGCTTG
2. GATCGGAAGAGCACACGTCTGAACTCCAGTCAC
3.                                  NNNNNN
4.                                        ATCTCGTATGCCGTCTTCTGCTTG
5.                                        TCGTATGCCGTCTTCTGCTTG
6. GATCGGAAGAGC
```

Figure 3.3. Structure of the Illumina P7 indexed adapter. (1)
The P7 adapter sequence; **(2)** the index read primer for multiplexed samples binds to the sequence 5' to the six-nucleotide barcode; **(3)** the variable six-nucleotide barcode (shown as Ns) is used for demultiplexing samples; **(4)** the sequence that is reverse complementary to the final 24 nucleotides of the P7 primer is used as the PCR primer for the enrichment PCR; **(5)** this sequence is reverse complementary to oligonucleotides immobilized on the flow cell; and **(6)** the first 12 nucleotides are complementary to the 3' nucleotides of the P5 primer.

3.1.5 Adapter fork

To generate the adapter fork, both adapters are combined with a buffer and the solution is heated and then allowed to slowly cool down. The reverse complementary sequences of the universal adapter and the indexed adapter will hybridize to each other, forming a Y-shaped "forked" structure. The annealed adapter has a T-overhang, because there is one non-matching base at the 3' end of the universal adapter. This T is connected to the previous base (a C) by a phosphothiorate (Ψ) bond, which is resistant to exonuclease activity. This T is necessary to bind to the 'A' overhang of the ligated fragments, and thereby encourage correct ligation. The 5' phosphate group on the bottom strand enables double-stranded ligation (Figure 3.4).

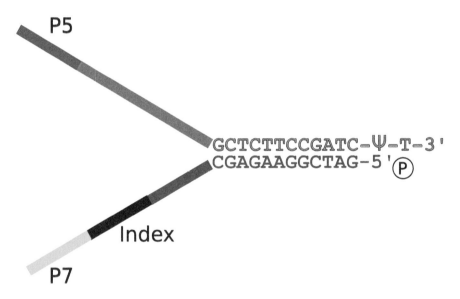

Figure 3.4. Y-shaped adapter. The Illumina forked or "Y-shaped" adapters are annealed to one another by a region of complementarity constituting 12 nucleotides, or about 20% of the length of each oligo.

3.1.6   Ligation reaction

The adapter forks are mixed with the end-repaired and A-tailed DNA fragments together with T4 DNA Ligase, which joins the DNA strands of the adapters and the DNA fragments by catalyzing the formation of a phosphodiester bond (Figure 3.5). In order to perform this reaction in an optimal fashion, the adapters and DNA fragments need to be combined with an appropriate molarity, usually an adapter:fragment ratio of about 10:1. If the adapter concentration is too high, the reaction will favor the formation of adapter dimers that can be difficult to remove and thus can dominate in the subsequent PCR amplification [160].

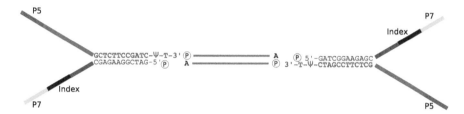

Figure 3.5. Illumina adapters and insert. Amplicon structure of a DNA fragment to which forked adapters have been ligated at both ends. Each strand consists of the P5 primer capture site (for attachment to the flow cell), a site for the read 1 sequencing primer, the insert (DNA fragment to be sequenced), a site for the read 2 sequencing primer, the index sequence (barcode), and the P7 capture site for attachment to the flowcell.

Following ligation, the products of the ligation reaction are purified and a size selection is performed to remove unligated adapters, as well as any adapters that may have ligated to one another ("adapter dimers"). For instance, for the HiSeq X, size selection is generally performed to achieve an insert size of around 350 bp for 2×150 bp reads. Originally, size selection was performed using agarose gels and cutting out a slice of gel at the desired size range. Agarose gel-based approaches have largely been replaced for most library preparation protocols with bead-based methods for size-selective binding of DNA fragments to the magnetic beads. This procedure for size selection has increased speed and throughput and lowered risks of cross-contamination.

3.1.7   Enrichment PCR

The purpose of enrichment PCR is to selectively enrich those DNA fragments that have adapter molecules on both ends and to amplify the total amount of DNA in the library. Enrichment PCR is performed with primers that anneal to the ends of the adapters (cf. line 5 in Figure 3.2 and line 4 in Figure 3.3). In general, depending on the amount of starting material the smallest possible number of PCR cycles is used to avoid skewing the representation of the library.

Prior to the first cycle of enrichment PCR, the ligated adapters still have "forks" on both ends. In the PCR, an initial denaturation step is performed. Note that each of the resulting single strands has a P5 sequence at its 5' end and a P7 sequence at its 3' end. Therefore, in the annealing step of the first cycle, only the PCR primer that is complementary to the P7 adapter can anneal. The extension step creates sequences that are reverse complementary to this template (Figure 3.6).

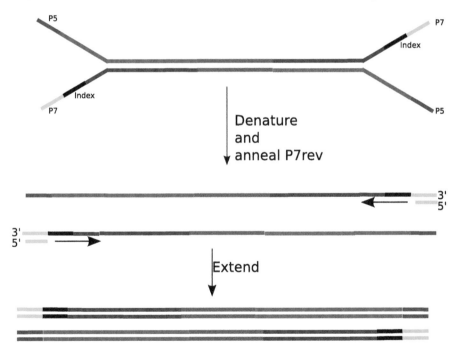

Figure 3.6. Enrichment PCR: First cycle. In the first cycle of enrichment PCR, only the P7 primer can anneal.

In the remaining cycles, the P5 primer (P5) can also bind. P5 is identical to the first 44 bp of the universal adapter (line 5 in Figure 3.2) and can thus bind to its reverse complement as generated by PCR (Figure 3.7).

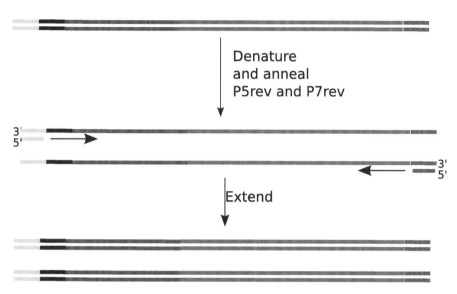

Figure 3.7. Enrichment PCR: Second and subsequent cycles. In the second and subsequent cycles of the enrichment PCR, both primers can bind.

After this step, the amount and purity of the DNA in the library is generally validated in order to quantify the amount and size distribution of the inserts. The most commonly used methods or devices for quantification are Nanodrop, Qubit, Bioanalyzer, and qPCR. At this point, samples with different barcodes (line 3 in Figure 3.3) can be pooled. Common artifacts that can arise during library PCR include duplicate reads as well as allele bias in amplification (see Chapter 6). Accurate quantification of the library is required in order to determine how much library to load onto the sequencer or to use in subsequent reactions such as exome enrichment.

3.2   FLOW-CELL PREPARATION

A flow cell is essentially a hollow glass slide. It can have one or multiple channels ("lanes"). For instance, the HiSeq flow cell has eight channels, and the MiSeq flow cell has only one. Each channel is randomly coated with a lawn of oligonucleotides that are complementary to the adapters. The Illumina HiSeq X Ten system introduced patterned flow cells with ordered wells designed to promote optimal spacing and uniform cluster sizes. Patterned flow cells have enabled increased cluster density compared to previous Illumina flow cells. The HiSeq 3000/4000 and the NovaSeq also utilize patterned flow cells.

Following the library preparation and cleanup steps, the PCR products are then denatured with NaOH. The single stranded DNA molecules are transfered to the flow cell. Here, they bind at random to oligos that are complementary to sequences in the adapters (cf. line 2 in Figure 3.2 and line 5 of Figure 3.3).

Then, an extension mix (buffer, dNTPs, DNA polymerase) is pumped into the channels of the flowcell, and the oligos on the flow cell are elongated using the ligated DNA fragments as a template, producing a double-stranded molecule, only one of whose strands is covalently attached to the flow cell. Formamide is added to denature these products, and the original fragment (which is not covalently attached to the flow cell) is washed away.

In the second step, the single-stranded DNA molecule bends and hybridizes to a second oligo on the flow cell surface. Extension is performed all the way back to the other oligo, forming a double-stranded "bridge", consisting of two covalently bound, single stranded (reverse complementary) copies of the original template (Figure 3.8).

Each cycle of the bridge amplification is performed at constant temperature (60°C). In most protocols, 35 cycles are performed:

1. Denaturation buffer (largely formamide) at 60°C is equivalent to the "denaturation step" in normal PCR.

2. Extension buffer (equivalent to the annealing step in normal PCR).

3. Extension mixture (equivalent to the extension step in normal PCR).

The result of the bridge amplification is a "colony" (cluster) of multiple bridges consisting of ∼1000 amplicons.

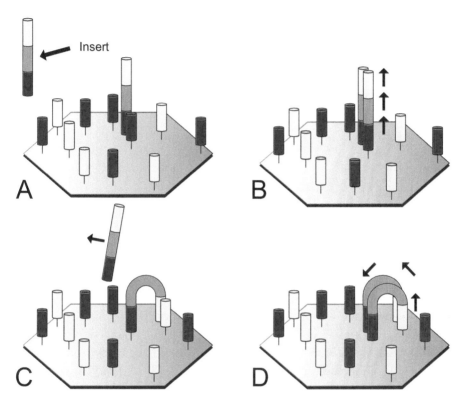

Figure 3.8. Bridge amplification. (A) A single molecule with insert and adapters attaches to a reverse complementary oligo on the flow cell surface; (B) the oligo on the flow cell is extended by the DNA polymerase using the attached molecule as a template. (C) Formamide is added to denature the amplification product, and the original molecule is washed away; in a subsequent step, the single-stranded fragment anneals to a second oligo on the flow cell surface, building a bridge (shown in the same panel for simplicity). (D) The second oligo is extended by DNA polymerase.

Following the colony PCR, each colony has amplicons in both orientations (i.e., `5'` to `3'` and `3'` to `5'`). However, the sequencing reaction can only be performed in one orientation at a time. Therefore, in order to sequence read 1, the reverse strand is cleaved and discarded, leaving a cluster with single-stranded forward strands. Now, the 3' ends of these strands as well as the reverse strand oligos on the flow cell are blocked to prevent unwanted DNA priming (Figure 3.9).

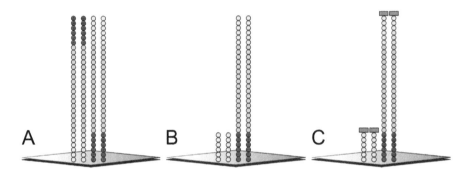

Figure 3.9. Flow cell preparation. (A) Following colony PCR, single-stranded amplicons representing both strands of the original template are attached to the flow cell. (B) Periodate is added to the flow cell and cleaves the extension product from the P5 (but not the P7) oligo. (C) Terminal transferase, a template-independent polymerase that catalyzes the addition of deoxynucleotides to the 3' hydroxyl terminus of DNA molecules, is added to block the 3' hydroxyl terminus of the oligos and the extension products so that they do not interfere with the subsequent sequencing reaction.

3.3   SEQUENCING BY SYNTHESIS

Sequencing by synthesis (SBS) is relatively simple to understand after all of the prep steps [276]. SBS is performed in cycles where one base at a time is incorporated. The sequencing primer corresponds to a region of the sequencing adapter. For single-end or for the first step of paired-end sequencing, the sequencing primer is shown in line 3 of Figure 3.2. The initial step of SBS involves the hybridization of the sequencing primer to the clusters produced by the bridge amplification. Where Sanger sequencing employs ddNTPs, Illumina's SBS procedure utilizes dNTPs with a 3'-OH blocking group ("terminator dNTPs"). In each cycle, all four terminator dNTPs are added simultaneously; each terminator dNTP has a reversible chemical block of the 3'-OH group as well as a unique fluorescent label. The terminator dNTPs can thus extend the sequencing primer by only one base according to the template. Following the one-base extension, the fluorescent dyes are excited by laser and the emission of signal from each cluster is recorded. The software of the sequencer can then identify which base was incorporated in each cluster by measuring the fluorescence intensity of the different fluorophores attached to the incorporated bases. Following this, the terminator dNTPs are unblocked at their 3'-OH group, and the next cycle is begun (Figure 3.10).

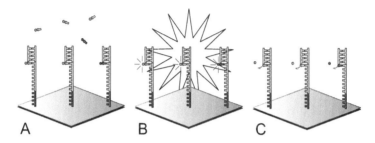

Figure 3.10. Sequencing by synthesis (SBS). SBS is performed one nucleotide at a time. **(A)** All four terminator dNTPs are simultaneously added to the flow cell. Each terminator dNTP has a reversible chemical block of the 3'-OH group and a unique fluorescent label. **(B)** Laser excitation of the fluorescent label and measurement of the fluorescence intensity for the 4 different fluorophores allows the incorporated nucleotide to be determined. **(C)** Unblocking of the terminator dNTPs, leaving behind a 3'-OH group, allows the next terminator dNTP to be incorporated into the growing strand in the next cycle.

One sequencing cycle is performed for each nucleotide in the read. Early Illumina devices produced sequence reads of 27–36 nucleotides in length [369]. Currently, the HiSeq devices routinely sequence up to 2×150 nucleotide reads, and the MiSeq yields 2×300 nucleotide reads.

Recent Illumina models (NextSeq and NovaSeq) use two-channel SBS that requires only two images to determine the four base calls; each cluster is imaged with red and green filters. A cluster observed only in red is interpreted as C, a cluster seen only with green as T, clusters observed in both red and green as A, and unlabeled clusters as G. Because only two channels are recorded instead of four, sequencing can be performed at a faster pace. Little published experience is available with regard to potential differences in quality with two- and four-channel SBS.

3.4 BASE CALLING AND QUALITY

A base-calling algorithm assigns (calls) bases and associated quality values for each position of the read. The quality of the sequenced bases is assessed by Illumina sequencers with a procedure called chastity filter. Illumina defines chastity as the ratio of the brightest base intensity divided by the sum of the brightest and second brightest base intensities. Intuitively, a perfect sequencing cycle would have a strong signal for one base and no signal for the other three bases [294].

$$\text{Chastity} = \frac{\text{Highest intensity}}{\text{Highest intensity} + \text{Next highest intensity}} \qquad (3.1)$$

By default, the purity of the signal from each cluster is examined over the first 25 cycles and calculated as the chastity value. At most one cycle may fall below the chastity threshold (default, 0.6), otherwise, the read will not pass the chastity filter (Figure 3.11). There are many potential reasons for poor quality signals. If there has been a PCR error in colony amplification such that a sizable proportion of the $\sim$1000 molecules in a cluster contains a different base at a certain position, then one may observe a strong signal for two bases—this is interpreted as a sign of poor quality. Phase errors occur if individual molecules do not incorporate a nucleotide in some cycle (e.g., because of incomplete removal of the 3' terminators, termed phasing) and then lag behind the other molecules, or if an individual molecule incorporates more than one nucleotide in a single cycle (e.g., because of incorporation of nucleotides without effective 3'-blocking, termed prephasing). This results in the loss of synchrony in the readout of the sequence copies of a cluster.

The proportion of sequences in each cluster which are affected by phasing and pre-phasing increases with cycle number, which is a major reason why the quality of Illumina reads tends to decline at high cycle numbers [196] (see Figure 6.4 for an example of a decline in quality toward the end of a read). Finally, overlapping colonies and impurities on the flow cell may interfere with the detection of the bases.

Illumina uses a proprietary (unpublished) algorithm to calculate the

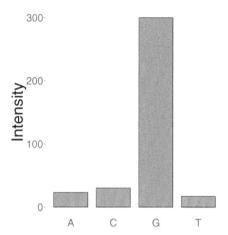

Figure 3.11. Illumina chastity. Example intensity values for one imaging cycle. The intensity values of the four bases are shown with arbitrary units. The G has the highest intensity (300) followed by the C (30), so the chastity is calculated as $300/(300 + 30) = 90.1\%$.

Phred score (see Chapter 5 for details on the Phred score), which is used to store an assessment for the error probability of a base call. The Phred score is computed based on intensity profiles (shifted purity: how much of signal is accounted for by the brightest channel?) and signal to noise ratios (signal overlap with the background: is the signal from the colony well delineated from the surrounding region of the flow cell?). The algorithm attempts to quantify the chastity of the strongest base signal, whether a signal for a given base call is much stronger than that of nearby bases, whether a spot representing a colony gets suspiciously dim during the course of sequencing (intensity decay), and whether the signal in the preceding and following cycles appears clean or not.[4]

3.5 SINGLE-END VERSUS PAIRED-END READS

As we have seen, the Illumina sequencing protocol involves the fragmentation of DNA into small fragments followed by the ligation of two different adapters on each end. Up to now, we have been implicitly talking about single-end (SE) sequencing, where we only sequence one end of the DNA fragment. To perform paired-end (PE) sequencing, the sequencing product from the first read is removed by denaturation

---

[4]Quality Predictors in RTA v1.12, Illumina technical note.

and washing steps. If mutiplexing of libraries is used, an index read primer is annealed and sequencing is performed in the same way as in a single-read run for a number of cycles corresponding to the length of the index. After this, the sequencing product is again removed by denaturation and washing steps.

Then the P7 strands (that are attached to the flow cell) are used to regenerate the complementary strand. An enzyme is used to cleave the extension product from the P7 oligo analogous to the removal of the P5 extension product (Figure 3.9). After the resynthesis of the complementary strand and washing away the previous template for read 1 and the index read, the read 2 sequencing primer is annealed, enabling sequencing of the other end of the insert in the same way as sequencing was performed for read 1 (Figure 3.12).

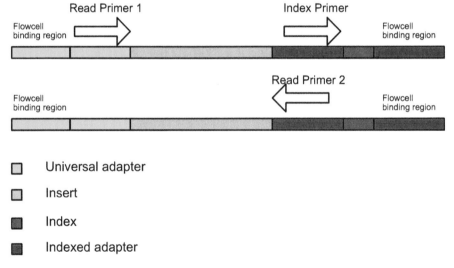

Figure 3.12. Paired-end sequencing. The figure shows the structure of a DNA fragment ("insert") that has been ligated to the P7 and P5 adapters. The sequencing primers are designed to be compatible with the adapters. Read primer 1 is used to sequence the insert initially. If multiplexing is used, then a separate primer is used to sequence the molecular barcode which is located in the P7 indexed primer. For paired-end sequencing, read primer 2 is used to sequence (the reverse complement of) the insert from the other direction.

It is useful to consider the relationship of the two reads to the original fragment. The word "insert" refers to the DNA fragment that is "inserted" between the two adapters. The two sequence reads cover the 5' and the 3' ends of the insert, but do not necessarily cover the entire fragment. If the combined length of read 1 and read 2 is less than the length of the insert, there will be an unknown gap between them (Figure 3.13A). If the combined length of the reads is longer than the length of the insert, then the reads will overlap. This allows the sequences to be "stitched" together into a longer single sequence, whereby the quality values of the overlapping bases are assessed twice, and the confidence in these base calls may be higher than with non-overlapping reads (Figure 3.13B). If the insert length is even smaller than the read length, then not only do the reads overlap, but they run over into the adapter sequences (Figure 3.13C). This situation is known as adapter contamination and is covered in more detail in Chapter 6.

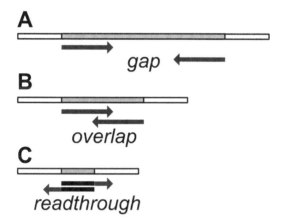

Figure 3.13. Insert size with paired-end sequencing. Depending on the relative size of the insert and the sequence reads, there may be a gap between read 1 and read 2 (**A**), the two sequence reads may overlap (**B**), or they may not only overlap but spill over into the adapter sequences (**C**).

3.5.1 Library size

The term library size is used to denote the average length of the fragments being sequenced. For Illumina paired-end, this could be anything from ~150 bp to 1200 bp, depending on what chemistry and library preparation method were used. The optimal library size depends on the application. For instance, if one is looking for single-nucleotide and other small variants in coding exons and is performing 2×100bp paired-end sequencing, then a 500 bp library size would probably be fine. But for 2×300bp paired-end sequencing, the reads would overlap, which may or may not be desired (cf. Figure 3.13). Although the library size (average fragment length) can be estimated with gel electrophoresis, an exact determination of the fragment lengths can be performed only by aligning all paired reads to the reference genome and measuring the distance from the ends of both reads.[5] This can be done using Picard Tools,[6] which is introduced in Chapter 8. The following command compares an aligned exome file (NIST7035_dedup.bam) to the reference genome (hs38DH.fa) and outputs a summary of the alignment metrics to a new file (insertmetrics.out).

```
$ java -jar picard.jar \
    CollectInsertSizeMetrics \
    I=NIST7035_dedup.bam \
    O=insertmetrics.out \
    H=insertSizeHistogram.pdf
```

This analysis showed a mean insert size of 188 nucleotides for an exome data set that will be described in detail in later chapters (Figure 3.14).

3.5.2 Major advantages of paired-end sequencing

Obviously, paired-end sequencing produces twice the total number of reads as an analogous single-end sequencing experiment (e.g., sequencing one flow-cell lane at 2×100 nt will generate twice as many sequenced nucleotides as sequencing the lane at 1×100 nt). The cost and effort involved in acquiring the sample and preparing the library are the same, although the sequencing itself costs more and takes longer. In most cases, however, paired-end versus single-read sequencing does not double the costs so the price per base is usually lower. Alignment of read pairs can be more accurate than alignment of single reads. For instance, if one end of the fragment is located in a highly repetitive sequence, it could be mapped to hundreds or more locations in the

---

[5]Assuming that the fragment does not originate from a region which harbors a structural variant, such as a deletion, an insertion or a translocation. These cases will be discussed in Chapter 19.

[6]https://broadinstitute.github.io/picard/

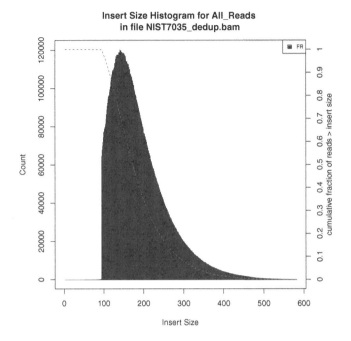

Figure 3.14. Insert size histogram. Insert size distribution of the NA12878 exome that will be described in Chapter 4. The figure was generated by Picard's CollectInsertSizeMetrics module.

genome, and would usually be discarded by aligners. If paired-end sequencing was performed and the other end of the fragment maps to a unique sequence located at an appropriate distance to an instance of the repetitive sequence, then both ends of the insert can be mapped (Figure 3.15).

Paired-end reads can be exploited for the detection of some types of structural variation, as we shall see in Chapter 19. We will see in Chapter 6 that PCR duplicates are a common artifact in Illumina data. Generally, duplicates are removed if the 5' end of the read has an identical genomic position. With paired-end reads, the 5' ends of both reads are examined to identify duplicates.

3.6 MOLECULAR BARCODING

As the output of NGS machines such as Illumina's HiSeq series continues to increase, a single flow cell may generate sufficient sequence

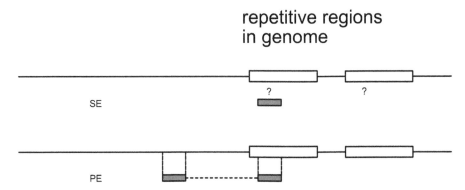

Figure 3.15. Mapping repetitive reads with PE sequences. Mapping reads in repetitive sequences can be made easier by paired-end sequencing because one read can act as an anchor for the other. In this example, the single-end read originates from a repetitive sequence that occurs multiple times in the genome (two occurrences are shown in this figure). The read will receive a mapping quality of zero, and depending on the mapping algorithm will either be discarded or will be assigned at random to one of the sequences (symbolized as question marks). If the same read is part of a read pair where the second read can be uniquely mapped at an appropriate distance from a particular instance of the repetitive sequence, then the first read can also be unambiguously mapped.

output for a large number of experiments. Multiplexing is a way of marking individual samples with a unique barcode, thereby allowing them to be mixed, sequenced on the same lane, and separated ("deconvoluted") following sequencing. In theory, there are n^4 different barcodes if the barcode length is n, but it is critical that the barcodes can be unambiguously differentiated from one another even in the case of sequencing errors so that in practice, sets of barcodes are used each of which are separated from all other barcodes by more than one difference [462]. The indexed adapter contains a 6 bp "barcode" sequence (line 3 in Figure 3.3) that can be used to combine several samples in the same lane but still be able to identify the provenance of the reads following sequencing. Several barcodes are shown in Table 3.1. With a few exceptions, the barcodes are separated from one another by a Hamming distance[7] of at least 3.

---

[7]The number of positions at which two sequences (in general, two strings) differ.

Table 3.1. Sequences of selected TruSeq barcodes.

Sequence	TruSeq name
ATCACG	TSBC01
CGATGT	TSBC02
TTAGGC	TSBC03
TGACCA	TSBC04
ACAGTG	TSBC05
. . .	. . .

There are two main approaches to barcoding. So-called inline barcodes are adjacent to the sample DNA and read from the same sequencing primer as part of the sequence read. In contrast, multiplex barcodes are read by a separate sequencing primer and reactions. The multiplex barcode is not adjacent to the insert sequence, but instead is located in the single-stranded portion of the indexed adapter. With multiplex barcoding, a separate primer is used to sequence the barcode (see Figures 3.12 and 3.16).

The index read is generated automatically by the Illumina sequencer. First, read 1 of the insert is generated using the read 1 sequencing primer. The read 1 sequencing product is removed, and the index sequencing primer is annealed to the same strand. Then, sequencing is carried out to determine the 6 bp sequence of the index. If single-end sequencing is being performed, then the experiment ends here. If paired-end sequencing is being performed, then the original template strand is used to regenerate the complementary strand in a second round of colony PCR after sequencing of the index read. Then, the read 2 sequencing primer is annealed, and an equal number of sequencing cycles are performed as for read 1. The Illumina software identifies the index sequence from each cluster, and outputs separate files for each index in each lane.

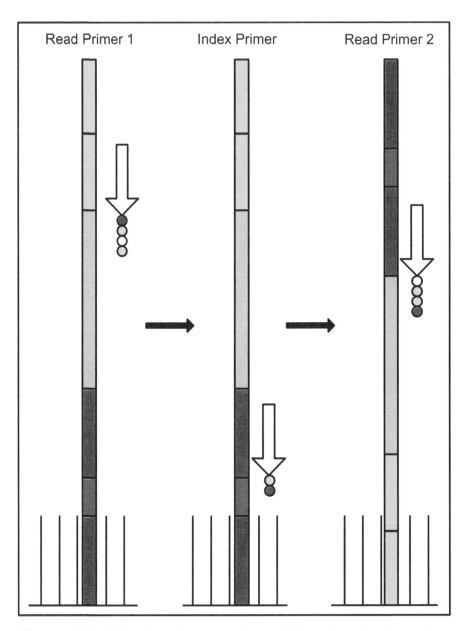

Figure 3.16. Paired-end sequencing with molecular barcodes.
Paired-end sequencing with molecular indexing involves three separate
sequencing primers and three sequencing reads.

3.7 MATE-PAIR SEQUENCING

There are two main types of sequencing libraries for the Illumina technology, fragment libraries or mate-pair libraries. Up to now, we have described the fragment library protocol, whereby adapters are ligated directly to DNA fragments to create a sequencing library; if the fragments are sequenced from both ends, we call it paired-end sequencing.

Mate-pair sequencing utilizes a different library preparation methodology that allows the pairs to be located much further apart than with paired-end sequencing. The insert size with mate-pair libraries is typically 2–5 kilobases or longer.

The mate pair protocol is summarized in Figure 3.17. In the first step the DNA is either fragmented and end-repaired using biotinylated nucleotides or the fragmentation is performed using a transposase enzyme that integrates a biotinylated adapter sequence into the DNA (biotin is a commonly used labeling reagent that will subsequently allow the DNA fragments to be purified using streptavidin beads). Second, each fragment is circularized by intramolecular ligation of their blunt ends. This means that the original ends of the fragment become located directly adjacent to one another and are identified in the circle by the biotin label. In the third step, the circularized DNA is fragmented by sonication or nebulization, and the biotinylated fragments are purified by affinity capture. Sequencing adapters (A1 and A2) are ligated to the ends of the captured fragments. Fourth, the fragments are hybridized to a flow cell, in which they are bridge amplified. The first sequence read is obtained with adapter A2 bound to the flow cell. Fifth, the complementary strand is synthesized and linearized with adapter A1 bound to the flow cell, and the second sequence read is obtained. Finally, mapping of the reads will show that the two sequence reads (arrows) will be directed outwards from the original fragment.

Mate-pair library preparation is significantly more expensive and time consuming than paired-end library preparation and requires considerably larger quantities of input material (Table 3.2). On the other hand, mate-pair sequencing has the advantage of spanning much larger fragment sizes and thus, allowing a better detection of structural variants, and improved coverage of repetitive regions.

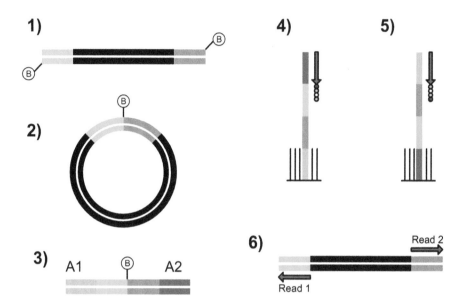

Figure 3.17. Mate-pair sequencing. Steps 1–6 are described in the text.

Table 3.2. Comparison of paired-end and mate-pair sequencing.

	Paired-end	Mate pair
Insert size	≈ 250 bp	2–20 kb
DNA	$\leq 1\ \mu$g	$\sim 4\ \mu$g
Lab work	Easier	Harder
Costs	Less	More

3.8 EXERCISES

Exercise 1

If paired-end sequencing is performed with 2×250 nt reads, is the insert size 500 nt?

Exercise 2

Explain why the identification of PCR duplicates is more reliable with paired-end sequencing than with single-end sequencing.

Exercise 3

The Hamming distance between two strings of equal length is the minimum number of substitutions to change one string into another. For instance, the Hamming distance between the Index 1 (i7) adapter sequences A704 (ACAAACGG) and A705 (ACCCAGCA) is five.

```
A704:   ACAAACGG
A705:   ACCCAGCA
```

Write a script that takes as input a list of molecular barcode sequences and outputs the pair of barcodes with the minimum Hamming distance. The barcode sequences can be obtained from the online Illumina documentation.

See hints and answers on page 499.

CHAPTER 4

Obtaining WES/WGS Data for This Book

T O take full advantage of this book, it is highly recommended that readers analyze real data with the tools and scripts that we present. However, medical whole-exome sequencing (WES) and whole-genome sequencing (WGS) data is identifiable, and it is necessary to restrict sharing of medical WES/WGS data to protect the privacy of research participants [192]. Thus, most medical WES/WGS data is not available for download without restrictions.

Therefore, we have illustrated many of the examples in this book with publicly available datasets, including the Corpasome, data from the Genome in a Bottle Consortium, and data from a brain cancer specimen. Except the tumor data, these datasets were not generated to address medical questions. Readers should feel free to analyze other datasets while working through the book, but please be conscious of restrictions on use of non-public data. For instance, hospitals forbid diagnostic data to be taken out of their firewall without specific permission to do so.

4.1 THE CORPASOME

The Corpasome refers to genomic data derived from the Corpas family. Manuel Corpas, who works as a genome bioinformatician, engaged on a personal journey to explore his own genome as well as that of several family members. The family hails from Málaga, Spain, and investigated exomes and some other data by means of direct-to-consumer (DTC) genomic testing. The Corpas family decided to make this information publicly available for the use of the greater scientific and genomic community [90, 91]. One of the authors of this book had the pleasure of collaborating with Manuel Corpas on the analysis of DTC genomic data from a family quartet [92]. The story of the Corpasome and the description of the reaction of the Corpas family to the results of exome sequencing makes for interesting reading and is highly recommended.

4.1.1 Downloading the Corpasome data

We will download the exome from Manuel Corpas himself that was published in the 2012 F1000Res article entitled *Low budget analysis of Direct-To-Consumer genomic testing familial data.*[1] To get the data, readers can navigate to the F1000Res article and download the data directly from the table called *Son exome files* in the Methods section of the paper. There are seven files available for download, and it is easiest to download all files at once (clicking on the `Download` button will download a zip archive called `92584.zip`). Unzipping the file (`$ unzip 92584.zip`) will generate six files:

- `Sons_exome_fastq_file_1.fq` (1.8G)

- `Sons_exome_fastq_file_2.fq` (1.8G)

- `Sons_exome_fastq_file_3.fq` (1.7G)

- `Sons_exome_fastq_file_4.fq` (1.7G)

- `Sons_Aligned_Bam_File.bam` (4.1G)

- `Sons_Index_of_BAM_file.bai` (5.8M)

- `Sons_VCF_file.vcf` (8.8M)

For the exercises in subsequent chapters in this book, we will use the FASTQ files, which contain a total of 32,116,828 reads (see Chapter 5).

4.2 THE GENOME IN A BOTTLE CONSORTIUM

The Genome in a Bottle Consortium (GIAB) is a public-private-academic consortium that aims to develop the technical infrastructure to enable WGS to become a tool for clinical practice. One of the means

---

[1]The article is freely available and can be found via PubMed at `https://www.ncbi.nlm.nih.gov/pubmed/24627758`.

toward achieving that goal is to provide reference data for use in analytical validation. For this purpose, the GIAB has released a number of datasets that are of very high quality. Information about these datasets is available at the GIAB Website[2] and the GIAB GitHub page.[3] As we will see throughout this book, the analysis of WES/WGS data, and in general of high-through NGS data, remains a difficult undertaking with many sources of error. Not only do the individual technological platforms such as Illumina, Life Technology, and Complete Genomics display different types of bias [365], but even the bioinformatics analysis pipelines do not entirely agree on the set of called variants from the same sample [317]. GIAB therefore deeply and multiply sequenced the NA12878 individual from the HapMap/1000Genomes project. Overall 11 WGS and 3 WES datasets were derived from five sequencing platforms, producing benchmark genotype calls that can be used to asses the accuracy of methods for variant calling and to understand sources of error and bias in bioinformatic algorithms [492].

4.2.1 Downloading the GIAB data

For this book, we will use data from an exome generated in this project that is available at the GIAB FTP site.[4] When you access the FTP site to download the data, navigate to the directory entitled `Garvan_NA12878_HG001_HiSeq_Exome`. The README file to be found in that directory explains that the exomes were sequenced at the Garvan Institute of Medical Research in Australia, using an Illumina HiSeq2500 with a Nextera Rapid Capture Exome and Expanded Exome kit. We will use the following two files for the exercises in subsequent chapters in this book:

- `NIST7035_TAAGGCGA_L001_R1_001.fastq.gz` (1.8G)

- `NIST7035_TAAGGCGA_L001_R2_001.fastq.gz` (1.9G)

Note that the complete exome data has three additional paired-end runs. We will not use them in this book in order to simplify the exercises, but if you would like to do so, additionally download the following files.

`NIST7035_TAAGGCGA_L002_R1_001.fastq.gz`

---

[2]http://www.genomeinabottle.org/ or http://jimb.stanford.edu/giab/
[3]https://github.com/genome-in-a-bottle
[4]ftp://ftp-trace.ncbi.nih.gov/giab/ftp/data/NA12878/

```
NIST7035_TAAGGCGA_L002_R2_001.fastq.gz
NIST7086_CGTACTAG_L001_R1_001.fastq.gz
NIST7086_CGTACTAG_L001_R2_001.fastq.gz
NIST7086_CGTACTAG_L002_R1_001.fastq.gz
NIST7086_CGTACTAG_L002_R2_001.fastq.gz
```

There are several ways of combining the files including concatenating the FASTQ files prior to the analysis, or performing the analysis separately on each set of paired-end sequences and then merging the resulting BAM files with SAMtools or BamTools.[5] None of the other analysis steps need to be changed.

4.3 NCBI'S SEQUENCE READ ARCHIVE AND THE EUROPEAN GENOME-PHENOME ARCHIVE

There are many other sources of publicly available WES and WGS data. The Sequence Read Archive (SRA) of the National Center for Biotechnology Information (NCBI) [206] and the European Genome-phenome Archive (EGA) [237] have large collections of data that can be downloaded for analysis.

Some of the sequence datasets in SRA require approval to download, but some cancer-related datasets are available without further approval. In Chapter 30, we will use one such dataset with a tumor-normal sample pair from an individual with an IDH1-mutated glioma, a severe brain cancer. Users may wish to substitute other datasets available at the SRA:

```
https://www.ncbi.nlm.nih.gov/sra/
```

The glioma sample pair was published by Park *et al.* [322] and can be found under the accession numbers SRX894623 (the tumor sample) and SRX894593 (the matched control sample). The samples have two sequencing runs each which might be merged but the sequencing depths of the single runs are sufficient for the exercises in this book. Hence, we only use the run SRR1825651 for the tumor and the run SRR1823052 for the normal control.

First, download and install the NCBI SRA Toolkit[6] that offers dedicated applications for working with SRA data. The Ubuntu

---

[5]https://github.com/pezmaster31/bamtools

[6]Some Linux distributions, including Debian, include it in their standard packages, otherwise it is available at: https://trace.ncbi.nlm.nih.gov/Traces/sra/sra.cgi?view=software

Linux 64 bit architecture version is a compressed tar archive (`sratoolkit.2.8.1-ubuntu64.tar.gz`) that contains a bin subdirectory with numerous individual applications, including one called `fastq-dump` that we will use to download the data.

The glioma data can then be downloaded with the SRA toolkit's `fastq-dump` utility as follows:

```
$ fastq-dump --split-files --gzip --skip-technical \
    --read-filter pass --origfmt --readids --clip \
    SRR1825651
```

This will download the reads for the tumor. The data of the control sample can be downloaded analogously by substituting the run ID SRR1823052.

The meaning of the `fastq-dump` options is

`--split-files` separates paired-end reads into left and right ends, and puts the forward and reverse reads (read 1 and read 2) in two separate FASTQ files. This format is required by alignment software such as BWA.

`--gzip` Compress the output using `gzip`.

`--skip-technical` Include only biological reads in case the "Illumina multiplex library construction protocol" was used for the sequencing. In this case the data would include undesired application reads and technical reads.

`--read-filter pass` Do not include reads which are useless (mostly composed only of N's).

`--origfmt` Causes the Definition line (first line of the four FASTQ lines of a read) to contain only the original sequence name.

`--readids` Append the read number after the spot ID as 'accession.spot.readid' on the Definition line. For paired-end reads, this appends the ID "1" onto the IDs of the reads from one file and the ID "2" onto the IDs of the reads from the second file, which is the format expected by downstream software such as BWA.

`--clip` Make sure tag sequences used for whole genome amplification and other similar tags are removed. Include this option, if you're not sure whether the sequences you are downloading might contain such tags or not.

`--outdir <path>` Optional output directory, default is the current working directory.

We have found that downloading files with the `fastq-dump` utility is not always reliable. An alternative is to use `wget` to retrieve the SRA files and then use `fastq-dump` locally to extract the reads in FASTQ format. In this case, the commands would be

```
$ wget https://ftp-trace.ncbi.nlm.nih.gov/sra/sra-instant/reads/\
    ByRun/sra/SRR/SRR182/SRR1825651/SRR1825651.sra
```

```
$ wget https://ftp-trace.ncbi.nlm.nih.gov/sra/sra-instant/reads/\
    ByRun/sra/SRR/SRR182/SRR1823052/SRR1823052.sra
```

Following the download, create FASTQ files from the SRA files with `fastq-dump` using the same options but specifying the local file name (e.g., `SRR1825651.sra`) instead of the run ID (e.g., `SRR1825651`):

```
$ fastq-dump --split-files --gzip --skip-technical \
    --read-filter pass --origfmt --readids --clip \
    SRR1825651.sra
```

These commands, which may take hours to complete, will create paired FASTQ files for the `SRR1825651` (tumor) and `SRR1823052` (normal) sequencing runs. If the pipeline was successful, you should see the following four files.

`SRR1823052_pass_1.fastq.gz` (4.3G)
`SRR1823052_pass_2.fastq.gz` (4.5G)
`SRR1825651_pass_1.fastq.gz` (4.2G)
`SRR1825651_pass_2.fastq.gz` (4.3G)

The SRA files (`SRR1825651.sra` and `SRR1823052.sra`) may be deleted following the generation of the FASTQ files.

4.4 1000 GENOMES EXOME DATA

For the investigation of *de novo* variants in Chapter 18 we downloaded WES data for a trio (mother, father, and daughter). The sequences were generated as a part of the 1000 Genomes project [111]. The data can be downloaded at the 1000G FTP site using `wget`:

```
$ wget --progress=bar \
    ftp://ftp.1000genomes.ebi.ac.uk/vol1/ftp/\
```

```
data_collections/1000_genomes_project/\
data/CEU/NA12891/exome_alignment/\
NA12891.alt_bwamem_GRCh38DH.20150826.CEU.exome.cram
```

The sequence data is compressed in the CRAM format that is described in Chapter 9. It can be converted to BAM format using the following command.

```
$ samtools view -bh \
    NA12891.alt_bwamem_GRCh38DH.20150826.CEU.exome.cram \
    -o NA12891.bam
```

The URL and the SAMtools command for the other two datasets are analogous. The father is NA12891, the mother NA12892, and the daughter is NA12878.

4.5   1000 GENOMES WGS DATA

We use a family trio from the 1000 Genomes Project to demonstrate *de novo* variant calling as well as the CNV callers CNVnator and DELLY. This is the Illumina platinum high coverage WGS dataset with the SRA Project Accession PRJNA186949. This project contains samples from the same three individuals as above (daughter: NA12878–SRR622457; father: NA12891–SRR622458; mother: NA12892–SRR622458). Download the FASTQ files with the following commands:

```
$ fastq-dump --split-files --gzip SRR622457
$ fastq-dump --split-files --gzip SRR622458
$ fastq-dump --split-files --gzip SRR622459
```

This may take several hours but will finally generate six FASTQ files, two for each sample. In order to perform the analysis with DELLY and CNVnator described in Chapter 19, follow the description in Chapter 8 on how to perform the alignment with BWA-MEM and postprocess the BAM files. You should wind up with the following files.

```
NA12878_dedup.bam
NA12891_dedup.bam
NA12892_dedup.bam
```

II

Raw Data Processing

CHAPTER 5

FASTQ Format

I T is extremely important to understand the main file formats used for the computational analysis of WES/WGS data, including mainly FASTQ, SAM/BAM, and VCF.[1] This book will devote chapters to each of these formats.

5.1 BCL FILES

Illumina sequencers perform image analysis and base calling, producing binary base call (*.bcl) files as their primary output. Newer Illumina models such as the HiSeq produce compressed BCL files (*.bcl.gz). Illumina's CASAVA (**C**onsensus **A**ssessment of **S**equence **A**nd **VA**riation) software was the major component of Illumina's sequencing analysis software platform, but was recently replaced by the BaseSpace platform. We will not attempt to provide a detailed description of the Illumina-specific software, but will briefly describe how it transforms the raw image data into FASTQ files.

The instrument control software of the Illumina sequencer performs image analysis, investigating the raw images to locate sequence clusters and outputting the sequencing intensity, the X and Y positions of the cluster on the flowcell tile, and an estimation of the noise of each cluster. Base calling, which is performed by the instrument control software's *Real Time Analysis* (RTA), uses the cluster intensities to output the sequence of bases read from each cluster together with a confidence level for each base. Additionally, the base calling procedure indicates whether a read passes filtering. The *.bcl files are writ-

---

[1] Additionally, pileup format will be introduced in Section 30.3.

ten by the instrument control software to the **BaseCalls** directory.[2] BCL is a vendor-specific format that is not used by other sequencing devices, and we will not go into more detail here. Most users of the Illumina sequencing technology use the BCL to FASTQ conversion software (**bcl2fastq** or more recently **bcl2fastq2**), which converts ***.bcl** files into ***.fastq.gz** files (compressed FASTQ files), and demultiplexes multiplexed samples (i.e., produces separate **fastq.gz** files for each barcoded sample in each lane).

5.2 FASTQ FILES

FASTQ is a file format for sharing sequencing read data combining both the sequence and an associated per base quality score. Its name derives from the venerable FASTA format, which is a text-based format for representing DNA and amino acid sequences that was originally published together with a software package for nucleic acid and protein alignment in 1985 [261]. The FASTA format is one of the most ubiquitous of all formats in bioinformatics. The first line starts with a > (greater-than) symbol, directly followed by a name or an unique sequence identifier, optionally followed by a description of the sequence. The following lines contain the sequence (usually with 70 or 80 characters to a line). One file can contain multiple sequence entries, but each must begin with a > line. For instance, the FASTA file for the Homo sapiens fibrillin 1 (*FBN1*) messenger RNA starts off as

```
>NM_000138.4 Homo sapiens fibrillin 1 (FBN1), mRNA
AGTATTTCTCTCGCGAGAAACCGCTGCGCGGACGATACTTGAAGAGGTGGGGAAAGGAGGGGGCTGCGGG
AGCCGCGGCAGAGACTGTGGGTGCCACAAGCGGACAGGAGCCACAGCTGGGACAGCTGCGAGCGGAGCCG
AGCAGTGGCTGTAGCGGCCACGACTGGGAGCAGCCGCCGCCGCCTCCTCGGGAGTCGGAGCCGCCGCTTC
(additional 165 lines)
```

The FASTQ format is an extension of the FASTA format that additionally stores a numeric quality score associated with each nucleotide in a sequence. Each read in the file is represented by four lines with the following information (without any length limit):

```
@<title and optional description>
```

---

[2]Details about the BCL file format, the sample sheet that is used to tell the Illumina software how to assign reads to samples and samples to projects, and information about barcoding can be found in the **bcl2fastq** User Guide and other documentation available at the Illumina Website (**https://www.illumina.com/**).

```
<sequence line>
+<optional repeat of title line>
<quality line>
```

The @ line comprises a free format field used primarily for the title, or identifier, of the reported sequence. Illumina has a typical naming convention for the title line which we will explain below. The second line contains the called bases, i.e., the sequence of the read. The + line follows. Although early versions of the FASTQ format included a repeat of the title line on this line, it has become standard for the line to comprise just the plus (+) sign, which substantially reduces the file size. Finally, the fourth line contains the ASCII-encoded Phred quality scores and must be of the same length as the sequence line. Thus each of the bases reported in the second line has a quality score reported in the corresponding position of the fourth line. Some previous version of FASTQ used different schemes for encoding quality values, but the community has settled on what has come to be known as the Sanger/Illumina 1.8+ format (Phred+33), and this is what we will describe here.[3]

An article on the FASTQ format published in 2010 provides a fuller explanation of the history of the quality score encoding [81].[4] An example of a read in FASTQ format is shown in Figure 5.1.

5.2.1 The Phred score

The Phred quality score is defined as

$$\boxed{Q = -10 \log_{10} p} \tag{5.1}$$

where p is the probability that the corresponding base call is **wrong** and Q is the Phred score (rounded to the closest integer value). The Phred quality score is thus a simple transformation of the error probability that represents a simple but reasonably space-efficient encoding (Table 5.1).

---

[3]The previous Illumina encoding reported quality scores using an offset of 64 instead of 33.

[4]Information on older encodings can also be found at: `https://en.wikipedia.org/wiki/FASTQ_format#Encoding`

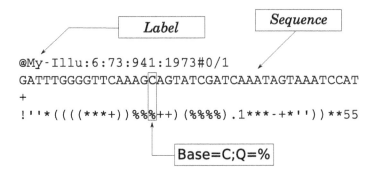

Figure 5.1. FASTQ. The FASTQ format has four lines per read. The first line starts with '@', followed by the label, or identifier. The second line contains the actual sequence of the read. The third line starts with '+', and usually has no other information (although a repeat of the label is allowed according to the format specifications). The fourth line contains the Phred quality (Q) scores encoded as ASCII characters. In this example, a box is drawn around a C base with the quality code of "%", which has the ASCII code 37, and thus represents a Phred score of 37-33=4, corresponding to an estimated error probability of 0.398. Thus, the C is a low-quality base call.

5.2.2   ASCII encoding of Phred scores

The transformation described in Equation 5.1 converts a probability to an integer value between 0 and 93. The values are not actually stored as a one- or two-digit number in the FASTQ file, but instead as a single character (char), again resulting in a substantial reduction of file size. Nonetheless, Phred qualities from 0 to 93 can represent a very broad range of error probabilities, ranging from 1.0 (100% error probability, or simply a wrong base) through to $10^{-9.3}$, which corresponds to an extremely accurate base call.

To store the Phred scores as characters, the quality scores are converted to ASCII characters. ASCII (American Standard Code for Information Interchange), is an early character encoding standard for representing characters in computers and other devices that was first published in 1963. ASCII codes 0–31 are unprintable—for instance, ASCII code 007 corresponds to a control code originally sent to ring an electromechanical bell on older systems, or to play a system warning sound on some computers. The first non-space printable character is ASCII code 33, and the last printable ASCII code is 126. Thus, FASTQ

Table 5.1. Base Quality and Accuracy.

Q_{Phred}	p	Accuracy
0	1	0%
10	10^{-1}	90%
20	10^{-2}	99%
30	10^{-3}	99.9%
40	10^{-4}	99.99%
50	10^{-5}	99.999%
60	10^{-6}	99.9999%
70	10^{-7}	99.99999%
80	10^{-8}	99.999999%
90	10^{-9}	99.9999999%
93	$10^{-9.3}$	99.99999995%

Note: Base quality Phred scores and their associated error probability (*p*) and base accuracy. A selection of values from the lowest (0) to the highest (93) Phred score representable in a FASTQ file is shown.

files use ASCII codes 33–126 to encode Phred qualities from 0 to 93 (Table 5.2).

5.2.3   Illumina FASTQ file naming scheme

Illumina uses a standard naming scheme for the FASTQ files. It is useful to understand how the scheme is structured. The general scheme for files generated by `bcl2fastq` is

```
<sample_name>_<barcode_sequence>_L<lane>\
  _R<read_number>_<set_number>.fastq.gz
```

Let's take a closer look at the following GIAB file from Chapter 4:

```
NIST7035_TAAGGCGA_L001_R1_001.fastq.gz
```

The components of this name are:

- `sample_name`: NIST7035. This is the sample name provided in the sample sheet for the sequencing run.

- `barcode_sequence`: TAAGGCGA. This is the nucleotide sequence of the molecular barcode used to tag the sample for multiplexing.

Table 5.2. Base Quality and ASCII Encoding.

ASCII character	Decimal value	Phred score
!	33	0
"	34	1
#	35	2
$	36	3
⋮	⋮	⋮
A	65	22
B	66	23
⋮	⋮	⋮
x	120	87
y	121	88
z	122	89
{	123	90
|	124	91
}	125	92
~	126	93

Note: Examples for the ASCII encoding of Phred scores. The Phred score can be determined by subtracting 33 from the ASCII character's decimal value.

- **lane**: 001. The lane number (1–8).

- **read_number**: 1. The read number for paired-end reads. R1 means read 1, and for a paired-end sequencing run, there is an additional file with R2 (read 2) whose name otherwise exactly matches the filename for read 1.

- **set_number**: 001. FASTQ files are set to have a maximum file size by the `--fastq-cluster-count` command line option of the `configureBclToFastq.pl` script that is part of the Illumina CASAVA software suite. If there is more data, then the data are divided into separate FASTQ files with the corresponding file size (specify 0 to ensure creation of a single FASTQ file). The different files corresponding to the same sample/barcode/lane are distinguished by the 0-padded 3-digit set number.

Certain Illumina sequencers use different FASTQ file naming schemes. Consult the Illumina documentation for details.

Note that for paired-end runs, the matched FASTQ files must have exactly the same number of reads and the reads must have the same order in both files. This can be checked using the UNIX commands zcat and wc. The cat command reads data from text files and outputs their contents to the command line interface, and the zcat command does the same with gzip-compressed files. The wc command counts words, lines, and characters in text files. By combining the commands as follows we see that each of the two files we downloaded has the same number of lines.

```
$ zcat NIST7035_TAAGGCGA_L001_R1_001.fastq.gz | wc -l
80812008
$ zcat NIST7035_TAAGGCGA_L001_R2_001.fastq.gz | wc -l
80812008
```

We can now divide the total number of lines by four to get the total number of reads (because each read takes up a total of four lines in the FASTQ file). Note that we cannot simply grep for lines that begin with @, because the ASCII symbol that corresponds to a Phred score of 31 is also @, and thus quality lines may also start with this character.

5.2.4 Illumina's FASTQ read naming scheme

Consider the following label, which was taken from a read from the NA12878 exome dataset described in Chapter 4.

```
@HWI-D00119:50:H7AP8ADXX:1:1101:2100:2202 1:N:0:TAAGGCGA
```

The syntax of the label lines corresponds to the following scheme:

```
@<instrument>:<run_number>:<flowcell_ID>:<lane>:<tile>:\
    <x-pos>:<y-pos>  <read>:<is_filtered>:\
    <control_number>:<index>
```

The information stored in the above label is summarized in Table 5.3. The first part of this label, up to the space, is used as a read name or identifier.

Table 5.3. Illumina sequence identifiers.

Item	Explanation
HWI-D00119	The unique instrument name
50	The run ID (this is the 50th time that this machine has been run)
H7AP8ADXX	The flow cell ID
1	The flow cell lane (1–8)
1101	Tile number within the flow cell lane
2100	X-coordinate of the cluster within the tile
2202	Y-coordinate of the cluster within the tile
1	Member of a pair (1 or 2; 2 can only be used for paired-end or mate pair sequencing)
N	Y: read violated the chastity filter (such reads may be filtered out or left in the FASTQ file); N: read did not violate the chastity filter
0	0 when none of the control bits are on, otherwise it is an even number. On HiSeq X and NextSeq systems, control specification is not performed and this number is always 0.
TAAGGCGA	index sequence (barcode)

Note: Each read label stores information in a standard scheme (The scheme has been used with CASAVA 1.8 and later). We have omitted the (optional) UMI (Unique Molecular Identifier) field because it is not used for exome or genome sequencing.

FURTHER READING

Cock PJ, Fields CJ, Goto N, Heuer ML, Rice PM (2010) The Sanger FASTQ file format for sequences with quality scores, and the Solexa/Illumina FASTQ variants. *Nucleic Acids Res* **38**:1767-71.

5.3 EXERCISES

Exercise 1

The definition of the Phred score in terms of the error probability was shown in Equation 5.1. Analogously, we can calculate the error probability from the Phred score as follows.

$$p = 10^{-Q/10} \tag{5.2}$$

Convert the ASCII symbols for the first five bases of the following read to Phred scores and calculate the average Phred score of these bases. Then use Equation 5.2 to calculate the average error probability.

```
@My-Illu:6:73:941:1973#0/1
GATTTGGGGTTCAAAGCAGTATCGATCAAATAGTAAATCCATTTGTTCAACTCACAGTTT
+
!''*((((*∗∗+))%%%++)(%%%).1∗∗∗-+∗''))∗∗55CCF>>>>>>CCCCCCC65
```

Exercise 2

Explain the information contained in the following read label.

```
@HWI-D00107:50:H6BP8ACWV:5:2204:10131:51624 2:N:0:AGGCAGAA
```

Exercise 3

Explain the meaning of the following read name.

```
@Machine42:1:FC7:7:19:4229:1044 1:N:0:TTAGGC
```

Exercise 4

What does the quality character > mean? What is the associated error probability?

Exercise 5

Imagine there is a 100 bp read each of whose bases has the quality score &. What is the expected number of errors for this read?

CHAPTER 6

Raw Data: Quality Control

P ROBABLY the most important law in bioinformatics is: "Garbage In, Garbage Out". Therefore, quality control (Q/C) is absolutely necessary for all stages of the analysis pipeline. Q/C for WES/WGS data can be separated into three stages: FASTQ (raw data), alignment, and variant calling [149]. In this chapter, we will look at Q/C for assessing the quality of the raw sequence data. Q/C serves as a quick screening tool for excluding data with serious quality issues and flagging data with questionable quality. The most important parameters to check are

▶ base quality

▶ nucleotide distribution

▶ GC content distribution

▶ duplication rate

▶ adapter sequence contamination.

The types of errors encountered in WES/WGS data vary depending on which sequencing technology was used. In this chapter, we focus on the Q/C analysis of Illumina data, which currently is the predominant technology for WES/WGS.

6.1 FASTQC

There are a number of tools that are widely used in the community for quality control. We will use a tool called FastQC, which you can download from the Babraham Bioinformatics Website.[1] At the end of the chapter, we will review other tools that are currently available for the quality control of whole-exome and whole-genome sequencing (WES/WGS) data.

At the time of this writing, the current version of FastQC was v0.11.5. FastQC is available as a binary (executable) file for MacOS, Windows, and Linux, or the source code can be downloaded. Follow the instructions provided at the Babraham Bioinformatics Website to install FastQC on your system. If desired, readers can perform the analysis on the Corpas or the GIAB exome data that was introduced in Chapter 4. To get started, we will use the graphical user interface (GUI) (Figure 6.1). Open FastQC and select the FASTQ files with the File|Open... menu item.

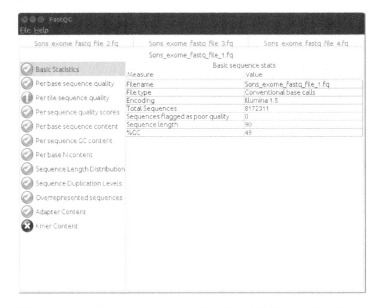

Figure 6.1. Screenshot of the FastQC application.

FastQC is powerful and easy to use. Each file being analyzed has its own tab, and there is a menu on the left with items such as **Basic statistics** and **Per base sequence quality** that show the results.

---

[1]http://www.bioinformatics.babraham.ac.uk/projects/fastqc/

We refer readers to the Babraham Bioinformatics Website for further information about how to use the program. Especially for scripts and pipelines it is preferable to use the command line interface of FastQC, which will be presented in Section 6.8.

6.2 BASE QUALITY

Base quality scores are the single most important parameter in the Q/C of raw sequencing data. A quality (Q) score is an estimation of the probability of error in base calling, with higher Phred scores corresponding to better qualities. The Phred algorithm was originally developed for automated Sanger sequencing, assessing peak shape and peak resolution of the sequence chromatograms [117]. Quality scores for Illumina data are also measured using the Phred scale explained in Chapter 5 (of course, a different algorithm is used to calculate the scores than with Sanger sequencing). The method used by Illumina to generate quality scores was discussed in Section 3.4.

In practice, quality values for current Illumina devices usually range from 0 to 60 (see Chapter 5 for details on the FASTQ format and Phred scores). As a rule of thumb, values of 30 or more (error probability less than 0.1%) are considered good, and values under 20 (error probability higher than 1%) are considered bad. Reads are judged by the distribution of quality scores across all the bases of the read. There is no simple rule that can always be applied to perform Q/C filtering of NGS data, but the general strategy is to discard reads whose overall quality is below a certain threshold. On the Illumina platform, base quality usually begins to decline as the cycle number increases, that is, toward the end of a read. The major reason for this are phase errors, which were discussed on page 34.

Cluster density is an important parameter for run quality. If too few DNA fragments undergo colony amplification on the flow cell, this results in lower data output. If too many DNA fragments are present, a situation termed "overclustering", image analysis problems may ensue because the sequencer cannot reliably find the boundary between neighboring clusters. The overall increase in brightness on the flow cell can make it difficult for the sequencer to find the correct focus. If neighboring clusters "bleed" into each other, the ratio of base intensity is decreased with consequent ambiguous base calling. Overclustering is typically manifested by a reduction in the percentage of clusters passing filter (%PF). Newer Illumina sequencers (e.g., HiSeq 4000, HiSeq

X Ten, NovaSeq) have patterned flow cells that allow higher densities with less sensitivity to overclustering.

As Illumina has introduced newer sequencing chemistry, the overall length and quality of the reads has increased substantially, and the decline in quality is observed at ever higher cycle numbers. In some cases, the base quality for the initial cycles are estimated to be lower than those in the middle sections of the read owing to the way Illumina's quality score algorithm estimates the quality score [149].

FastQC and most other WES/WGS Q/C software present the quality report as a base Q-score versus cycle number plot, whereby the background color indicates FastQC's ranges for high-quality calls (green), calls of reasonable quality (orange), and calls of poor quality (red) (Figure 6.2).

6.2.1 Base quality and paired-end reads

For paired-end reads, the median base quality score may be lower for the reverse read, probably related to the longer period of time the template was on the sequencer prior to sequencing. Figure 6.3 shows that the quality of the reverse read of the Genome in a Bottle Consortium (GIAB) NA12878 exome dataset is slightly worse than that of the forward read. Figure 6.4 shows another set of forward and reverse reads, in which the reverse read displays markedly worse quality than the forward read.

Finally, note that FastQC automatically determines which Phred encoding method was used for the quality scores in both datasets. For the Corpas exome data in Figure 6.2, Illumina 1.5 encoding was used, and in the other datasets, Sanger/Illumina 1.8+ encoding was used.

6.3 NUCLEOTIDE DISTRIBUTION

In exome data, one expects that there should be no difference in the average proportion of bases at different positions of the read. That is, the nucleotide distribution (A, T, C, G) across cycles across all reads generally should remain relatively stable, except for minor fluctuations at the end of the read. However, some types of library preparation produce a biased sequence composition, normally at the start of the read. For instance, cDNA synthesis using random hexamer priming induces biases in the nucleotide composition at the beginning of transcriptome sequencing reads [157]. This is not necessarily the case with

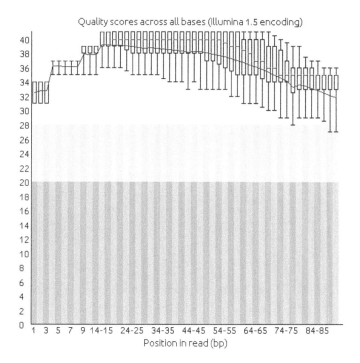

Figure 6.2. Base quality. FastQC base quality report (`Per base sequence quality`) for the Corpas exome data. The central red line of each box of the boxplot represents the median base quality value at a given read position. For instance, the median Phred quality score for the first base over all reads was 34, and the median quality of the bases at position 15 was 40. The yellow box represents the inter-quartile range (25–75%), and the upper and lower whiskers represent the 10% and 90% points. The blue line shows the mean quality. The overall base quality of the data can be regarded as good.

WES or WGS data, but libraries fragmented using transposases inherit an intrinsic bias in the positions at which reads start. This effect can be seen in Figure 6.5, which shows the FastQC evaluation of the sequence content of the Corpas exome, which (presumably) was fragmented using ultrasonication and that of GIAB NA12878, which was fragmented using Illumina's Nextera system, a transposon-based method for production of DNA fragment libraries [279]. It can be seen that the Nextera library shows a sequence bias in the early cycles. It has been our experience that this phenomenon has no noticeable effect on downstream analysis. Other reasons for deviations from uniformity

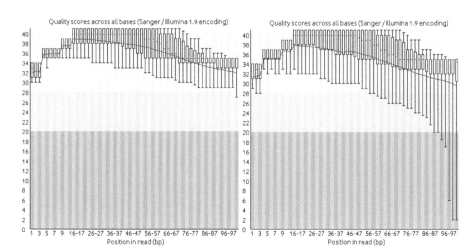

(a) (b)

Figure 6.3. FastQC Per base sequence quality for the (a) forward and (b) reverse reads of the GIAB NA12878 exome dataset. The reverse strand reads have slightly worse overall quality than the forward strand reads as manifested by the broader interquantile range in the later cycles. Still, the overall quality of both reads may be considered as good.

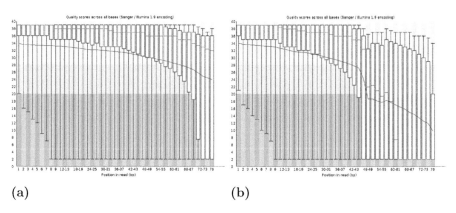

(a) (b)

Figure 6.4. FastQC Per base sequence quality for the (a) forward and (b) reverse reads of the SRR098359 dataset. The decline in base quality in the reverse reads is more noticeable than in the exomes shown in Figure 6.3.

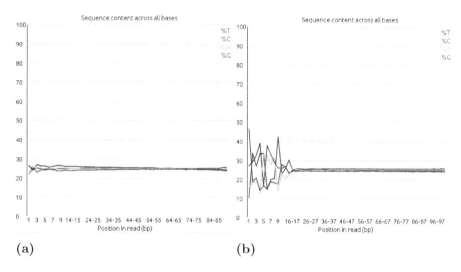

Figure 6.5. Per base sequence content. FastQC `Per base sequence content` for (a) the Corpas exome and (b) the NA12878 exome dataset. The sequence content is nearly uniform in the Corpas dataset, but shows fluctuations at the beginning of the reads of the NA12878 exome.

of per base sequence content include overrepresented sequences such as adapters.

6.4 GC CONTENT DISTRIBUTION

The total percentage of GC content sequenced can also be used as a quality control parameter. The percentage of GC in the genome varies across species and across the regions of each genome. For (human) exome regions, the GC content is about 49–51%, while for whole-genome sequencing, the GC content is around 41%. Figure 6.6 shows a comparison of an exome and a genome dataset from the same individual. It can be seen that the GC distribution is shifted toward higher ranges in the exome dataset. An abnormal GC content percentage, say, more than 10% deviation from normal range, can indicate contamination [149].

6.5 DUPLICATION RATE

Library preparation involves a number of steps that begin with the fragmentation of DNA by ultrasonication or other means, the ligation

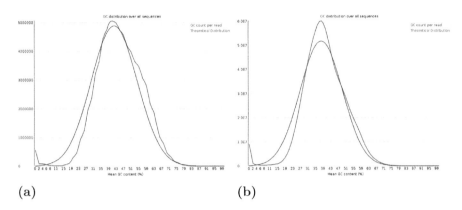

(a)   (b)

Figure 6.6. FastQC `Per sequence GC content` distribution. (a) Exome (SRR098401) and (b) genome (SRR622457) of the 1000 Genomes individual NA12878. The mode of the exome data is approximately 41% and that of the genome data 38%. In many cases, the observed shift is larger than this.

of adapters to both ends of the fragment, and PCR amplification of the ligated fragments prior to addition of fragments to the flow cell, bridge amplification, and sequencing-by-synthesis. The PCR amplification step essentially is intentionally creating duplicates of the original fragments in order to have sufficient quantities of DNA for the subsequent steps. Therefore, PCR duplication is inevitable in Illumina sequencing, and the key question is whether there is an excessive amount of PCR duplication in the final results.

Duplicate reads can arise from other sources in addition to PCR duplicates, including sequencing artifacts such as poly-A and poly-N reads, noise in cluster detection, and from genomic DNA shearing at the same location in different molecules [354]. Duplicate reads can be observed if multiple independent inserts arise from identical genomic positions during library construction. This effect may be seen especially at higher depth sequencing experiments because the chance of obtaining identical fragments increases with the number of fragments that are produced [488]. Another cause of duplicate reads is the incorrect detection of a single amplification cluster as multiple clusters by the optical sensor of the sequencing instrument. These duplication artifacts are referred to as optical duplicates. Finally, paralogous gene sequences with high degrees of sequence identity may occasionally be a cause for apparent duplication.

The PCR amplification step can introduce significant biases due to preferential amplification of shorter molecules, molecules without extremely high or low GC composition, and other molecules that are preferentially amplified for unknown reasons. This effect causes the sequencing to be a non-random sampling of the source genome, and is particularly problematic if any single molecule experiences a PCR error early in amplification as this error is propagated and sampled many times during sequencing [104].

Therefore, removing duplicate reads (reads of the same length and sequence identity) is a widely-employed practice to correct this bias when analyzing NGS data, following the underlying assumption that PCR amplification is responsible for most of the read duplication [488].

The `Sequence Duplication Levels` panel of FastQC plots the number of times the same sequence is seen in a 200,000 read subset of the input data. In a diverse library most sequences will occur only once in the final set. In most libraries you would expect to see not more than >10% duplicate rates. Duplication rates above 30% are often considered to be very high. Figure 6.7 shows the overall amount of duplication in the Corpas exome, with its 32,116,328 single-end reads, and the NA12878 exome dataset, with its 20,203,002 paired-end reads. It can be seen that the overall level of duplication is higher in the NA12878 dataset. Differences have been noted between library preparation methodologies [395], and as mentioned before the NA12878 sample was prepared using a transposase-based system. In general, duplicate reads should be removed from WES/WGS data prior to downstream analysis, as we will see later on.[2]

6.6 K-MER CONTENT

Consider what happens if the insert sequence is shorter than the read length. If the read length, L_R, is longer than the insert size, L_I, then the read produced by the sequencing machine will include $L_A = L_R - L_I$ nucleotides from the adapter sequence (Figure 6.8). Depending on the library preparation protocol, adapter fragments will be present in the 3' end of the read and possibly also the 5' end. If these fragments ("adapter contamination") are not removed, they can lead to either

---

[2]In most cases duplicate reads are not actually removed but only marked as such, so that downstream analysis tools can ignore them but no original sequencing data is lost.

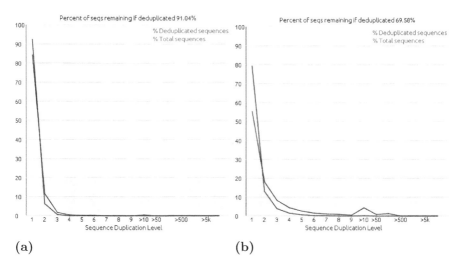

(a)   (b)

Figure 6.7. Duplication level. FastQC plot of `Sequence Duplication Levels` in (a) the Corpas exome and (b) the reverse reads of the GIAB NA12878 exome dataset.

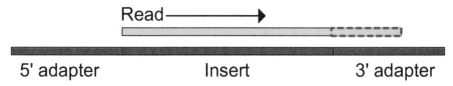

Figure 6.8. Adapter contamination. Illumina read running into the 3' adapter. If the length of the read is longer than the length of the specific insert, then the 3' end of the read will originate from the adapter sequence on the other side (for example, the read length is 100 and the insert was only 80 nucleotides long, then the last 20 base pairs, as indicated here with a dashed red line, will be adapter contamination).

missed alignments because the sequenced construct does not match the reference genome sequence or, if the read is mapped to the reference genome, to a misleading increase in the number of mismatches at the end of the mapping.

The `Kmer Content` module of FastQC measures the number of each 7-mer at each position in the sequencing library. FastQC uses a binomial test to assess the coverage at all positions. It then reports k-mers with statically significant positional enrichment. The analysis of the Corpas exome file reveals several overrepresented 7-mer sequences, including **AGATCGG** and several related 7-mer sequences: **ATCGGAA, GATCGGA, TAGATCG, AAGATCG**, etc., with observed-to-expected count ratios of up to 63.6. We see some positional enrichment toward the end of the reads (Figure 6.9).

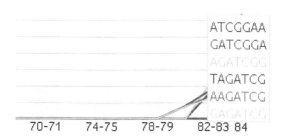

Figure 6.9. The `Kmer content` module of FastQC. Here, we see several adapter-related k-mers that are enriched toward the end of the reads, which is a sign of adapter contamination. This plot shows an excerpt of the `Kmer content` window of FastQC for the Corpas exome data.

Considering the following Illumina Universal adapter:

```
5'-AAT GAT ACG GCG ACC ACC GAG ATC TAC ACT \
   CTT TCC CTA CAC GAC GCT CTT CCG ATC T-3'
```

The reverse complement sequence of the last 7 nucleotides of the adapter is AGATCGG – exactly the same as the overrepresented k-mer! Therefore, we can infer that a substantial number of inserts in the Corpas exome are shorter than the read length, leading to an overrepresentation of adapter-related k-mers at the 3' end of the reads. This is an indication that adapter clipping should be performed (see Chapter 7).

Not all enriched k-mer sequences originate from adapter sequences, but in general significant enrichment of k-mers in WES/WGS data suggests the presence of an artifact and the cause of the enrichment should be sought before proceeding with further analysis.

6.7 PER TILE SEQUENCE QUALITY

Beside the sample quality itself also the sequencing process may have impact on the quality of the reads. Issues such as bubbles passing the flow cell or damaged (smudged or scratched) flow cells may result in loss of information or a drop in quality. This can be seen in the heatmap (Figure 6.10) that represents the base qualities for all the tiles of the flow cell, where cold colors indicate an average base quality and warm colors a substantial drop in quality.

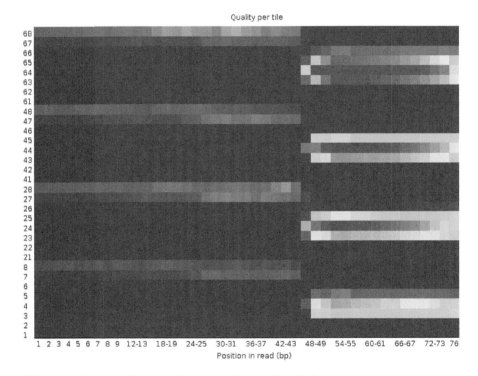

Figure 6.10. Per tile quality. FastQC `Per tile sequence quality` plot for the reverse read of sample SRR098359. The plot shows the deviation from the average quality for each tile, with hotter colors showing that a tile had worse qualities than other tiles for that base. A good plot should be blue all over. In this example, it is apparent that some tiles display poor overall quality.

6.8   RUNNING FASTQC FROM THE COMMAND LINE

FastQC can be run from the command line as a part of larger software pipelines. The basic command is simple.

```
$ fastqc file.fa.gz
```

Entering `fastqc -h` will cause the program to print a list of options, including the number of threads, non-standard adapter sequences, etc.

6.9   OTHER Q/C TOOLS

FastQC is probably the most popular Q/C tool for next-generation sequencing data, but additional useful Q/C applications include the NGS QC Toolkit [323], the FASTX Toolkit,[3] HTQC [471], and QC3 [150]. We will explore additional Q/C applications in subsequent chapters about quality control for WES/WGS alignments and for variant calling Q/C.

FURTHER READING

Guo Y, Ye F, Sheng Q, Clark T, Samuels DC (2013) Three-stage quality control strategies for DNA re-sequencing data. *Briefings in Bioinformatics* **15**(6):879–889.

Zhou X, Rokas A (2014) Prevention, diagnosis and treatment of high-throughput sequencing data pathologies. *Molecular Ecology* **23**(7):1679–1700.

6.10   EXERCISES

Exercise 1

Assuming that Q ranges from 0 to 41, what is the range of the error rates that are reported for a base?

Exercise 2

Write a script or program to find the average quality values of a WES FASTQ file in the same way FastQC does, that is, averaged for each read position.

Your code should input a FASTQ file and go through it line by line.

---

[3]`http://hannonlab.cshl.edu/fastx_toolkit/`

It should determine which lines represent quality strings, and calculate the Phred quality scores corresponding to their ASCII code (see Chapter 5 for information about the FASTQ format and Phred scores). Keep track of the values in an array or vector, and calculate the average over all quality strings. Output the average values for each read position as well as the mean value over all bases.

Exercise 3

Extend your code to calculate the average per read GC content as well as "N" content for the sequences in the file.

Exercise 4

Now extend your code to look for overrepresented k-mers. In Python this can be achieved using a Dictionary data structure (C++: map, Java: HashMap, Perl: hash) in which the keys are the k-mers and the values are the counts. Output the top ten k-mers and estimate the expected count for the k-mers based on the GC content and the assumption of independence of individual k-mer positions. You may use k=4,5, or 6 for your experiment.

See hints and answers on page 499.

CHAPTER 7

Trimming

R EMOVAL of low-quality segments of NGS reads, referred to as "trimming", is a common strategy to mitigate sequencing errors and artifacts in Next-Generation Sequencing data analysis [489]. With Illumina data, the two most commonly encountered causes of low-quality segments in reads are (i) the tendency for reduced quality at the 3' end of reads owing to phase errors, and (ii) adapter contamination (see Chapter 6). Trimming refers to the process of removing low-quality or artefactual sequences or portions of sequences prior to downstream analysis — essentially by surgically eliminating only low quality regions [102].

7.1 TRIMMING

Trimming has been shown to improve the overall data quality and to enable better results in downstream analysis [102, 489]. On the other hand, excessive trimming may reduce the quality of downstream results [272]. Therefore, bioinformatics processing needs to address two related questions: (i) Should a given data set be trimmed? and (ii) How stringently should the dataset be trimmed?

The results of the Q/C analysis with FastQC may suggest the need for trimming, for instance, if the later cycles show a low mean quality score or if there is evidence of adapter contamination. We will demonstrate and explain this process using the application called Trimmomatic [40]. We note that the approaches toward trimming differ between tools and also between different NGS applications. The trimming strategy needs to be adjusted according to whether single-end or paired-end sequencing was performed. We have found that Trimmomatic performs well for WES/WGS data, but many other trimming applications have been published, including ngsShoRT [65] and AdapterRemoval [389].

7.1.1 Quality trimming

Quality Trimming applies a filter to segments of reads based on their Phred quality values. As explained in chapter 6, with Illumina reads, the quality of the data begins to get worse towards the 3' end of the read. Therefore, a typical approach towards quality filtering is to assess the quality of bases and determine where to truncate the read, retaining the 5' portion, and discarding the lower quality 3' portion. Trimmomatic applies a sliding window approach that examines the average quality of a set of contiguous bases by sliding a window over the read starting at the 5' end and trimming if the quality falls below a threshold. This approach is more robust than trimming starting at a single poor quality base because it prevents an isolated poor quality base from causing the removal of subsequent high-quality data [40].

7.1.2 Adapter trimming

The issues surrounding adapter contamination were introduced in Section 6.6 (see Figure 6.8). Adapter trimming, also known as adapter clipping, removes adapter sequences from NGS reads. Programs such as Trimmomatic therefore need to have a file with a list of adapter sequences (or other potentially contaminating sequences). Trimmomatic uses the term "technical sequences" to refer to adapter sequences and polymerase chain reaction (PCR) primers, but, for simplicity , in the following we will use the word "adapter". In the simple mode (which is most useful for single-end reads), each read is scanned from the 5' end to the 3' end to determine if any of the user-provided adapter sequences are present. If the adapter overlaps with the 5' end of the read, then the entire read is discarded. Otherwise, the 3' terminus of the read is discarded starting from the first overlapping nucleotide.

There is no generally accepted operating protocol for adapter removal, but we recommend that users examine the raw FASTQ data with FastQC as described in the previous chapter, and if signs of adapter contamination are found (see Figure 6.9 in Chapter 6), Trimmomatic should be applied with the appropriate adapter sequences. Quality trimming should be applied especially if the overall quality is poor towards the 3' end of reads, although downstream alignment and variant calls are able to cope with a certain amount of poor quality bases and reads.

Trimmomatic has a "palindrome mode" that is optimized for the detection of "adapter read-through". When "read-through" occurs, both reads in a pair will comprise the same sequence (in reverse complementary orientation) followed by contaminating sequence from the "opposite" adapter (see Figure 3.13C).

7.2 TRIMMOMATIC

Trimmomatic is a Java executable that can be downloaded from the Usadel lab homepage [40].[1] You will therefore need to have a Java runtime environment on your computer to run it. Users can download the binary version (which contains the executable `jar` file, the adapters directory, and the license), or the source code, which can be compiled using `$ ant dist` (the executable will be created in the directory `dist/jar`).

Trimming software typically works in stages. With Trimmomatic, the user needs to indicate the stages and the order of the stages. The following five stages represent a good starting point, but may need to be altered depending on the specific experiment.

1. Remove adapters

2. Remove leading low quality or N bases (below quality 3)

3. Remove trailing low quality or N bases (below quality 3)

4. Scan the read with a 4-base wide sliding window, trimming when the average quality per base drops below 15 (i.e., truncate the 3 portion of the read starting at the first base that causes the quality within the sliding window to drop below the threshold; this is the last base of the window, as can be seen in Figure 7.1)

5. Discard reads whose length after trimming is 36 bases or less

In order to remove adapters, users need to know which adapters were used for the exome library preparation. See the information provided in Chapter 2 for background information on various kinds of adapter sequences. Trimmomatic comes with a directory called

---

[1]`http://www.usadellab.org/cms/`

"adapter" that contains the sequences of the most commonly used Illumina adapters. Some older data will have used TruSeq2 adapters (e.g., with the Genome Analyzer II machines) and most newer data will use TruSeq3 or Nextera adapters (e.g., HiSeq and MiSeq machines). Trimmomatic comes with the following single-end (SE) and paired-end (PE) adapter sequences (in the adapter subdirectory of the Trimmomatic distribution):

```
NexteraPE-PE.fa
TruSeq2-PE.fa
TruSeq2-SE.fa
TruSeq3-PE-2.fa
TruSeq3-PE.fa
TruSeq3-SE.fa
```

7.2.1 Illumina Clip

The ILLUMINACLIP argument is as follows:

```
ILLUMINACLIP:<fastaWithAdaptersEtc>:<seed mismatches>:\
    <palindrome clip threshold>:<simple clip threshold>
```

To understand this, it is important to understand how Trimmomatic works. It uses a two step strategy to search for matches between adapters and reads. Trimmomatic initially searches for matches between short sections (up to 16 bp) of the adapters in each possible position of the read. The parameter seed mismatches controls the maximum number of mismatches allowed between the adapter sequence and a subsequence of the read to still be considered a match. The match calculated by the full alignment must exceed the simple clip threshold in order for trimming to be performed.

The full alignment score is calculated by increasing the alignment score by 0.6 for each matching base and by reducing the alignment score by Q/10 for each mismatched base (where Q is the Phred encoded quality score of the mismatched base). A perfect match of a sequence with a length of n bases is thus $n \times 0.6$, which is about 7 for a 12 base perfect match and about 15 for a 25-base perfect match. Therefore, values of between 7–15 are recommended for this parameter. The palindrome clip threshold specifies how accurate the match between two 'adapter ligated' reads must be for PE palindrome read alignment (cf. Figure 3.13C).

Consider the following example.

```
ILLUMINACLIP:./adapters/TruSeq2-SE.fa:2:30:10
```

This will cause Trimmomatic to use the TruSeq2 single-end adapters for its analysis, allowing up to 2 seed mismatches, with a clip threshold of ten. The format of the ILLUMINACLIP argument is the same for paired-end and single-end runs, except that if the dataset being analyzed consists of single-end reads, then the palindrome clip threshold is ignored. For paired-end reads, this argument would cause a match accuracy threshold of 30 is to be used.

7.2.2   Trimmomatic for single-end reads

We will perform trimming on the first of the four Corpas exome files (see Chapter 4). Recall from the previous chapter that this data set employs an older quality encoding (Illumina 1.5, see Figure 6.2) that corresponds to Phred+64 encoding (Chapter 5). We therefore use the -phred64 flag.

```
$ java -jar trimmomatic-0.36.jar SE \
    -phred64  \
    -threads 2 \
    -trimlog son.log \
    Sons_exome_fastq_file_1.fq \
    trimmed_output.fq \
    ILLUMINACLIP:./adapters/TruSeq2-SE.fa:2:30:10 \
    LEADING:3 \
    TRAILING:3 \
    SLIDINGWINDOW:4:15 \
    MINLEN:36 \
    TOPHRED33
```

Let us consider what these options mean.

SE
>   Single end reads.

-phred64
>   Quality scores in the FASTQ file were encoded with Phred+64.

-threads 2
>   The number of threads to be used by Trimmomatic.

-trimlog
> Write a log to the indicated file.

Sons_exome_fastq_file_1.fq and **trimmed_output.fq**
> The input and output files are indicated at this point in the command line. These files are required to be in FASTQ format and may be compressed (**gz**).

ILLUMINACLIP:./adapters/TruSeq2-SE.fa:2:30:10
> The location of the file with the Illumina adapters is given, followed by seed mismatches, palindrome clip threshold, simple clip threshold, i.e., we allow up to 2 mismatches to the adapter sequence, and require a score of at least 10 for the alignment between any adapter sequence against a read. The value of 30 is for the palindrome clip threshold, but that is not used in SE mode.

LEADING:3 and TRAILING:3
> Specifies the minimum quality required to keep a leading (5') or trailing (3') base (here, a minimum Phred score of 3 is indicated).

SLIDINGWINDOW:4:14
> Window size of 4, minimum mean quality in window 14.

MINLEN:36
> Discard all sequences that are smaller than 36 base pairs after the other trimming operations.

TOPHRED33
> Convert quality scores to Phred+33 in the output file.

7.2.3 Trimmomatic for paired-end reads

We will demonstrate the use of Trimmomatic for paired-end sequencing reads on the NA12878 exome data described in Chapter 4. This paired-end dataset was sequenced using the Illumina Nextera Rapid Capture Exome kit. The NA12878 dataset is more modern than the Corpas exome, and so the quality values are already in Phred+33 encoding and do not need to be converted.

```
$ java -jar trimmomatic-0.36.jar PE \
  -threads 4 \
  -trimlog NIST7035.log \
  NIST7035_TAAGGCGA_L001_R1_001.fastq.gz \
```

```
NIST7035_TAAGGCGA_L001_R2_001.fastq.gz \
NIST7035_trimmed_R1_paired.fastq.gz \
NIST7035_trimmed_R1_unpaired.fastq.gz \
NIST7035_trimmed_R2_paired.fastq.gz  \
NIST7035_trimmed_R2_unpaired.fastq.gz \
ILLUMINACLIP:./adapters/NexteraPE-PE.fa:2:30:10 \
LEADING:3 \
TRAILING:3 \
SLIDINGWINDOW:4:15 \
MINLEN:36
```

The command for PE data is similar to that for SE. The command indicates paired-end mode (PE), and the single input file is replaced by two input files with the forward and reverse reads. The next four arguments represent the names of the output files: paired and unpaired forward reads as well as paired and unpaired reverse reads. Only read pairs of which both reads are retained can be written to the paired output files. If one of the two reads is dropped because it falls below the minimum length, the other read is written to the corresponding unpaired file. This is necessary to guarantee the correct number and order of read pairs in the two output files for paired reads.

7.3   IS TRIMMING REALLY NECESSARY?

As with most things in life, there is no free lunch in bioinformatics. The trimming step is time consuming, and generally produces four files with the paired and unpaired reads, which adds to the complexity of the analysis. With WES or WGS data, a separate trimming step may not be absolutely necessary — there are different opinions in the community. Variant calling tools such as GATK take base quality into account, and the overall quality of data from modern Illumina sequencers is high. Aligners such as BWA are able to soft-clip portions of sequences that do not correspond to the reference (see Section 9.4). Soft-clipping contaminating adapter sequences that do not match the reference sequence has similar effects as adapter clipping. However, even if trimming is not performed, it is advisable to perform quality-control on the sequence data, and trimming should be considered especially if there are concerns about adapter contamination and low quality bases.

In production environments where large volumes of sequencing data

```
@HWI-D00119:50:H7AP8ADXX:1:1101:2100:2202
ATACTGTTGT...........CATGCCAACCTGGCAACCA
+
@@@DDDDB:F...........5:A,5:4(80(2<?B@B<?
```

```
@HWI-D00119:50:H7AP8ADXX:1:1101:2100:2202
ATACTGTTGT...........CATGCCAACC
+
@@@DDDDB:F...........5:A,5:4(80
```

Figure 7.1. Quality trimming. A read is shown before and after trimming by trimmomatic. The nine last bases were trimmed because of a low average quality in a sliding window (average score of less than 15 in a sliding window of 4 bases: `SLIDINGWINDOW:4:15`). Consider the average scores of several windows of quality characters (this can be done using the Unix command `echo -n "5:4(" | od -A n -t d1` to get the scores for `5:4(`, i.e., 53, 58, 52, 40. Subtracting 33 from each number and taking the average gives us and average score of 17.75, which is above threshold. However, the average score of `(80(` is only 13, which is below threshold and triggers the trimming of the read as shown in the illustration. Trimming starts at the 3' base of the first below-threshold window, which in our example corresponds to the "T" of the `ACCT` with quality string `(80(`. Note that the middle portion of the sequence and the corresponding quality string have been omitted for better legibility.

need to be analyzed, tools are generally connected into pipelines to avoid unnecessary I/O. In our current pipeline, we use bwakit to trim adapters with **trimadap**, mark duplicates with **samblaster**, and sort the final alignment with **SAMtools** (see Chapter 8). We do not perform a separate quality trimming step.

A recent article evaluates many published tools for read trimming [102].

FURTHER READING

Bolger AM, Lohse M, Usadel B (2014) Trimmomatic: a flexible trimmer for Illumina sequence data. *Bioinformatics* **30**:2114–2120.

Del Fabbro C, Scalabrin S, Morgante M, Giorgi FM (2013) An extensive evaluation of read trimming effects on Illumina NGS data analysis. *PLoS One* **8**(12):e85024.

7.4 EXERCISES

Exercise 1

Write a program or script to "guess" the encoding of a FASTQ file. Recall that there is a different offset between the ASCII code and the Phred quality score for the earlier version of the FASTQ format and the current version (33 or 64), which leads to a different range of ASCII characters being observed in the file. You can test your script on the Corpas exome data (which has Illumina 1.5 format, with an offset of 64) and the newer Genome in a Bottle Consortium exome data (which uses the current Sanger encoding with an offset of 33).

Exercise 2

Write a program or script that performs sliding-window analysis of quality strings and outputs the position of the first window with a below-threshold average quality. Test your results by comparing your script with the output of the Unix **od** utility as shown in the legend to Figure 7.1.

Exercise 3

Examine the following two Trimmomatic commands.

```
$ java -jar trimmomatic-0.32.jar SE \
```

```
-phred33 \
sample.fastq \
aclipped.fq.gz \
ILLUMINACLIP:TruSeq2-SE.fa:2:30:10 \
MINLEN:25
```

and

```
$ java -jar trimmomatic-0.32.jar SE \
   -phred33 \
   sample.fastq \
   aclippedtrim.fq.gz \
   ILLUMINACLIP:TruSeq2-SE.fa:2:30:10 \
   LEADING:3 \
   TRAILING:3 \
   SLIDINGWINDOW:4:15 \
   MINLEN:25
```

Explain what each command does and how they differ from one another.

Exercise 4

Use Trimmomatic to process the Corpas exome files. Write a script to calculate the average quality values of the reads before and after trimming based on the information in the Trimmomatic log files. What do you observe?

Exercise 5

Write a program or script to parse the logfile created by Trimmomatic. The logfile has one line for each read with the following fields:

▶ The read name

▶ The surviving sequence length

▶ The number of bases trimmed from the start (5' end) of the read

▶ The position of the last surviving base in the original read

▶ The number of bases trimmed from the end (3') of the read

For instance, the following lines show that the first three reads were not trimmed at all but the last four bases were trimmed from the fourth read.

```
FCB021RACXX:4:1208:14092:195285#CAGATCAT 90 0 90 0
FCB021RACXX:4:1208:14092:195285#CAGATCAT 90 0 90 0
FCD044UACXX:4:1202:2462:65186#CAGATCAT 90 0 90 0
FCB021RACXX:4:1308:13583:89169#CAGATCAT 86 0 86 4
(...)
```

Write a script that counts the number of reads with a given number of bases removed from the start and from the end of the reads, and output the counts of reads in each category. Your script should produce output in the following format:

```
Bases removed from start of reads

Number of sequences with 0 bases removed from start: 33848
Number of sequences with 5 bases removed from start: 7
(...)

Bases removed from end of reads

Number of sequences with 0 bases removed from end: 32500
Number of sequences with 1 bases removed from end: 292
(...)
Number of sequences with 52 bases removed from end: 7
Number of sequences with 53 bases removed from end: 8
Number of sequences with 54 bases removed from end: 5
```

III

Alignment

CHAPTER 8

Alignment: Mapping Reads to the Reference Genome

R EFERENCE-BASED ASSEMBLY follows the goal of finding the *differences* between an individual's genome and the reference genome, by mapping reads to the reference genome and then performing variant calling to identify the differences. Reference-based assembly, which we will refer to simply as alignment, thus differs from genome assembly, which is the process of puzzling together a complete genome sequence of an organism for which shotgun sequencing has been performed. *De novo* assembly algorithms with de Bruijn graphs, which we will not discuss in this book, are computationally demanding and error-prone and therefore have not been used much in WES/WGS diagnostics or disease gene discovery programs. In contrast, reference-based assembly algorithms are relatively fast and accurate, because they exploit knowledge about the human reference genome to guide alignment of NGS reads.

In this chapter, we will explain the foundations of reference-based alignment and show how to use BWA-MEM for this purpose.

8.1 THE GENOME REFERENCE(S)

Clearly, a high quality human genome assembly is essential for medical genomics and translational research. The human reference assembly is now in its 38th edition, and is one of the highest quality mammalian assemblies available [74]. The assembly declared to be "finished" in 2004 [180] actually still contained over 300 gaps in the euchromatic portion of the genome, and numerous other errors. The Genome Reference Consortium (GRC), which was founded in 2007, aims to improve the human reference genome assembly (as well as those of mouse, zebrafish, and chicken).

At the time of this writing (early 2017), the human genetics and genomics community is transitioning from the GRCh37 genome assembly to the GRCh38 assembly. We will discuss the relevance of the new human genome build (GRCh38) for WES/WGS in Section 8.7. Different institutions have produced slightly different versions of the GRCh37 and GRCh38 genome assemblies. The University of California Santa Cruz (UCSC) genome browser[1] referred to the GRCh37 build as hg19, and the Broad Institute produced a b37 genome build with a few minor differences. The current build is GRCh38 (referred to by UCSC as hg38). The various resources sometimes also use different identifiers to refer to the chromosomes. For example, the most common way of referring to chromosome 22 is

```
>chr22
```

but one also sees chromosomes named only by the number

```
>22
```

or by accession number (GenBank ID)

```
>CM000684.2
```

It is a common source of errors to use one way of referring to the chromosomes for a database or tool that is expecting that chromosomes are referred to in another way (computers have no way of knowing that 22 and chr22 are referring to the same thing without some form of mapping!).

---

[1]https://genome.ucsc.edu/

8.1.1 Obtaining the reference genome

There are many ways of getting the human reference genome files.[2] An archive containing the current version of the human genome assembly can be downloaded from the UCSC Genome Browser:[3]

```
hg38.fullAnalysisSet.chroms.tar.gz
```

We used the following commands to unpack the contents of this archive and to concatenate them into a single FASTA file called hs38DH.fa.

```
$ tar xvzf hg38.fullAnalysisSet.chroms.tar.gz
$ cd hg38.fullAnalysisSet.chroms
$ cat *.fa > hs38DH.fa
```

Following this, the only file we need is hs38DH.fa, and the other files can be deleted if desired.

Reference genome files such as hs38DH.fa contain not only the primary assembly (the sequences for the "standard" chromosomes chr1, chr2,..., chr22, chrX, chrY, chrM) but also a number of other types of sequence data:

Random contigs Unlocalized sequences that are known to originate from specific chromosomes, but whose exact location within the chromosomes is not known (e.g., chr9_KI270720v1_random).

ChrUn Unplaced sequences that are known to originate from the human genome, but which cannot be confidently placed on a specific chromosome.

---

[2]However, websites come and go. In order to create the examples for this book, we originally downloaded the genome file using the script run-gen-ref from the bwakit (https://github.com/lh3/bwa/tree/master/bwakit), which is provided by Heng Li, the author of BWA [252]. Recently, however, one of the URLs used by run-gen-ref to download the genomes has changed, and so the script does not work without modification. To use the bwa-kit script, change the value of the variable url38 to url38="ftp://ftp.ncbi.nlm.nih.gov/genomes/all/GCA/000/001/405/GCA_000001405.15_GRCh38/seqs_for_alignment_pipelines.ucsc_ids/GCA_000001405.15_GRCh38_full_analysis_set.fna.gz". If readers have any difficulties obtaining the files needed to work through the exercises in this book, they are advised to consult the book's GitHub site (see page xvii) for up-to-date information.

[3]http://hgdownload-test.cse.ucsc.edu/goldenPath/hg38/bigZips/analysisSet/

EBV Epstein-Barr virus sequence, representing the genome of Human herpes virus 4 type 1, the cause of infectious mononucleosis. This disease results in fever, sore throat, and enlarged cervical lymph nodes. About 98% of adults have been exposed to the virus by the age of 40 years. Since the virus remains latent in the body after infection, it is very common to find EBV sequences when performing human genome sequencing.

HLA Human leukocyte antigen (HLA) sequences representing selected alleles of the HLA A, B, C, DQA1, DQB1, and DRB1 loci. The HLA loci encode the major histocompatibility complex (MHC) proteins and are highly variable in the population.

Decoy sequences A major motivation for including the "decoy" sequences in the reference genome is that if a sample actually contains a genomic segment that is not in the reference assembly, aligners may spend a lot of CPU time trying to find a good match, or worse, aligners may wrongly assign the reads with a low mapping quality to segments with similar sequences in the reference genome. If the segment matches to one of the decoy contigs and the decoy is included in the reference, then the segment will quickly be assigned to the decoy and prevent the aligner from uselessly searching for other matches. Thus, the decoy sequences are a pragmatic solution to this, contain EBV and human sequences that in effect "catch" reads that would otherwise map with low quality to other regions in the reference and lead to unnecessary computation and avoid false variant calls related to false mappings.

Alternate contigs Alternative sequence paths in regions with complex structural variation in the form of additional locus sequences (see Section 8.7).

Table 8.1 presents a summary of all the categories of sequences available for the hg38 human genome assembly. The `hs38DH.fa` file generated as described above does not contain all of the sequences but is sufficient for many purposes including the exercises in this book.[4]

Not all tools and annotation databases have been updated yet, and thus we will use the previous build of the human genome, GRCh37

---

[4]The table presents the full list produced by bwakit (see footnote 2).

Table 8.1. Individual sequences available for the hg38 reference genome assembly.

Type	Example	Nr.
Standard	chr22	25
Random	chrY_KI270740v1_random	42
chrUn	chrUn_GL000218v1	127
chrUn decoy	chrUn_KN707606v1_decoy	2841
alt contigs	chr3_KI270778v1_alt	261
EBV	–	1
HLA	HLA-C*01:08	525
Total		3366

for a few of the examples in the book.[5] One can find a copy of the hs37d5.fa.gz file at

```
ftp://ftp-trace.ncbi.nih.gov/1000genomes/ftp/technical/
  reference/phase2_reference_assembly_sequence/
```

After downloading the file, we unpacked it with **gunzip**.

```
$ gunzip hs37d5.fa.gz
```

In this case, the file already contains all of the human genome b37 primary assembly plus the decoy sequences in one file.

8.1.2 The Genome Reference Consortium

Alternatively, one can use the genome build of the Genome Reference Consortium (GRC). This build may be more up to date, but the disadvantage is that accession numbers rather than the "plain" chromosome names are used, and users may need to convert those names prior to downstream analysis. To download the sequence data for GRCh38, readers can go to the Website of the Genome Reference Consortium[6] and access the latest minor release (patch 10, GRCh38.p10, at the time of this writing) via the FTP site.[7] Download the following file

---

[5]GRCh37 is also known as hg19.

[6]https://www.ncbi.nlm.nih.gov/grc/human

[7]The next minor assembly update (GRCh38.p11) is planned to be released in summer 2017.

`GCA_000001405.25_GRCh38.p10_genomic.fna.gz`

Note that the suffix fna stands for "fasta nucleic acid". Extracting the archive file with `gunzip` will generate the file

`GCA_000001405.25_GRCh38.p10_genomic.fna`

We will not use this sequence further in this book.

8.1.3 Coordinate conversion using liftOver

New sequencing projects will most likely use the current genome assembly (hg38/GRCh38) as a reference. However, at the time of this writing, several useful resources for the bioinformatic analysis of WES/WGS data are still only available for the previous assembly (hg19/GRCh37). The GRCh38 assembly has not only corrected single nucleotide errors but also updated the overall structure of the genome assembly by changes such as gap filling. This means that the genomic positions of variants are more often than not different between the two assemblies.

To address this issue, coordinates can be converted from one assembly to the other using UCSC's `liftOver` tool.[8] For this purpose, `liftOver` requires a data file ("chain file") that defines the specific coordinate changes from one genome assembly to the other.[9]

Using `liftOver` to convert the coordinates of an input file in either BED or GFF/GTF format is fairly easy:

```
$ liftOver <in.bed> hg19ToHg38.over.chain.gz \
    <out.bed> <out_unmapped.bed>
```

The command shown above converts positions in the BED file `in.bed` from hg19 to hg38. Any positions that cannot be mapped are recorded in `out_unmapped.bed`. The following, analogous command works for GFF files.

```
$ liftOver -gff <in.gff> hg19ToHg38.over.chain.gz \
    <out.gff> <out_unmapped.gff>
```

---

[8]Available at `http://hgdownload.soe.ucsc.edu/downloads.html#utilities_downloads`

[9]Data files defining changes from hg19 to other genome assemblies and from hg38 to other assemblies can be found at `http://hgdownload.soe.ucsc.edu/goldenPath/hg19/liftOver/` and `http://hgdownload.soe.ucsc.edu/goldenPath/hg38/liftOver/`.

Alternatively, one can also use UCSC's liftOver webtool[10] which, however, works only with input in BED format. For converting the variant coordinates specified in VCF files, Picard (see Section 10.1) can also be used.

```
$ java -jar picard.jar LiftoverVcf \
  R=hs37d5.fa \
  CHAIN=hg19ToHg38.over.chain.gz \
  I=<in.vcf> \
  O=<out.vcf> \
  REJECT=<out.unmapped.vcf>
```

where option R is used to specify the reference genome in FASTA format, here hs37d5.fa for hg19. NCBI's Genome Remapping Service webtool[11] is another alternative for converting VCF files. A useful tool for remapping the coordinates in input files of various other types is CrossMap[12] which supports SAM/BAM, BED, GFF, VCF, BigWig, Wiggle and bedGraph. For example:

```
$ CrossMap.py vcf hg19ToHg38.over.chain.gz <in.vcf> \
    <ref_genome.fa> <out.vcf>
```

Finally, for the conversion of coordinates specified in GRanges objects in R, we recommend the R/Bioconductor workflow for liftOver.[13]

It is rarely possible to remap all coordinates, and thus in most cases some proportion of the variant positions cannot be mapped.

8.2 BURROWS–WHEELER TRANSFORM

Most alignment algorithms construct auxiliary data structures, called indices, for the read sequences or the reference sequence, or sometimes both, in order to accelerate the alignment process [255]. Many current alignment algorithms rely on indices related to the Burrows–Wheeler Transformation (BWT) in order to localize reads to the reference genome. The BWT applies a reversible transformation to a block of input text. The transformation does not itself compress the data, but reorders it to make it easier to compress with simple algorithms such

---

[10]https://genome.ucsc.edu/cgi-bin/hgLiftOver
[11]https://www.ncbi.nlm.nih.gov/genome/tools/remap
[12]http://crossmap.sourceforge.net/
[13]https://www.bioconductor.org/help/workflows/liftOver/

as move-to-front coding [54]. The BWT is the basis for the widely used bzip2 compression, and is also used by many of the read mapping algorithms in common use today, including BWA [252, 253], SOAP2 [259], and Bowtie 2 [236]. The BWT leads to a block-sorted data structure that is well suited to searching short strings in a larger text. In conjunction with the FM-index [121] and some other auxiliary data structures, BWA exploits the BWT to enable search with time linear in the length of the search string within a larger string (such as the genome sequence). We will not attempt a full explanation of the algorithms here, which would require several chapters.[14]

8.3 BWA

Heng Li and Richard Durbin implemented an alignment package based on backward search with the Burrows–Wheeler transform called the Burrows–Wheeler Alignment tool, or BWA for short [252, 253]. BWA has become one of the most widely used aligners for Illumina data, and more recently the same authors published BWA-MEM, which has a range of new features and is applicable to a wide range of sequence lengths from 70 bp to a few megabases [251].[15]

The project homepage is located at `https://github.com/lh3/bwa`. Precompiled releases can be downloaded from the `releases` directory of the GitHub page. Download the latest version of BWA (we have used `bwa-0.7.15.tar.bz2` in this book). If you downloaded the source code, unpack the software and compile it with the following commands.

```
$ tar jxf bwa-0.7.15.tar.bz2
$ cd bwa-0.7.15
$ make
```

This will produce an executable called `bwa`.

---

[14]The book's GitHub page (see page xvii) contains lecture slides with an explanation of the BWT and the FM-index as they are used in BWA.

[15]Since BWA-MEM was developed for reads of ≥ 70 bases, for shorter reads it is advisable to use the standard BWA algorithm. Both are provided by the same tool.

8.3.1 Indexing the reference genome

In order to perform an alignment, BWA first needs to construct the BWT and the FM-index for the reference genome. This is done with the following command.

```
$ bwa index -a bwtsw hs38DH.fa
```

The generation of the index took a little over 45 minutes on a consumer desktop, and produces the files listed in Table 8.2. Note that previous versions of BWA generated nine files, because the forward and reverse stranded indices were written to separate files. The index files are used as input for the read mapping algorithm described below.

Table 8.2. BWA index files

File	size	explanation
hs38DH.fa.amb	22K	A text file that records the locations of ambiguous ("amb"), non-ACGT nucleotides (e.g., N) in the reference
hs38DH.fa.ann	69K	A text file that records the name and length of the reference sequences
hs38DH.fa.bwt	3.1G	The BWT of the input sequence
hs38DH.fa.pac	773 M	The packaged sequence (in which four nucleotides are encoded as one byte)
hs38DH.fa.sa	1.6G	The suffix array index

8.4 READ GROUPS

Downstream analysis by Picard and GATK will require read group(s) that represent the provenance of the data and can be specified when performing read alignment (as we will see shortly). The read group defines the set (group) of reads that correspond to the same sample and were sequenced in the same run and same lane of a flow cell. This is important because algorithms such as the variant caller GATK need to know whether reads were sequenced together on the same lane or not (otherwise, it would be impossible to correct for lane-specific biases). Other downstream applications need to know whether the reads come from one or many biological samples. If available, detailed information about the read groups is stored in the header of the SAM alignment file produced by BWA, and each read in the file is assigned to one of the read groups.

The read group information should be composed in the following format

```
@RG\tID:group\tSM:sample\tPL:illumina\tLB:lib\tPU:unit
```

The "\t" stands for the tab character but should be written as \t rather than an actual tab. The meaning of the fields is as follows

- ID: The read group identifier. This tag must be unique because it identifies the read group each read belongs to. The full name of the ID will be written into the @RG line of the header of the SAM file, and will be referred to in the RG:Z tag of each record (Z stands for a printable "string", e.g. "RG:Z:group". See Chapter 9 for details on the SAM format). Ideally, the read group IDs are composed using the flow cell, lane name and number, making them a globally unique identifier across all sequencing data in the world (see also the section on Illumina's FASTQ read naming scheme in Chapter 5).

- SM: The name of the sample. GATK treats all read groups with the same SM as containing data for the same sample. The value of the SM field is also the name that is used for the sample column in the VCF file after variant calling (see Chapter 13).

- PL: The platform or technology used to produce the reads. Valid values are ILLUMINA, SOLID, LS454, HELICOS and PACBIO.

- LB: DNA preparation library identifier. The Picard module MarkDuplicates uses this information to recognize duplicates in case the same DNA library was sequenced on multiple lanes (see Section 10.1.4).

- PU: Platform unit. The PU holds three types of information and is formated as {FLOWCELL_BARCODE}.{LANE}.{SAMPLE_BARCODE}. FLOWCELL_BARCODE refers to the unique identifier for a particular flow cell. LANE indicates the lane of the flow cell and the SAMPLE_BARCODE is a sample/library-specific identifier. The PU is not required by GATK, but takes precedence over ID for base recalibration if it is present (see Section 10.4).

The files that we are using are from a single sample (NA12878), and a single lane, the platform is the Illumina HiSeq2500. The flow cell ID is H7AP8ADXX, the lane is 1 (remember the _L001 in the file name, indicating lane 1), and the sample barcode is TAAGGCGA (see Table 5.3 for information about how to find this information in standard Illumina read labels). We will therefore construct our read group as

```
@RG\tID:rg1\tSM:NA12878\tPL:illumina\tLB:lib1\tPU:H7AP8ADXX:1:TAAGGCGA
```

Alternatively, it is possible to use Picard to add read group information to the BAM file if desired using the `AddOrReplaceReadGroups` function. We will introduce both the BAM file and the Picard tool shortly. Finally, note that the read group does not always have to include all the information listed above.

8.5 READ MAPPING WITH BWA-MEM

We will now perform alignment with the files of paired reads obtained from Trimmomatic. Use the following command.

```
$ bwa mem -t 8 \
  -R '<read group info>' \
  hs38DH.fa \
  NIST7035_trimmed_R1_paired.fastq.gz  \
  NIST7035_trimmed_R2_paired.fastq.gz > NIST7035_aln.sam
```

We have abbreviated the read group info shown in the previous section as `<read group info>` for legibility. The `-t 8` flag causes BWA to use 8 threads. The output is directed to a new file named `NIST7035_aln.sam`. We will examine the format of SAM files in the following chapter, but for now suffice it to say that the files contain a header with general information about the alignment followed by information about each aligned read.[16] SAM files are plain text files that tent to take up a lot of disk space. However, the SAM output of BWA can be converted into a compressed version, the BAM file, using SAMtools [254]:

```
samtools view -Sb NIST7035_aln.sam > NIST7035_aln.bam
```

The option `S` and `b` indicate that the input is a SAM file and the output should be a BAM file.

---

[16]The SAM/BAM format is discussed in Chapter 9.

In practice, we can save disk space and avoid writing the intermediate SAM file by piping[17] the standard output of **bwa** directly as standard input into **samtools**:

```
$ bwa mem -t 8 -R '<read group info>' hs38DH.fa \
  NIST7035_trimmed_R1_paired.fastq.gz  \
  NIST7035_trimmed_R2_paired.fastq.gz | \
  samtools view -Sb - > NIST7035_aln.bam
```

In Chapter 7, we performed adapter and quality trimming of the reads, and also obtained two files with reads that no longer have a pair after trimming. It would be possible to align these reads and merge the BAM files of all of the data. This step would add complexity and time to the analysis pipeline, and there are alternate strategies that we discussed in the final section of Chapter 7. Both SAMtools and Picard can be used to merge BAM files. We will leave this step as an exercise for the reader.

8.6   SORTING AND INDEXING

Many downstream tools require the reads in the BAM file to be sorted according to chromosome and chromosomal position. Furthermore, many tools require the sorted BAM file to be indexed for analysis or visualization. We will cover these procedures in Chapter 10.

8.7   THE GRCH38 GENOME ASSEMBLY

The initial assemblies of the human genome were represented as a consensus haploid sequence for each chromosome. This consensus sequence was the best attainable consensus sequence for the human genome, and was called the golden path. Variants were represented with reference to the corresponding position of the golden-path assembly. However, this representation of the human genome is too simplistic and does not sufficiently take common large-scale structural variation into account. It has now been widely recognized that it is not possible to adequately represent genomic regions with substantial structural allelic diversity using a single consensus sequence for the human genome [75]. In fact, several chromosomal regions display a sufficiently high degree

---

[17]"Piping" uses the vertical bar or pipe symbol ('|') to pass the standard output from one tool to another.

of variability that they cannot be adequately represented by a single sequence, but instead require alternate sequences in form of alternate locus scaffolds (or alternate loci) [74].

The Genome Reference Consortium (GRC) therefore introduced a new graph-like assembly model with alternative sequence paths in regions with complex structural variation in the form of additional locus sequences. While the previous genome assembly, GRCh37, included three regions with nine alternate locus sequences, the GRCh38 assembly, which was released in December 2013, has a total of 178 regions with 261 alternate loci (Figure 8.1). An alternate locus is a sequence that is an alternate representation of a genomic region in a largely haploid assembly. Thus, alternate loci are provided for genomic regions that show substantial variability in the population and are embedded in an otherwise haploid representation of the genome.

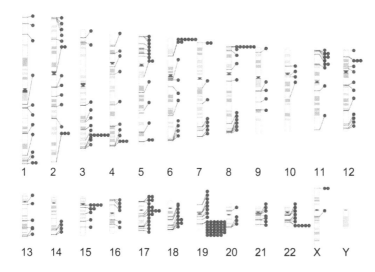

Figure 8.1. Alternate loci in the GRCh38 Genome Assembly. Genomic regions in the GRCh38 genome assembly with alternate locus scaffolds (alternate loci). The regions range from 33,439 to 5,081,216 nt in length (mean 344,634 nt, median 169,569 nt), with most regions being between 100 and 200 kilobases. The illustration was produced using PhenoGram [461].

We recently showed that stretches of sequences that are largely but not entirely identical between the primary assembly and an alternate locus can result in multiple variant calls against regions of the primary assembly. In WGS analysis, this results in characteristic and recognizable patterns of variant calls at positions that we term alignable scaffold-discrepant positions (ASDPs; Figure 8.2).

A

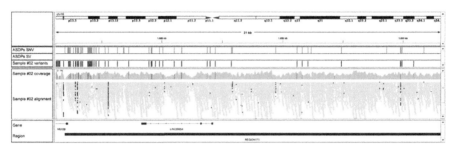

B

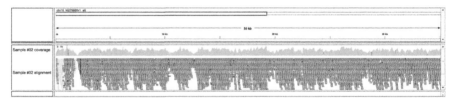

Figure 8.2. REGION173 vs. KI270856.1. (A) Numerous homozygous variants in REGION173 (1,871,114–1,903,127 on chromosome 16) were called that probably relate to misalignment of the reads to REGION173 on chromosome 16 rather than to the alternate locus KI270856.1 **(B)** The corresponding region on the alternate locus KI270856.1 displays far fewer discrepancies with aligned reads. The sequenced individual can be inferred to be homozygous for KI270856.1. See [183] for details on the analysis.

Our results, that we will not further summarize here, suggested that there are many challenges and opportunities for variant calling with the GRCh38 genome assembly model [183].

FURTHER READING

Li H, Homer N (2010). A survey of sequence alignment algorithms for next-generation sequencing. *Brief Bioinform* 11:473-483.

Jäger M, Schubach M, Zemojtel T, Reinert K, Church DM, Robinson PN (2016) Alternate-locus aware variant calling in whole genome sequencing. *Genome Med* **8**:130.

8.8 EXERCISES

Exercise 1

In this chapter, we have applied BWA-MEM to paired-end (PE) sequencing data. In this exercise, you can explore its use for single-end reads.

Check the online manual of BWA (`man bwa`) and find out how the aligner is used for single-end (SE) data. Use it to align the two FASTQ files with unpaired reads obtained from Trimmomatic (Section 7.2.3), and create the corresponding BAM files:

```
NIST7035_trimmed_R1_unpaired.fastq.gz
NIST7035_trimmed_R2_unpaired.fastq.gz
```

Exercise 2

BAM files can be merged using SAMtools or Picard. Find out how this works and test it by merging the two single-end BAM files created in Exercise 1 with the paired-end BAM file of the sample (`NIST7035_aln.bam`) to create a merged BAM file `NIST7035_aln_merged.bam`. (In the following chapters, you can substitute this file for `NIST7035_aln.bam` if you like, but obviously the results will be different.)

Exercise 3

The final exercise for this chapter is probably the simplest in the entire book: compare the file sizes of the SAM file obtained from BWA-MEM (`NIST7035_aln.sam`) and the BAM file after compression via SAMtools (`NIST7035_aln.bam`). What is the approximate rate of compression? For this reason, it is best to avoid writing an intermediate SAM file.

CHAPTER 9

SAM/BAM Format

Peter Robinson and Peter Hansen

S EQUENCE Alignment/Map (SAM) format is a tab-delimited text format that aims to be a universal format for storing alignments of NGS reads to a reference genome. BAM (Binary Alignment/Map) is the compressed (binary) version of SAM. The SAM/BAM format can store alignments from all major aligners, such as Bowtie 2 [236] or BWA [252, 253], supports indexing for quick searches and viewing (see Chapter 17), and is widely supported by variant calling software. There is an extensive online documentation of the format, and we will not attempt to cover all of the details here. Instead, with the aid of small toy examples, we will give intuitive explanations of the various components of the format. Furthermore, we will show how to analyze the characteristics of alignments represented by SAM/BAM files. For this purpose, we will mainly use SAMtools [254], although many of the procedures can also be performed with Picard, Sambamba [421], or SAMBLASTER [119].

9.1 SAM BASICS: SINGLE-END READS

We will illustrate the basic SAM format using a toy example in which five reads of length 10 are mapped to a reference sequence of length 40 (Figure 9.1). SAM files consist of an optional header section followed by an alignment section. Header lines start with '@', while alignment lines do not. In the alignment section of the SAM file, each line represents

Table 9.1. Mandatory fields of the SAM Format.

Col	Field	Description	Example
1	QNAME	Query template NAME	read_1
2	FLAG	Bitwise FLAG	0
3	RNAME	Reference sequence NAME	chrE
4	POS	Left-most mapping POSition (1-based)	11
5	MAPQ	MAPping Quality	37
6	CIGAR	CIGAR string	10M
7	RNEXT	Ref. name of the mate or NEXT read	*
8	PNEXT	Position of the mate or NEXT read	0
9	TLEN	Observed Template LENgth	0
10	SEQ	Segment SEQuence	ACGCATACTG
11	QUAL	Base QUALity string	DIGAFHHBCA

Note: Each line in the alignment section of a SAM file comprises 11 mandatory fields.

an alignment for one read and comprises at least 11 mandatory fields (Table 9.1).

The SAM/BAM format is intended to be as generic as possible, and uses the word *template* to refer (in the current context) to DNA fragments/inserts. The SAM format uses the word *segment* to refer to a contiguous sequence or subsequence; for instance, the two reads of a read pair can be referred to as two segments. The field QNAME stores the name of the query sequence, usually a read,[1] RNAME stores the name of the reference sequence, usually a chromosome (e.g., chr7), and POS denotes the position on the chromosome to which the read has been aligned. We will discuss the FLAG field in detail in Section 9.6. The field MAPQ represents the mapping quality assigned by the aligner and reflects the confidence with which the read could be mapped to the indicated position. The CIGAR string provides a compact representation of the alignment that will be explained below. The fields RNEXT, PNEXT and TLEN are only used for paired-end reads. Therefore, for single-end reads, RNEXT is set to "*" for *not-applicable* and PNEXT and TLEN are set to 0. Finally, the field SEQ contains the nucleotide sequence of the read and the field QUAL shows the base quality scores for each position.

---

[1]Segments having the same QNAME are regarded as originating from the same template, i.e., reads from a read pair usually have the same QNAME.

```
@SQ SN:chrE LN:40
read_1  0 chrE  9 37    10M * 0 0 TAACGCATAC JJJJIGIHIJ
read_2  0 chrE 10 37    10M * 0 0 AATGCATACT JIJIIIIHGD
read_3  0 chrE 11 37 4M2I4M * 0 0 ACGCAAATAC IIIIIIHHFH
read_4  0 chrE 13 37 4M1D6M * 0 0 GCATCTGTTG JHHJIGIHIJ
read_5 16 chrE 21 37    10M * 0 0 TTGTCCCGGA HFHHIIIIII
```

(a) SAM file

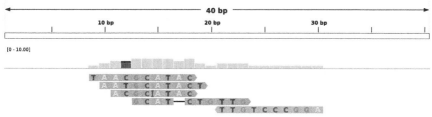

(b) Integrative Genomics Viewer (IGV)

Figure 9.1. SAM format basics: SE sequences. Toy example for single-end read mappings represented in **(a)** SAM format and **(b)** in the Integrative Genomics Viewer (IGV) [429] (Chapter 17). Five reads of length 10 are mapped to a reference sequence of length 40. The first read matches perfectly and starts at position 9. The second read starts at position 10 and has one mismatch. The third read has an insertion of two As, which is encoded by the CIGAR string 4M2I4M (the insertion is shown as a vertical bar in IGV). The fourth read has a deletion, encoded by the CIGAR string 4M1D6M. Unlike the first four reads, which map to the forward strand, the fifth read maps to the reverse strand. In the SAM file this is encoded by the bitwise flag 0 for the forward strand and 16 for the reverse strand. In IGV, reads can optionally be colored by strand: here, reads mapping to the forward strand are shown in red and reads mapping to the reverse strand are shown in blue.

9.2   SAM BASICS: PAIRED-END READS

In SAM files, read pairs are represented on two lines. There are two fields that are used to refer to the reference sequence and position of the "next" aligned segment of the read pair, RNEXT and PNEXT. If both members of the pair map to the same chromosome, as is almost always the case, RNEXT is set to '=' and PNEXT is set to the start position of the paired read (Figure 9.2). The TLEN field (template length) shows the distance between the left-most and right-most mapped base of a pair, if both are mapped to the same reference. In such cases TLEN of the leftmost segment is positive and TLEN of the right-most segment is negative, but their absolute value is the same for both members of the pair and corresponds to the inferred insert size. If both members of the pair are properly mapped to the forward and reverse strand, this is indicated by bitwise flags of 99 and 147, or 83 and 163. The bitflags for paired-end reads indicate whether both members of a pair are mapped or not, or if only one read of the pair is mapped, or if reads are mapped in correct orientation. Bitwise flags are explained in detail in Section 9.6.

9.3 CIGAR STRINGS

A CIGAR[2] string is comprised of a series of operation lengths plus the operations that describe how exactly a read has been aligned to the reference sequence (Figure 9.3). The three most important CIGAR operations are M (Match/Mismatch), I (Insertion), and D (Deletion), but there are six additional operations, some of which will be discussed below. All CIGAR operations are summarized in Table 9.2. A CIGAR string can be thought of as a series of operation lengths plus the CIGAR operation, which allows an abstract and compact representation of arbitrary alignments. For instance, consider the following alignment, in which the query sequence (Q) is aligned to the reference sequence (R) beginning at position 7 of the reference sequence.

```
AGCATGTTAGATAA--GATAGCTGG  R
      || |||||  |||| ||||
------TTGGATAAAGGATA-CTGG  Q
```

There are 8 initial matches, then an insertion of 2 nucleotides in the query, then four additional matches, then a deletion of one base in the query, and then four more matches. We can represent this as 8M2I4M1D4M. Note that the third base of the query sequence (G) is actually different from the base of the reference (A) to which it has

---

[2]CIGAR stands for Concise Idiosyncratic Gapped Alignment Report.

```
@SQ SN:chrE LN:40
read_1  99 chrE  2 42 10M = 27  35 CTCGTTATAA JJJJIGIHIJ
read_2  99 chrE 11 42 10M = 30  29 ACGCATACTG JHHJIGIHIJ
read_1 147 chrE 27 42 10M =  2 -35 CGGATGGAAA DGHIIIIJIJ
read_2 147 chrE 30 42 10M = 11 -29 ATGGAAAACC HFHHIIIIII
```

(a) SAM file

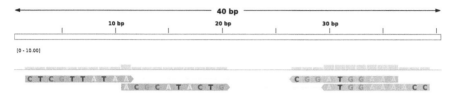

(b) Integrative Genomics Viewer (IGV)

Figure 9.2. SAM format basics: PE sequences. Toy example for paired-end read mappings represented in (a) SAM format and (b) in IGV. Two pairs consisting of four reads of length 10 are mapped to a reference sequence of length 40. Note that members of the same pair have the same QNAME and map to different strands. Read 1 is mapped to the forward strand at position 2 and forms a pair with the read mapped to the reverse strand at position 27. RNEXT is set to =, because the second read of the pair is mapped to the same reference, i.e., chrE, and PNEXT is set to 27, because the reverse read is mapped to this position (the reverse read is aligned to positions 27–36). Therefore, TLEN is set to 35, because this is the distance between the left-most and right-most mapped base of the two reads. The TLEN entry for the reverse read is shown with a minus sign.

Different bitwise flags are used for paired-end reads than for single-end reads. The flags 99 and 147 indicate that both members of the pair were mapped properly to the forward and the reverse strand.

been aligned. Nonetheless this is an alignment match (M), although it is not a sequence match.

9.4 CLIPPED READS

Major mapping tools such as BWA[252] try to map a portion of a read if they cannot map the full-length read to the reference genome. In these

Table 9.2. CIGAR operations

Op	Description
M	alignment match (sequence match or mismatch)
I	insertion (additional non-reference base)
D	deletion (reference base missing in the read)
N	skipped region from the reference
S	soft clipping (clipped sequences still present in SEQ)
H	hard clipping (clipped sequences not present in SEQ)
P	padding (silent deletion from padded reference)
=	sequence match
X	sequence mismatch

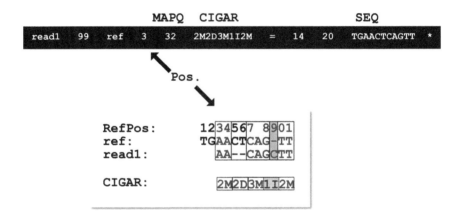

Figure 9.3. CIGAR format. The line represents an alignment of read1 beginning at position 3 of the reference (*ref*). The alignment consists of two matched nucleotides followed by a deletion of 2 nucleotides, three matching nucleotides, and insertion of one nucleotide, and two matching nucleotides. For legibility the base quality string was replaced by an asterisk (see text).

cases, the unalignable portion can be excluded from the alignment by a process called clipping. With soft-clipping, which is shown in the CIGAR string using an "S", the clipped sequence bases aren't removed from the SEQ string, but are not used by variant callers and are not shown in viewers such as IGV. Hard-clipping (H) is similar to soft

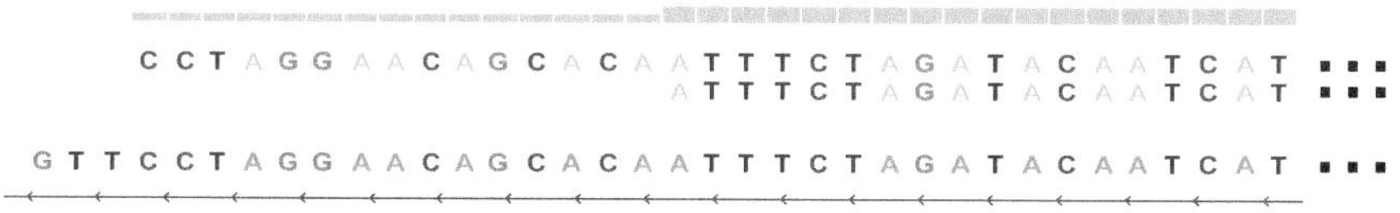

```
C C T A G G A A C A G C A C A A T T T C T A G A T A C A A T C A T ▪ ▪ ▪
                      A T T T C T A G A T A C A A T C A T ▪ ▪ ▪

G T T C C T A G G A A C A G C A C A A T T T C T A G A T A C A A T C A T ▪ ▪ ▪
```

(a)
```
Ref:    GTTCCTAGGAACAGCACAATTTCTAGATACAATCAT
Read1:     CCTAGGAACAGCACAATTTCTAGATACAATCAT
Read2:       ggtcacatgattgtATTTCTAGATACAATCAT
```

(b)

Figure 9.4. Soft clipping (a) An IGV screenshot of the 5' end of an alignment of two reads to the reference. The second read was soft-clipped; it was aligned with REF=chr4, POS=16034528, and a CIGAR of 14S87M. The first 14 nucleotides were thus soft-clipped. They were not removed from the sequence in the SAM file, but are not shown in IGV. **(b)** The corresponding alignment of Read1 and Read2 with Ref. The soft-clipped (unalignable) bases of read2 are shown in lower case.

clipping (S) but differs in that the hard-clipped subsequence is not present in the alignment record (Figure 9.4).[3]

9.5 MAPPING QUALITY

The fifth field of a SAM file reports the mapping quality of the read (see also Chapter 8). The MAPQ score (MAPping Quality) reflects the probability that the read is aligned to the wrong position in the genome. Like the base quality score described in Chapter 5, it is a Phred-scaled posterior probability that the mapping position indicated by the aligner is incorrect

$$MAPQ = -10 \log_{10} P(\text{mapping position wrong}) \qquad (9.1)$$

The MAPQ is rounded to the nearest integer. For instance, one in a thousand reads with a MAPQ of 30 would be predicted to be mis-

---

[3]BWA utilizes soft clipping for the primary alignment so that the original raw data can be regenerated from the BAM file if needed. This isn't necessary for the secondary alignments, and therefore BWA uses hard clipping to conserve disk space.

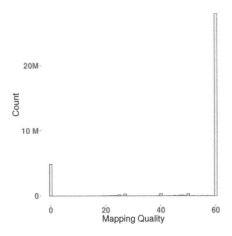

Figure 9.5. Mapping quality. Distribution of MAPQ (mapping quality) values for aligned reads from the NA12878 exome.

aligned. A MAPQ value of 255 indicates that the mapping quality is not available.

One of the major challenges for read alignment algorithms is the existence of repetitive sequences in the human genome such that a short read may align equally well to two or more positions. In this case, BWA assigns the read a MAPQ of zero and picks one of the positions at random. As can be seen in Figure 9.5, BWA-MEM assigned most reads a mapping quality score of 60, with the second most common score being zero, and it assigned a much smaller number of reads intermediate scores. Readers should be aware of the fact that different aligners use different methods for calculating mapping quality, and the MAPQ scores obtained by different aligners are not generally directly comparable.

9.6 THE SAM BITFLAG

The second field of each record in a SAM file represents a bitfield with values from 12 bitflags (Table 9.3).

A bitfield is used to store a set of Boolean (Yes/No) values in a compact fashion. In our case, we want to store 12 Yes/No attributes about each read. In theory, we could store twelve separate char values (one byte each), but if we note that the 12 values can be stored as the individual bits of a bitfield, we can save a substantial amount of

Table 9.3. SAM Format: Bitflags.

Bit (hex)	Bit (dec)	Description
0x1	1	Template has multiple segments (multiple reads, usually a read pair)
0x2	2	Each segment of the template is properly aligned according to the aligner
0x4	4	Segment is unmapped
0x8	8	Next segment in the template is unmapped
0x10	16	SEQ is reverse complemented
0x20	32	SEQ of the next segment in the template is reverse complemented
0x40	64	First segment in the template
0x80	128	Last segment in the template
0x100	256	Secondary alignment
0x200	512	Segment does not pass quality controls
0x400	1024	Segment is a PCR or optical duplicate
0x800	2048	Supplementary alignment

Note: The current SAM format uses 12 bitflags, each of which can have a value of 1 (Yes, True) or 0 (No, False). The bitflags can be combined independently of one another, and are displayed in SAM files as the corresponding decimal integer value.

storage. Table 9.3 displays the twelve bitflags, and Figure 9.6 shows an example of how the bitflags are represented as a decimal integer value. In order to understand how the bitfields are represented, it is helpful to recall how the positional notation of numbers works. In familiar decimal notation, the number 453 stands for $4 \times 10^2 + 5 \times 10^1 + 3 \times 10^0 = 400 + 50 + 3 = 453$. In binary notation, a number such as 101101 stands for $1 \times 2^5 + 0 \times 2^4 + 1 \times 2^3 + 1 \times 2^2 + 0 \times 1^1 + 1 \times 2^0 = 32 + 8 + 4 + 1 = 45$. It is often convenient to use hexadecimal ("hex") notation, where the base is not 10 or 2 but 16; the digits are represented by 0–9 followed by A,B,C,D,E, and F. To be fully comfortable with SAM files and using programs such as SAMtools to work with them, it is helpful to understand how to convert between these three number representations.

For example, a value of 99 indicates that the read has multiple segments in sequencing (usually this refers to a paired-end or mate-

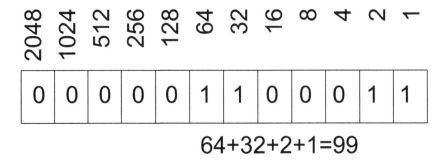

Figure 9.6. In this example, the bitflags 0x1 (1 in decimal notation), 0x2 (2), 0x20 (32), and 0x40 (64) are set. The decimal representation as displayed in the SAM file is thus $64+32+2+1=99$

pair read; $0x1=1\times16^0=1$), that each segment of the read could be mapped properly (correct mapping of both reads of the read pair, $0x2=2\times16^0=2$), that the mate read is mapped to the reverse strand ($0x20=2\times16^1=32$), and that the read being referred to in the current SAM record is the first segment in the template (the first read in the pair, $0x40=4\times16^1=64$) (Figure 9.6 and Table 9.3).

Bit operations can be used to filter reads in a SAM file. For instance, the SAM format requires that for each read one and only one of the associated lines satisfies

```
FLAG & 0x900 == 0
```

This line is called the *primary line* of the read. The hexadecimal value 0x900 is equal to 0x800 plus 0x100, meaning that the bit flags for supplementary alignment (0x800) and secondary alignment (0x100) are set. Recall that & stands for the bitwise logical AND operation on a pair of bits. If both bits are 1, the result of the operation is 1, otherwise 0. For example

```
        1010
AND     1100
        =1000
```

Thus, if a read is annotated either as a secondary alignment[4] or a

---

[4]If a read maps to multiple locations due to repetitive sequences or other reasons, then one of these alignments would be marked primary and all others will have the secondary alignment bitflag set.

supplementary alignment,[5] then we will obtain `FLAG & 0x900 != 0`.[6] The SAM format states that an arbitrary number of lines may represent secondary or supplementary alignments, but that only one line can be the primary (representative) alignment.

SAMtools can filter alignments to output only those aligned reads that match certain criteria, including overlapping with regions indicated in a BED file (`-L`), belonging to certain read groups (`-r`, `-R`), originating from a certain library, or having a certain minimum number of CIGAR bases consuming the query sequence.[7] The option `-f <INT>` will only output alignments with all bits set in `INT` present in the `FLAG` field (`INT` can be specified as a decimal number or as a hexadecimal number by beginning with '0x'). This performs an AND operation on all of the bits set in `INT`, meaning that sequences will only be shown if at least the corresponding bit positions are set to 1 in `FLAG`. The option `-F <INT>` does not output alignments with any bits set in `INT` present in the `FLAG` field. The `-F` option thus skips any read for which `FLAG & INT != 0`. So, to output only primary reads, we can use the following command.

```
$ samtools view -F 0x900 NIST7035_aln.bam
```

The reads that are output with this command had a total of 38 different values for the bitfield, none of which included the bits 0x100 or 0x800. The following command will extract all of the secondary alignment reads.

```
$ samtools view -f 0x100 NIST7035_aln.bam
```

The `flagstat` function of SAMtools provides a summary of the number of records corresponding to each of the bit flags of Table 9.3. For instance, the evaluation of our example dataset shows the following.

```
$ samtools flagstat NIST7035_aln37.bam
35210329 + 0 in total (QC-passed reads + QC-failed reads)
```

---

[5]According to the SAM specification, a chimeric alignment is an alignment of a read that cannot be represented as a linear alignment. Usually this means that a read consists of multiple segments that align to different parts of the genome. The segments do not have large overlaps. One of the segments is considered to have the representative alignment, and the others are called supplementary and have the supplementary alignment flag set.

[6]The result could be 0x100, 0x800 or 0x900.

[7]The `M/D/N/=/X` operators are said to consume reference bases.

```
0 + 0 duplicates
35210007 + 0 mapped (100.00%:-nan%)
35210329 + 0 paired in sequencing
17605154 + 0 read1
17605175 + 0 read2
34991352 + 0 properly paired (99.38%:-nan%)
35209746 + 0 with itself and mate mapped
261 + 0 singletons (0.00%:-nan%)
19085 + 0 with mate mapped to a different chr
9718 + 0 with mate mapped to a different chr (mapQ>=5)
```

9.7 SAM TAGS: OPTIONAL FIELDS

An arbitrary number of optional fields may follow the 11 mandatory fields. All optional fields follow the `TAG:TYPE:VALUE` format where `TAG` is a two-character string, `TYPE` is one of six data types (Table 9.4), and `VALUE` is the actual value.

Table 9.4. SAM format: data types for the optional tags

Type	Description
A	Single character
Z	String
i	Signed 32-bit integer
f	Single-precision float (real number)
H	Hexadecimal number string
B	General array

Different aligners use different optional fields. Readers can find full documentation of common fields at the SAMtools Website,[8] and aligners such as BWA also provide documentation about the fields they use. Here, we will present several selected fields to give readers a feeling of how these fields are used in practice.

9.7.1 The NM field

The `NM` tag, which is predefined, takes an integer value (`i`). The `NM` field indicates the edit distance to the reference, including ambiguous

---

[8]`https://samtools.github.io/hts-specs/SAMtags.pdf`

bases[9] but excluding clipping. Recall that the CIGAR string 101M indicates that a 101 nt long read was aligned to the genome without gaps. However, it does not make any statement about whether the alignment contained mismatching positions (sequence mismatches). A CIGAR string of 101M and the tag NM:i:0, however, matches perfectly (edit distance of zero). A read with a CIGAR string of 101M and the tag NM:i:1 has one mismatched base (edit distance of one).

9.7.2 The MD field

The MD field provides additional, reference-centered, information about the alignment. It is a string for mismatching positions that allows one to infer where single-nucleotide variants (SNVs) and deletions are located without looking at the reference. It is easiest to explain how the MD field works using an example. One of the reads in the NA12878 exome was mapped to chr1:21989502 with a mapping quality of zero (it maps equally well to another position on chromosome 1), and a CIGAR string of 101M. Consider its MD tag:

```
MD:Z:2C77A4G4A5C4
```

This means that there are two sequence matches (beginning at the position POS=21989502 on the reference chr1), followed by a mismatch with a C in the reference, 77 matches, a mismatched A, 4 more matches, a mismatched G, 4 matches, another mismatched A, 5 matches, a mismatched C, and finally 4 matches. Compare the alignment of the read with the corresponding region of chromosome 1:

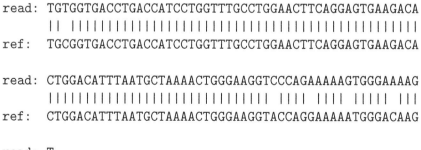

```
read:  TGTGGTGACCTGACCATCCTGGTTTGCCTGGAACTTCAGGAGTGAAGACA
       ||  ||||||||||||||||||||||||||||||||||||||||||||||
ref:   TGCGGTGACCTGACCATCCTGGTTTGCCTGGAACTTCAGGAGTGAAGACA

read:  CTGGACATTTAATGCTAAAACTGGGAAGGTCCCAGAAAAAGTGGGAAAAG
       ||||||||||||||||||||||||||||| |||| |||| ||||| |||
ref:   CTGGACATTTAATGCTAAAACTGGGAAGGTACCAGGAAAAATGGGACAAG

read:  T
       |
ref:   T
```

---

[9] "N" bases

Let us now consider an example with a deletion. This is a read mapped to chr1:31504512 with a CIGAR string of `11M5D60M` (total length of mapped read 83 nucleotides) and a tag of `MD:Z:11^TTTTG6G23G29`. The CIGAR string tells us that the read is aligned for the first 11 bases, then has a deletion of 5 nucleotides, and again is aligned for 60 bases. The CIGAR string does not tells us what bases were deleted, nor does it tell us whether the aligned bases are matches or mismatches. The `MD` string tells us that the first 11 bases perfectly match, that the bases TTTTG were deleted (this is denoted by the caret character ^ followed by the deleted bases, `^TTTTG`), the following 6 bases matched followed by a mismatched G in the reference, 23 matching bases followed again by a mismatched G in the reference, followed by 29 matching bases.

```
read:  TTGGGCAAGTT.....TTTTTTTTTTTTTTTTTTTTTTTTGAGACAGAG
       |||||||||||     ||||||  ||||||||||||||||||||||| |||
ref:   TTGGGCAAGTTTTTTGTTTTTTTGTTTTTTTTTTTTTTTTTTTGAGACGGAG

read:  TCTCTCTCTGTTGCCCGGGCTGGAGT
       ||||||||||||||||||||||||||
ref:   TCTCTCTCTGTTGCCCGGGCTGGAGT
```

If there are multiple adjacent mismatches, then a 0 is used. For instance

```
read:  CGATACGGGGAC
       |   |||  ||||
ref:   CACTACTCGGAC
```

This would produce the CIGAR 12M (twelve aligned positions without insertions or deletions) and the `MD` string `MD:Z:1A0C3T0C4`, indicating that between the mismatched A and C and T and C there are no (0) matching bases. In case the first or last base is a sequence mismatch, it is also preceded or followed by a 0 (e.g., `MD:Z:0A100` or `MD:Z:100A0`).

Note that insertions aren't specified in the `MD` field because it is reference-centered and an insertion doesn't constitute a loss of information about the reference. Moreover, the inserted bases can be inferred from the `SEQ` field together with the CIGAR string of a read. The CIGAR string `30M1I70M`, for example, corresponds to `MD:Z:100` if all aligned bases (`M`) show sequence matches. The read has a total of 101 bases but only the 100 reference bases to which it aligns are described in the `MD` tag.

The MD field must be compatible with the CIGAR string (excluding insertions).

9.7.3 The RG field

The RG field indicates the read group of the read, e.g., RG:Z:rg1. See Chapter 8 for information about read groups.

9.7.4 The AS field

The AS field indicates the alignment score generated by the aligner. For instance, AS:i:84 indicates that BWA-MEM assigned the read an alignment score of 84.

9.7.5 End-user reserved fields

The fields X?:?, Y?:?, and Z?:? are reserved for end-users. This means that alignment programs such as BWA are allowed to define their own fields whose tags begin with the letters X, Y, or Z. BWA has defined a number of tags that begin with X. Note that the different BWA programs use different combinations of tags. For instance, the tag XA shows alternative alignments using the format: chr,pos,CIGAR,NM; for each alternative alignment. For example,

```
XA:Z:chr1,+13074589,101M,3;chr1,-13152100,101M,3;chr1, \
-12882405,101M,3;chr1,-12827800,101M,4; \
chr1_KI270766v1_alt,-97694,98M3S,3;
```

This tag shows that the read could also be mapped to alternative locations. The first alternative hit was on chromosome 1 at position 13,074,589, it was aligned without indels (CIGAR 101M), and the edit distance (NM) was three. A negative position indicates that the alternative alignment lies on the reverse strand. Other optional tags of BWA include XS, which shows the suboptimal alignment score.

9.8 BAM AND CRAM

The binary (compressed) version of SAM is a BAM file, which is the main file type used for storing and analyzing read alignments. BAM is compressed in the BGZF (Blocked GNU Zip Format) format. In practice, BAM files are indexed to allow fast retrieval of alignments that overlap specified regions (without the search needing to go through each of the reads one by one). As we will see in Chapter 10, BAM files need to be sorted according to chromosomal position before they can be indexed.

The BAM format is able to compress aligned reads by about 50–80% of their original size, but with the steady growth in the amount of data being produced, even that is not sustainable and attempts are being made to increase the degree of compression even further. CRAM files are alignment files like BAM files, but use additional compression approaches including reference-based compression to improve the efficiency [42, 171]. CRAM can compress BAM files in a lossless fashion (meaning that one can retrieve 100% of the information from the original BAM file after conversion from BAM to CRAM and back to BAM). Alternatively, it is possible to convert a BAM file to a CRAM file in a way that the quality scores are binned (saving additional space) or even discarded entirely if desired. We showed how to use SAMtools to convert a CRAM file to a BAM file on page 53. At the time of this writing, CRAM was undergoing steady development, and documentation was available at the Website of the European Bioinformatics Institute.[10]

FURTHER READING

The SAM/BAM Format Specification Working Group (2016) Sequence Alignment/Map Format Specification. `https://samtools.github.io/hts-specs/SAMv1.pdf` (latest version at the time of this writing: April 28, 2016).

9.9 EXERCISES

Exercise 1

With the above knowledge in hand, let us perform some exercises to deepen our knowledge of the BAM format. The first exercise will deal with the bitflag. We saw that the bitflags 99 and 147 were assigned to our paired-end reads. What do they mean? Write out these numbers in binary and hexadecimal representation and consult Table 9.3.

Exercise 2

What does the bitflag 16 mean?

---

[10]`http://www.ebi.ac.uk/ena/software/cram-toolkit`

Exercise 3

Determine the CIGAR strings, 1-based positions (POS) and the NM and MD tags for the following alignments:

(a)

```
ref:   CCATACT-GAACTGACTAAC
          ||| ||| || ||
read:     ACTAGAA-TGGCT
```

(b)

```
ref:   ATCCCCTCATCC-GCCTGCTCCTTCTCACATG
          |  | ||||| |||||||||||    |||||
read:    CTACGCATCCGGCCTGCTCCTT---ACATG
```

Exercise 4

Write down the alignment for the following read

```
read:   CATTCATACTGAA
```

to the reference genome

```
ref:   AGCATTACTACTAAATTT
```

where POS is 3 and the CIGAR string is 4M1D1M1I7M.

Exercise 5

If the CIGAR string is 50M and we have MD:Z:4A45 what is NM?

Exercise 6

If the CIGAR string is 6M1D23M and we have MD:Z:1T4^T1C21 what is NM?

Exercise 7

Write down a function, method, or program in a programming language of your choice that returns the CIGAR representation of a given alignment (using only the operations M, I and D). For instance, in the C programming language the prototype could look like:

```
char *cigarize(const char *reference, const char *read);
```
The arguments `reference` and `read` are null-byte terminated strings of the same length that have been previously aligned, with gaps in the sequences indicated using spaces or dashes.

Exercise 8

Use CRAM tools to compress a BAM file. Follow the instructions at the CRAM Website.[11] Determine how efficient compression is with various settings for lossless versus lossy compression.

See hints and answers on page 500.

---

[11]http://www.ebi.ac.uk/ena/software/cram-toolkit

Postprocessing the Alignment

W^{E} have seen how to align reads with BWA-MEM in Chapter 8, and we have explored how aligned reads are specified in SAM/BAM format in chapter 9. However, the raw alignments we obtain from BWA-MEM are not yet ready for downstream analysis and require several postprocessing steps. In this chapter, we will use the Picard Tools suite to validate the BAM file, sort the reads according to their genomic location, mark or remove duplicate reads, and index the BAM file.

We will then begin to explore the Genome Analysis Toolkit (GATK) [104, 285]. GATK is probably the most widely used variant caller, but it is a large software package that also has major functionalities for further improving the alignment and the quality scores of the BAM file prior to final variant calling.

The postprocessing of the alignment we will discuss in the following sections is often also referred to as "post-alignment" processing.

10.1 PICARD TOOLS

The Picard Tools suite is designed to work with the GATK variant calling software which we will examine later in this chapter and in Chapter 12. Both Picard and GATK are developed at the Broad Institute. The Picard suite contains numerous tools for manipulating NGS data and formats such as SAM/BAM and VCF, and focuses on post-alignment processing tasks required to get raw alignments ready for variant calling with GATK. The software can be downloaded at the Picard GitHub page.[1] We have used version 2.8.0 in this book, and downloaded the executable jar file (`picard.jar`). You will require Java version 1.8 or later to run Picard. To test whether Picard can be run on your system, enter the following command, which will also give you detailed information about the tools contained in the Picard suite.

```
$ java -jar picard.jar -h
```

Picard contains numerous tools, each of which can be invoked with a command in the following scheme.

```
$ java jvm-args -jar picard.jar PicardToolName \
  OPTION1=value1 \
  OPTION2=value2...
```

The Java Virtual Machine arguments (`jvm-args`) control aspects such as memory usage. For instance, `-Xmx8g -Xms2g` would start the Java virtual machine with 2GB memory and allow it to use a maximum of 8GB memory.

10.1.1   Validating the SAM/BAM file

Occasionally, downstream tools such as GATK will emit difficult-to-understand error messages that may be related to misformated SAM/BAM files. It can be useful to run the `ValidateSamFile` utility of Picard, which is able to check both SAM and BAM files.

```
$ java -jar picard.jar ValidateSamFile \
  INPUT=NIST7035_aln.bam \
  MODE=SUMMARY
```

When run in this fashion, `ValidateSamFile` will emit several lines of information. If the file is correctly formated, Picard will then emit the diagnosis: "No errors found".

10.1.2   Sorting and marking duplicates, and indexing the BAM file

In Chapter 6, we explained how PCR duplicates can originate during library preparation. PCR duplicates can lead to many kinds of bias in downstream applications, but are perhaps particularly important for

---

[1]`https://broadinstitute.github.io/picard/`

variant calling. PCR duplicates can lead to incorrect coverage assessments and erroneous variant calls [354] (Figure 10.1). This is because variant calling algorithms assume that all reads are independent, and if one particular fragment underwent substantial amplification such that say 90% of all reads covering a certain true, heterozygous variant originate from this fragment, then the variant might be called as homozygous instead of heterozygous (because only about 50% of the remaining reads, or a total of 5% of all reads, would have the reference sequence at this position). For this reason, it is highly recommended to remove duplicate reads prior to further downstream analysis. Even though some of the duplicate reads might be "real" (in the sense of two independent fragments that happen to start at the same position), in practice, the overwhelming majority of duplicate reads are PCR duplicates that would bias the result.

To do so, we will use Picard to sort the BAM file and then to mark duplicates.

10.1.3   Picard: SortSam

We will use the following command to sort the aligned reads of the BAM file that was produced using BWA-MEM and SAMtools (see Section 8.5). Sorting is performed according to genomic coordinates.

```
$ java -jar picard.jar SortSam \
    INPUT=NIST7035_aln.bam \
    OUTPUT=NIST7035_sorted.bam \
    SORT_ORDER=coordinate
```

This command creates a sorted BAM file, NIST7035_sorted.bam.

10.1.4   Picard: MarkDuplicates

Picard can mark, and optionally remove, duplicate reads from *sorted* BAM files. The MarkDuplicates tool works by comparing sequences with identical 5' positions (for single reads if single-end sequencing was performed, and for both reads if paired-end sequencing was performed). Reads with identical 5' positions and identical sequences are marked as duplicates.

```
$ java -jar picard.jar MarkDuplicates \
    INPUT=NIST7035_sorted.bam \
```

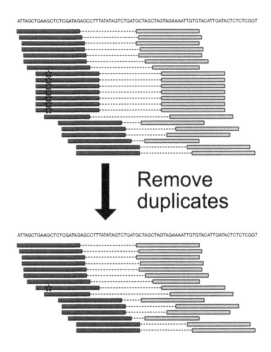

Figure 10.1. PCR duplicates. In this example, paired-end reads are shown as blue (forward read) and green (reverse read) rectangles connected by a dashed line. The red star on a subset of reads symbolizes a PCR error that has led to the incorporation of one erroneous base. PCR amplification of an individual molecule with the erroneous base leads to overrepresentation of the corresponding variant in the alignment. Removal of the duplicates removes all but one instance of the error. Downstream variant calling algorithms might be led to call a (false positive) variant if duplicates were not removed.

```
OUTPUT=NIST7035_dedup.bam \
METRICS_FILE=NIST7035.metrics
```

This creates a file called `NIST7035_dedup.bam` with the same contents as the input file, except that any duplicate reads are marked as such. Essentially, this means that the `0x400` bit is set in the bitfield of SAM records determined to be duplicates (see Section 9.6 for an explanation of bitfields).[2] We can check the result of MarkDuplicates by the following command, which uses SAMtools to view a BAM file as a SAM file and display only those records with the bitflag bit 0x400:

```
$ samtools view -f 0x400 NIST7035_dedup.bam
```

If we use the -c option to count matching entries without printing them, we see that there are a total of 2,134,576 duplicates.

```
$ samtools view -c -f 0x400 NIST7035_dedup.bam
2134576
```

As expected, performing the same command on the input file (NIST7035_sorted.bam) fails to reveal any duplicates. We check the total number of sequence lines in the BAM file using the following command.

```
$ samtools view -c -F 0x100 NIST7035_dedup.bam
35203956
```

The -F 0x100 flag[3] filters out alignments that are not the primary alignment. If we do not use this filter, then SAMtools will output the total number of mapped alignments, including those marked as secondary (see Chapter 9).

With these numbers, the overall rate of duplicate reads is calculated as 6.1%, which is an acceptable amount. A look into the NIST7035.metrics file produced by MarkDuplicates confirms this estimate of the duplication rate and shows additional information.

If we run MarkDuplicates with the TAGGING_POLICY=ALL option, we can get a readout of the number of duplicates judged to be PCR duplication artifacts compared to the number of optical duplicates, i.e., duplicate reads from a single amplification cluster that were incorrectly detected as multiple clusters by the optical sensor of the sequencing instrument.

```
$ java -jar picard.jar MarkDuplicates \
    INPUT=NIST7035_sorted.bam \
```

---

[2]With the additional option REMOVE_DUPLICATES=true duplicate reads can also be completely dropped. This reduces the file size of the output, but generally it is advisable not to discard original data.

[3]Equivalently, -F 256

```
OUTPUT=NIST7035_dedup.bam \
METRICS_FILE=NIST7035.metrics \
TAGGING_POLICY=All
```

This will cause the records of duplicated reads to be annotated with values for the DT tag as either library/PCR-generated duplicates (LB), or sequencing-platform artifact duplicates (SQ). It is instructive to look at an example of an optical duplicate in more detail (Table 10.1).

Table 10.1. Optical duplicates

Tile	X	Y	pos (f)	pos (r)	SQ ?
2105	7227	36599	245155301	245155319	No
2105	7238	36610	245155301	245155319	Yes
2115	19853	57971	245155319	245155349	No
2206	3972	38490	245155319	245155386	No

Note: Four read pairs that map to chr1:245,155,319. The second read was marked as an optical duplicate of the first because the start and end positions are identical, the reads come from the same tile, and the X and Y coordinates within the tile are near to one another. The other two reads are from different tiles and also have distinct start and end coordinates. The 'SQ ?' column shows whether MarkDuplicates marked the read as an optical duplicated (DT:Z:SQ in the BAM file).

10.1.5   Picard: BuildBamIndex

```
$ java -jar picard.jar BuildBamIndex \
  INPUT=NIST7035_dedup.bam
```

This creates an index file called NIST7035_dedup.bai, which is ready to be used in the variant calling workflow of the next chapter. Note that Picard will also create a BAM index if the option CREATE_INDEX=true is added to any command that writes a coordinate-sorted BAM file.

Alternatively, a sorted BAM file can also be indexed using SAMtools:

```
$ samtools index NIST7035_dedup.bam
```

Now, the BAM file is ready for fine-tuning by indel realignment and recalibrating the base qualities with GATK.

10.2 GATK

GATK has excellent and comprehensive documentation covering a very wide range of use cases. Because of the copious range of functions, GATK is an extremely powerful and flexible tool for variant calling, but this comes at the price of a steep learning curve. There is no single GATK command that is good for all situations, and potential users of GATK are well advised to study the documentation carefully and to check and validate their results. In this book, we have chosen one path through the variant calling process and aim to provide a thorough and intuitive explanation of the procedures and their effect on the read data. There are two main papers about GATK, the first of which describes the Map-Reduce framework [285], and the second of which covers the base quality score recalibrator, indel realigner, SNP calling, variant quality score recalibrator and their application to deep whole genome, whole exome, and low-pass multi-sample calling [104].[4]

The GATK best practices workflows provide step-by-step recommendations for a number of scenarios, including the calling of single nucleotide variants (SNVs) and small indels [438]. In this book, we will be discussing many but not all components of the best-practices pipeline, and will put emphasis on explaining how individual tools, methods, and approaches work rather than how to set up the final pipeline, which in any case needs to be adapted to local resources, requirements, and scientific or medical goals. Readers should also be aware of the fact that the GATK team is very actively developing GATK, and useful new features may be introduced with new versions.

To download GATK, go to the GATK website,[5] and find the download site in the menu. You can obtain GATK for free under an academic or not-for-profit license. The current version of GATK is 3.7,[6] and the software can be uncompressed as follows

```
$ tar jxf GenomeAnalysisTK-3.7.tar.bz2
```

All the GATK tools are called using the same basic command structure. Here's a simple example that counts the number of sequence reads in a BAM file:

---

[4]GATK offers a number of additional modules such as MuTect and MuTect2, which are specifically dedicated to calling somatic variants [76], a topic we will explore in Chapter 30.

[5]https://software.broadinstitute.org/gatk/

[0]As this book was going to press, a new, open-source 4.0 version of GATK was released.

```
$ java -jar GenomeAnalysisTK.jar \
    -T CountReads \
    -R example_reference.fasta \
    -I example_reads.bam
```

The -jar argument invokes the GATK engine itself, and the -T argument tells it which tool you want to run. Some arguments like -R for the *genome reference* and -I for the *input file* are passed to the GATK engine and can be used with all the tools. Most tools also take additional arguments that are specific to their function. GATK options also have long forms; for instance -I is equivalent to --input_file.[7]

Chapter 8 reflected GATK recommendations for read mapping and other processing steps for generating an initial alignment (BAM) file. While it would be possible to perform variant calling directly with these BAM files, experience has shown that indel realignment and base-quality recalibration help to correct for certain technical biases, improve the quality of the data, and lead to a more accurate final variant call set.

10.3 REALIGNMENT

The alignment of reads by BWA-MEM is done read-by-read, and may tend to accumulate erroneous single nucleotide variant (SNV) calls near true insertions or deletions (indels) due to misalignment, mainly because alignment algorithms penalize mismatches less than gaps. The IndelRealigner module of GATK performs a second pass over the BAM file and corrects some of the errors by performing a local realignment of reads around candidate indels.

Thanks to recent developments of GATK, a separate indel realignment step is no longer necessary for variant discovery since an equivalent functionality is now included in the HaplotypeCaller and the MuTect2 variant caller (which performs a haplotype assembly step). However, we will present the indel realignment step separately here in order to explain why it is useful. Also, if a variant caller other than GATK HaplotypeCaller or MuTect2 is used, it may still be beneficial to perform the indel realignment step with GATK.

The reason for which realignment may be necessary can be appre-

---

[7]We will not provide full listings of all of the options available for each of the GATK tools, which are copious, but full documentation is available at the GATK website.

ciated in Figures 10.2 and 10.3. Each read is aligned separately by read mappers; If an insertion or deletion is located towards the very beginning or end of a read, the aligner may favor alignments with mismatches or soft-clips instead of opening a gap in either the read or the reference sequence. The local realignment process considers all of the reads that span a given position. By combining the evidence from all of the reads, the realigner may find a high-scoring consensus that supports the presence of an indel event.

To perform realignment, we need to make use of not only the GATK executable, but also the coordinate-sorted and indexed BAM alignment data together with the reference sequence. GATK requires an index file and a dictionary file to accompany the reference sequence. The Picard tool `CreateSequenceDictionary` creates a sequence dictionary file (with ".dict" extension) from a reference sequence provided in FASTA format, which is required by many processing and analysis tools.

```
$ java -jar picard.jar CreateSequenceDictionary \
    R=hs38DH.fa \
    O=hs38DH.dict
```

The sequence dictionary has the format of a SAM header and contains entries for the sequence name (SN), length (LN), M5 (the MD5 hash of the sequence) and UR (the URI of the sequence, which in this case is the file-system path on the computer of one of the authors) for each component of the reference sequence. For instance, the line for chromosome 8 is:

```
@SQ    SN:chr8 LN:145138636 \
    M5:c67955b5f7815a9a1edfaa15893d3616 \
    UR:file:/home/peter/data/genome/hs38DH.fa
```

The FASTA index file, which allows the reference genome to be efficiently accessed, can be generated with the following SAMtools command:

```
$ samtools faidx hs38DH.fa
```

The `IndelRealigner` module can use a VCF file containing known indels that serve as a kind of "gold-standard" that GATK will use to improve the performance of local realignment. We downloaded the file `Mills_and_1000G_gold_standard.indels.hg38.vcf.gz`   from the

(a)

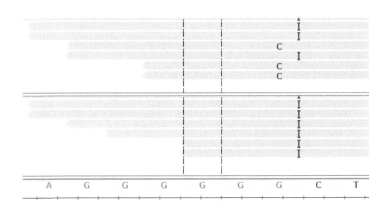

(b)

	Raw	Realigned
Actual	AGGGGGGCCT	AGGGGGGCCT
Reference	AGGGGGG-CT	AGGGGGG-CT
Read1	AGGGGGGCCT	AGGGGGGCCT
Read2	AGGGGGGCCT	AGGGGGGCCT
Read3	-GGGGGC-CT	--GGGGGCCT
Read4	-GGGGGGCCT	-GGGGGGCCT
Read5	---GGGC-CT	----GGGCCT
Read6	---GGGC-CT	----GGGCCT

Figure 10.2. Realigned insertion. In this example, there is an insertion of a cytosine residue directly 3' to a stretch of six guanine residues. That is, the insertion changes the sequence AGGGGGGCT to AGGGGGGCCT. (a) The upper panel shows an IGV screenshot of the alignment prior to the realignment step, and the lower panel shows the same reads following the GATK local realignment step. In the upper panel, we see that a missense change was identified in three reads, because the initial local aligner gave a better score to an alignment with one mismatch than to an alignment with one insertion. The full view would show 25 other reads with the insertion, in which the affected position was located towards the center of the read, so that not aligning the remaining reads with the insertion would lead to numerous mismatches in other parts of the alignment. The local realigner in essence used this information to correct the information of the three reads, so now all the reads show the insertion. (b) Excerpt of the corresponding raw and realigned alignments that correspond to the aligned reads shown in Panel (a).

GATK Website.[8] Also download the index (.tbi) for this file to the same directory.

Finally, we can run GATK

```
$ java -jar GenomeAnalysisTK.jar \
  -T RealignerTargetCreator \
  -R hs38DH.fa \
  -known Mills_and_1000G_gold_standard.indels.hg38.vcf.gz \
  -I NIST7035_dedup.bam \
  -o NIST7035_targetcreator.intervals
```

This command outputs a file with the sites of existent and potential indels. The first five lines of the NIST7035_targetcreator.intervals file are as follows. Note that the intervals are denoted using 1-based coordinates.

```
CM000663.2:14398-14400
CM000663.2:14718-14720
CM000663.2:15240-15241
CM000663.2:16819
CM000663.2:129011-129013
```

We can now perform the realignment using the following command to run the IndelRealigner module of GATK with the intervals we created in the previous step.

```
$ java -Xmx8G -Djava.io.tmpdir=/tmp -jar \
  GenomeAnalysisTK.jar \
  -T IndelRealigner \
  -R hs38DH.fa \
  -targetIntervals NIST7035_targetcreator.intervals \
  -known Mills_and_1000G_gold_standard.indels.hg38.vcf.gz \
  -I  NIST7035_dedup.bam \
  -o  NIST7035_indelrealigner.bam
```

The output file, NIST7035_indelrealigner.bam, is a coordinate-sorted and indexed BAM file that contains the same records as the input BAM file NIST7035_dedup.bam, but with changes to realigned

---

[8]To obtain the VCF file and its index, follow the instructions at https://software.broadinstitute.org/gatk/download/bundle. Similar files are also available for the GRCh37 genome build.

(a)

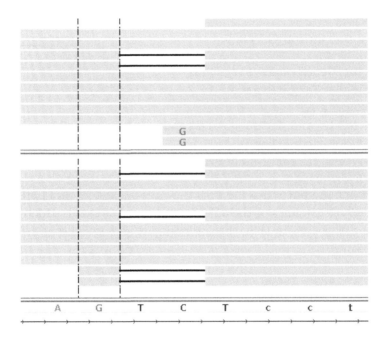

(b)

	Raw	*Realigned*
Actual	AG..TCCT	AG..TCCT
Reference	AGTCTCCT	AGTCTCCT
Read1	AG..TCCT	AG..TCCT
Read2	AG..TCCT	AG..TCCT
Read3	...GTCCT	.G..TCCT
Read4	...GTCCT	.G..TCCT

Figure 10.3. Realigned deletion. In this example, there is a deletion of a TC dinucleotide. That is, the deletion changes the sequence AGTCTCCT to AGTCCT. (a) The upper panel shows an IGV screenshot of the alignment prior to the realignment step, and the lower panel shows the same reads following the GATK local realignment step. (b) Excerpt of the corresponding raw and realigned alignments that correspond to the aligned reads shown in Panel (a). Note that in contrast to Figure 10.2, the deletion is heterozygous and not homozygous, so only about half of the reads are affected.

records and their mates. The changes to records are indicated by the OC tag to mark the original CIGAR string.

An example of a realignment performed by GATK is shown in Figures 10.2 and 10.3. As an example of the output of the realigner, note that in one of the reads shown in Figure 10.3, the realignment changed the CIGAR string from 101 bases aligned without insertions or deletions (OC:Z:101M) to 1M2D100M.

10.4 BASE QUALITY SCORE RECALIBRATION

Next-generation sequencers provide estimates of base quality for each sequenced base as Phred-scaled quality scores that reflect the likelihood that the base call is erroneous. We provided an overview of the data attributes used by Illumina to estimate the quality scores in Chapter 3, but the actual algorithm used is not publicly available. The reported quality scores may be inaccurate as the result of systematic biases. Base Quality Score Recalibration (BQSR) uses empirical data to identify and characterize systematic errors made by the sequencer during the estimation of quality scores. Base-quality scores are critical for variant calling algorithms, which weigh bases by confidence in order to make a final call of reference or alternate base at each position. The BQSR procedure examines four covariates:

1. Lane

2. Originally reported quality

3. Machine cycle (position in the read)

4. Sequence context (preceding and subsequent base)

It then estimates the influence of each of these covariates by counting the number of errors and observations. An important observation is that the great majority of variants in any sequenced human genome will have been previously observed. Therefore, if we call variants in a genome and remove the known variants from further consideration, the remaining variants will be highly enriched in sequencing errors. Therefore, the BQSR procedure makes the following key assumption (which strictly speaking is not true, because WES/WGS samples can have never before seen, i.e., unknown, variants; however, the assumption is true often enough to be useful for the purpose of recalibration):

>   Any sequence mismatch compared to the reference
>   genome is an error — except if it corresponds to a known
>   variant.

If we have a sufficient quantity of data, we can estimate an empirical error probability given the covariates (e.g., lane, sequence context, machine cycle) at the site by calculating the proportion of errors at all sites with the same covariates:

$$p(\text{error}) = \frac{\text{Number of mismatches to reference}}{\text{Number of observations}} \tag{10.1}$$

To perform BQSR on our exome, we will first download the latest version of the dbSNP known polymorphism data, `All_20161122.vcf.gz` (this file has variants called against the GRCh38 genome assembly, and can be downloaded from the dbSNP FTP site.[9] We will first need to index this VCF file, which can be performed using tabix [250]. Tabix is part of the htslib package,[10] which can be uncompressed and built as follows:

```
$ tar jxf htslib-1.3.2.tar.bz2
$ cd htslib/
$ ./configure
$ make
```

This will generate several executables including tabix. If desired, the executables can be installed system wide by `$ make install`. Assuming tabix is available in the PATH, the index file can now be created with the following command.

```
$ tabix -p vcf All_20161122.vcf.gz
```

This command produces the index file `All_20161122.vcf.gz.tbi`. Now we can use GATK to perform BQSR.

```
$ java -jar GenomeAnalysisTK.jar \
    -T BaseRecalibrator \
    -R hs38DH.fa \
    -I NIST7035_indelrealigner.bam \
    -knownSites All_20161122.vcf.gz \
    -o NIST-recal.table
```

---

[9]ftp://ftp.ncbi.nih.gov/snp/organisms/human_9606/VCF/
[10]http://www.htslib.org/download/

This command uses the GATK module `BaseRecalibrator` with the reference sequence `hs38DH` and the known sites `All_20161122.vcf.gz` to analyze the base quality scores as a function of several covariates in the input file `NIST7035_indelrealigner.bam` and to write the results to the file `NIST-recal.table`. This file will contain many tables with the result of the BQSR analysis.

We can now use the recalibration tables to create a recalibrated BAM file. The GATK `PrintReads` module is used.

```
$ java -jar GenomeAnalysisTK.jar \
   -T PrintReads \
   -R hs38DH.fa \
   -I NIST7035_indelrealigner.bam \
   -BQSR NIST-recal.table \
   -o NIST7035_recal.bam
```

It is possible to generate plots to demonstrate the effect of the BQSR procedure. To do so, we first need to generate a second pass recalibration table in order to analyze the biases in the recalibrated data (optimally, little or no bias will be left, but we would like to graphically compare the bias observed before and after). The command is analogous to the first `BaseRecalibrator` command except that we pass the initial recalibration table with the `-BQSR NIST-recal.table` option.

```
$ java -jar GenomeAnalysisTK.jar \
    -T BaseRecalibrator \
    -R hs38DH.fa \
    -I NIST7035_recal.bam \
    -knownSites All_20161122.vcf.gz \
    -BQSR NIST-recal.table  \
    -o NIST-secondpass.table
```

This run will output a second pass recalibration table file called `NIST-secondpass.table`. We can then use both recalibration tables to generate plots.

```
$ java -jar GenomeAnalysisTK.jar \
    -T AnalyzeCovariates \
    -R hs38DH.fa \
    -before NIST-recal.table  \
```

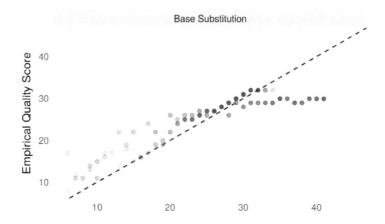

Figure 10.4. An excerpt of the BQSR plot produced by GATK's `AnalyzeCovariates` module. The pink dots represent data points before recalibration, and the blue dots after recalibration. It can be seen that there is a modest improvement in the agreement of the reported and empirical quality scores after recalibration. In practice, the effect of BQSR tends to be larger the more data is used (e.g., with WGS as opposed to WES data).

```
-after NIST-secondpass.table \
-plots BQSR.pdf
```

The file `BQSR.pdf` contains a summary of the recalibration data and several before/after plots. Figure 10.4 shows a scatter plot of the Phred-encoded quality scores originally reported by the Illumina software against the empirical quality scores calculated as explained above. The pink data points are from the pre-recalibration BAM file, and the blue data points are from the recalibrated BAM file. The recalibration procedure reduced the root mean square error (RMSE) in the reported quality scores from 5.21 to 3.21.

It should be noted that there are multiple variant calling pipelines, not all of which use BQSR or analogous steps. We will present an overview of some popular variant callers in Chapter 12. Whether or not BQSR is useful may depend on the particular mapper and variant caller used as well as on the overall sequence depth and level of divergence [263, 310, 431].

FURTHER READING

Van der Auwera GA, Carneiro MO, Hartl C, Poplin R, Del Angel G, Levy-Moonshine A, Jordan T, et al. (2013) From FastQ data to high confidence variant calls: the Genome Analysis Toolkit best practices pipeline. *Curr. Protoc. Bioinformatics* **43**:11.10.1–33.

10.5 EXERCISES

Exercise 1

We have presented SAMtools and a number of Picard applications for postprocessing the alignment file. Many other tools have become available in the community that perform the same tasks more efficiently. Sambamba exploits multi-core processing to reduce execution time [421].[11] You can confirm this if you like by performing one of the steps shown in this chapter with Picard and Sambamba and comparing the processing time using the Unix `time` command. For instance, the following command will measure how long it takes to execute Picard's `ValidateSamFile` application

```
$ time java -jar picard.jar ValidateSamFile \
    INPUT=NIST7035_aln.bam \
    MODE=SUMMARY
```

Although multi-core processing is certainly useful, since Sambamba generally runs about 3-4 times faster than Picard, further gains in efficiency can be obtained by using pipes to reduce disk space and time spent with file I/O. Recall that the pipe (|) passes the output of one command to another for further processing. For example, `ps aux` can be used to see a list of active processes, and `ps aux | grep firefox` will check if one of those processes is Firefox. This has a number of advantages — here, when performing read alignment we can avoid writing the unsorted BAM file to disk, and then having to read the same file from disk again when we perform the next step. Try the following command and use the Unix `time` command to determine how much faster it runs than the comparable Picard command.

```
$ bwa mem -t 8 \
  -R '<read group info>' hs38DH.fa \
  NIST7035_trimmed_R1_paired.fastq.gz  \
```

---

[11]Precompiled versions of Sambamba can be downloaded from the project's GitHub page: `https://github.com/lomereiter/sambamba/releases/`.

```
NIST7035_trimmed_R2_paired.fastq.gz | \
sambamba view -f bam -l 0 -S /dev/stdin | \
sambamba sort -m 4GB -o NIST7035_SB.bam /dev/stdin
```

By default, Sambamba uses up to 2GB of RAM for sorting, but the memory limit can be increased with the -m option as shown above.

Exercise 2

SAMBLASTER is an application that can perform fast duplicate marking [119]. It supports I/O piping similar to Sambamba. In this exercise, chain together the following commands.

```
$ bwa mem -t 8 \
  -R '<read group info>' hs38DH.fa \
  NIST7035_trimmed_R1_paired.fastq.gz  \
  NIST7035_trimmed_R2_paired.fastq.gz | \
  samblaster | \
  samtools view -Sb - > NIST7035_SB.bam
```

Use the Unix time command to determine how much time is saved compared with performing each command separately. In "real-life" exome and genome analysis pipelines, one often sees efficient tools such as Sambamba and SAMBLASTER used together with I/O pipes to conserve disk space and save processing time.

Exercise 3

Positions in the BAM file that actually undergo realignment by the GATK realignment procedure can be identified by examining the OC tag (see page 141). Try to find additional examples where realignment improved the final alignment of insertions or deletions in the NIST7035_recal.bam file, or use another WES/WGS sample of your choice. One way of finding lines with an OC tag is to combine SAMtools and grep.

```
$ samtools view NIST7035_indelrealigner.bam | grep -w OC
```

To examine the candidates, you can load the original and the realigned BAM file into IGV (see Chapter 17 for instructions on how to use IGV; it is instructive to view both alignment files in IGV simultaneously).

Exercise 4

In this chapter, we used SAMtools to determine how many reads in the NA12878 exome were PCR or optical duplicates. As an alternative approach, use the following awk expression

```
awk 'and($2,0x0400)'
```

combined with Unix tools to count lines. Note that the "and" function performs a logical AND operation on 0x0400 and the bit string represented in the FLAG field (in the second column, i.e., $2).

CHAPTER 11

Alignment Data: Quality Control

D EPTH AND COVERAGE are two key parameters by which to judge the quality of a sequencing run and the read analysis. An assessment of these parameters makes up the second key step in WES/WGS quality control (Q/C), and will be the major subject of this chapter. The words depth and coverage are used in different, and potentially confusing, ways in the literature, and therefore we will begin this chapter by defining how we are going to use these words in this book.

The empirical coverage of any given base of the reference genome in a sequencing experiment is defined as the number of reads whose mapping position "covers" (overlaps with) the base in question. The word "depth" (or the phrase "depth of coverage") is often used synonymously with "coverage".

It should be intuitively obvious that the more reads there are, the more certain we can be about the genotype at any given position. For instance, if some position is covered by just two reads, both of which have an alternate base (ALT) at the position, can we really be certain that the sample wasn't heterozygous for reference (REF) and ALT at this position and we just by chance happened to observe two ALT sequences? On the other hand, if we observe 49 ALT and 53 REF bases at some position, we will usually be able to confidently assert that the genotype is heterozygous. High sequencing costs are the major limiting factor that prevents researchers from simply sequencing more, and thus planning NGS experiments is always a compromise between sequencing costs and sequencing depth. The relationship between the total number of reads sequenced and the coverage is not linear. One of the reasons for this is that a higher duplication rate is seen at very high sequencing depths.

We have just spoken about the sequencing depth at specific positions, but for planning experiments, we are also interested in the expected coverage, which is the average number of times that each nucleotide is expected to be covered given the read length and the number of reads that have been sequenced. A simple model is sufficient to estimate the expected coverage in whole-genome sequencing.[1]

- G: genome size (e.g., 3.2×10^9 nucleotides for humans).

- L: read length (e.g., 100 nucleotides for a typical Illumina run).

- N: total number of reads.

It is now easy to calculate that the total number of sequenced bases (n_b) is

$$n_b = N \times L \tag{11.1}$$

Similarly, the average coverage per base (λ) is

$$\lambda = \frac{n_b}{G} \tag{11.2}$$

The Poisson distribution is commonly used to model read counts. Given an average or expected number of counts per position, the Poisson distribution can be used to calculate the probability that a particular position in the genome will be covered by k reads if all N reads are randomly distributed across the genome.

$$P(k) = \frac{\lambda^k e^{-\lambda}}{k!} \tag{11.3}$$

For instance, let us imagine we perform human genome sequencing on an Illumina HiSeq2500 machine in rapid run mode using one lane of a single flow cell and obtain 75 million read pairs with a read length of 2×100 nucleotides. We can easily calculate the average coverage as

- G: 3.2×10^9 bp

- L: 100 nt

---

[1]This simplified model ignores various sources of error such as ambiguous mappings.

- N: $2 \times 7.5 \times 10^7$

- $\lambda = 4.6875$

We can now estimate the probability that a given base will not be covered by calculating the probability of observing a coverage of zero given an expected coverage of 4.6875.

$$P(X = 0) = \frac{\lambda^0 e^{-\lambda}}{0!} = e^{-\lambda} = e^{-4.6875} \qquad (11.4)$$

Therefore, the probability of seeing at least one read at a given position is

$$P(X > 0) = 1 - P(X = 0) = 1 - e^{-4.6875} = 0.991 \qquad (11.5)$$

The probability of having a coverage of 1 or more is about 99%. The human genome project sequenced the human genome to an average coverage of about 6-fold with the goal of having every nucleotide of the human genome sequenced at least once. There are two major reasons why this is not sufficient for current WES/WGS resequencing projects, whose goal is to characterize variants rather than discover the reference sequence. First, as we will examine in the next chapter, a sufficient depth is required to call variants confidently (as a rule of thumb, a depth of 10 reads is a suggested threshold for homozygous variants and 20 for heterozygous variants, but variant calling additionally uses base and mapping qualities to assess reliability). Second, coverage in actual experiments is far from uniform, being affected by multiple factors such as GC content and mappability.[2]

With exome sequencing (WES), a third important factor is the enrichment of the target regions. WES enrichment kits focus on protein coding exons and a few other disease-relevant sequences such as microRNA genes. There are currently three major exome sequencing capture kits on the market: Illumina TrueSeq, Agilent SureSelect and NimbleGen SeqCap EZ. The capture regions for the exome capture kits range from 37.6 to 62.1 million base pairs. Capture efficiency is one of the most important quality control parameters for exome sequencing or other targeted sequencing (i.e., the percentage of reads that map to one of the targeted regions, sometimes called "on-target ratio"). Previous studies have shown that capture efficiencies between 40 and 70% are typical for exome sequencing.

---

[2]We will discuss both GC content and mappability in the context of copy number variants where they play a major role (see Section 19.4).

With WES, we are typically only interested in the coverage of the enriched regions. While the average coverage of these regions—often called "on-target coverage"—is of interest, we are also interested in the breadth of coverage. That is, what percentage of the targeted region of the genome was covered by at least X reads (Figure 11.1). This is important because the capture efficiency is far from uniform across the targeted regions. In a typical WES experiment, some regions will be highly covered, well beyond the average coverage of the enriched regions, while other regions will be barely covered even in an experiment with a substantial total read count.

11.1 COVERAGE ANALYSIS

Coverage analysis tells us how well the exome target-enrichment process actually worked. Did the target capture approach enrich for the baited genomic region, i.e., are the targeted regions enriched for aligned reads compared to non-targeted regions? What is the average coverage? A certain minimal coverage is required for accurate genotyping. Although there are no absolute rules, a minimum threshold of at least 20–25-fold coverage[3] is recommended to call heterozygous variants [194, 311, 457]. Studies have shown that most functionally relevant variants in the human exonic regions could be detected at an overall on-target sequencing depth of 120X, and that the proportion of regions covered at 25X or more plateaus after an overall depth of 120X is reached [194] (cf. Figure 11.1). Readers should be aware that these numbers are dependent on the technology used and greater demands on the technical performance of enrichment methods are likely to be placed as the technology used for enrichment and WES/WGS continues to improve.

A plot of the actual sequencing depth over the target regions is used to determine the proportion of the targets that has enough reads so we can be (relatively) sure we have called the correct genotype.

The manufacturers of exome capture kits provide a BED file that describes either the targets or the baits. A target is a region of the genome which is targeted for sequencing. In exome capture, the targets are typically the coding exons but may additionally include 5' and 3' untranslated regions (UTRs), splice sites, and promoter regions. A bait is a sequence designed to capture the targets; typically, baits are short

---

[3]We also say "20–25X coverage".

oligos that are used to hybridize to the input DNA. Note that baits and targets are different concepts. When evaluating a capture strategy, we are typically going to be concerned with how well the baits overlap with our targets.

We will show how to perform coverage analysis using the NA12878 exome. This exome was enriched with the Nextera Rapid Capture Exome and Expanded Exome kit. The manufacturer of this kit, Illumina, provides a file for download (`Nextera Rapid Capture Exome Targeted Regions Manifest v1.2`) in BED Format that specifies the targeted regions. An initial difficulty that we have with this is that the file uses genomic coordinates of the previous genome assembly GRCh37. The human genomics community has been slow to move to the newest genome build, and many of the resources that have been used are still only available for GRCh37 at the time of this writing. In practice, this means that many analysis pipelines still rely partially or entirely on the GRCh37 assembly. Although it would be possible to convert the genomic coordinates of the targeted regions to the GRCh38 build we have been using for most of this book,[4] there are sequences that cannot be automatically converted, and so for better comparability with published coverage analyses, we will demonstrate the coverage analysis with an alignment to the GRCh37 genome. We generated the BAM file using the hs37d5 reference mentioned on page 99 followed by duplicate removal as explained on page 131. We will show the commands we used but will refer readers to the earlier chapters for explanations.

We first create a BWA index and align the trimmed reads to the GRCh37 genome.[5]

```
$ bwa index -a bwtsw hs37d5.fa
$ bwa mem -t 8 \
  -R '@RG1' \
  hs37d5.fa \
  NIST7035_trimmed_R1_paired.fastq.gz  \
  NIST7035_trimmed_R2_paired.fastq.gz | \
  samtools view -Sb - > NIST7035_aln37.bam
```

We now sort the reads and remove duplicates before analyzing the coverage.

---

[4]See section 8.1.3.

[5]We abbreviate the read group information to @RG1 so that it will fit on the line. See Chapter 8 for more details on read groups.

```
$ java -jar picard.jar SortSam \
    I=NIST7035_aln37.bam \
    O=NIST7035_sorted37.bam \
    SORT_ORDER=coordinate
```

We will add the `REMOVE_DUPLICATES=true` option to the MarkDuplicates command to remove duplicate reads from the output BAM file.

```
$ java -jar picard.jar MarkDuplicates \
    INPUT=NIST7035_sorted37.bam \
    OUTPUT=NIST7035_dedup37.bam \
    REMOVE_DUPLICATES=true \
    METRICS_FILE=NIST7035_37.metrics
```

We can now use the output file `NIST7035_dedup37.bam` to calculate coverage metrics. First we will need to explain the BED format.

11.1.1 BED format

The Browser Extensible Data (BED) format is a simple tab-delimited format for annotation of genomic features. The BED format provides a flexible way to define the data lines that are displayed in an annotation track. BED lines have three required fields and nine additional optional fields. The number of fields per line must be consistent throughout any single set of data in an annotation track. The order of the optional fields is fixed: lower-numbered fields must always be populated if higher-numbered fields are used.

The first three fields are required:

chrom: The name of the chromosome on which the genome feature exists, e.g., "chr3".

start: The zero-based starting position of the feature on the chromosome (Note that some other commonly used formats such as VCF use one-based numbering, and conversion may be needed depending on the application).

end: The ending position of the feature in the chromosome. In contrast to the start position, the end position in each BED feature is one-based.[6]

BED format has another nine optional fields that are not needed for coverage analysis.[7]

11.1.2 Coverage analysis with BEDTools

BEDTools allows genomic intervals from multiple files in BAM, BED, VCF, and other formats to be intersected, merged, counted, complemented, and shuffled [347]. BEDTools has a wide range of functionalities and excellent online documentation.[8] We will use BEDTools for coverage analysis. The latest version of BEDTools can be downloaded from its GitHub page;[9] in this book, we have used version 2.26.0.

Use the following commands to generate the executables.

```
$ tar xvfz bedtools-2.26.0.tar.gz
$ cd bedtools2/
$ make
```

This will generate a number of executable programs in the `bin` subdirectory.

We will now download the target definition file from the Illumina Website:[10]

```
nexterarapidcapture_exome_targetedregions_v1.2.bed
```

According to the release notes, the targeted region encompasses about 45 Mb. This file has the targeted regions in 3-column BED format.

```
chr1    12098    12258
chr1    12553    12721
chr1    13331    13701
chr1    30334    30503
(...)
```

---

[6]Equivalently, and preferably for some algorithms and some string-handling functions of C or Java, this is one position after the end of the feature in zero-based numbering.

[7]Full documentation is available at the UCSC website: `http://genome.ucsc.edu/FAQ/FAQformat.html#format1`

[8]`http://bedtools.readthedocs.io/en/latest/`

[9]`https://github.com/arq5x/bedtools2/releases`

[10]`http://support.illumina.com/downloads.html`

Note that we abbreviate the name of the Nextera BED file in the following command in order to fit the command into one line.

The Nextera file uses the symbol "chr1" for chromosome 1 (and so on), but the `NIST7035_dedup37.bam` file does not use the abbreviation "chr" (e.g., chromosome 1 is simply "1"), like the reference genome FASTA file to which the reads were mapped.

We therefore use `sed` to remove the string "chr".

```
$ sed "s/^chr//g" nextera.bed > nextera.nochr.bed
$ head -n 3 nextera.nochr.bed
1    12098    12258
1    12553    12721
1    13331    13701
```

The BEDTools command for generating the coverage data is[11]

```
$ bedtools coverage \
    -hist \
    -a nextera.nochr.bed \
    -b NIST7035_dedup37.bam > NIST.bed.cov
```

BEDTools coverage computes both the depth and breadth of coverage of features (target regions) in the BED file with the features (reads) in the BAM file. The `-hist` option causes BEDTools to report a histogram of coverage for each target region as well as a summary histogram for **all** targets. The output is tab-delimited, and reports each feature in the BED file. Consider the first three lines of the file generated by the above command.

```
head -n 3 NIST.bed.cov
1    12098    12258    102    1     160    0.0062500
1    12098    12258    103    2     160    0.0125000
1    12098    12258    104    10    160    0.0625000
```

The first three columns report an exon on chromosome 1 located between position 12098 and 12258 (this corresponds to an entry in `nextera.nochr.bed`). The fourth column indicates the depth of coverage (e.g., 102 reads for the first line). The fifth column shows the

---

[11]For older versions of BEDTools (<2.24) use the following command line:
```
$ bedtools coverage -hist -abam NIST7035_dedup37.bam \
-b nextera.nochr.bed > NIST.bed.cov
```

number of bases at the indicated depth (for the three lines shown here, 1 base was covered by 102 reads, 2 bases by 103 reads, and 10 bases by 104 reads). The final two columns display the length of the feature (here: 160 nucleotides) and the fraction covered at the indicated depth (for instance the fraction of bases in this feature covered at a depth of 103 is $2/160=0.0125000$).

The output file concludes with a series of lines that begin with "all" that give the overall summary

```
$ grep ^all NIST.bed.cov > NIST.all.cov
$ cat NIST.all.cov
all    0      3840711    45326818    0.0847337
all    1      937808     45326818    0.0206899
all    2      797240     45326818    0.0175887
(...)
all    211    1395       45326818    0.0000308
(...)
all    1740   1          45326818    0.0000000
```

Here, the target region has a total of 45,326,818 individual positions. 3,840,711 were not covered at all (proportion $0.0847337 = 3840711/45326818$). 937,808 were covered by one read, 797,240 were covered by 2 reads, 1395 were covered by 211 reads, and so on. The maximum coverage of 1740 was attained by one position.

11.2   A COVERAGE PLOT IN R

We can use the results of the BEDTools analysis to make a coverage plot in R (Figure 11.1). The code used to make the plot is shown here. It is easy to extend this plot to show multiple exomes at once.

```
cover <- read.table("NIST.all.cov")
cov_cumul <- 1-cumsum(cover[,5])
plot(cover[1:200, 2], cov_cumul[1:200], type='l',
xlab="Depth",
ylab="Fraction of capture target bases >= depth",
ylim=c(0,1.0),
main="Target Region Coverage")
abline(v = 20, col = "gray60")
abline(v = 50, col = "gray60")
abline(v = 80, col = "gray60")
```

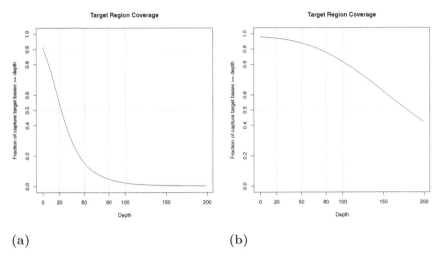

(a)   (b)

Figure 11.1. Exome coverage plots. (a) Coverage plot of the NA12878 sample. **(b)** Coverage plot of an in-house sample that was enriched with the Agilent SureSelect v6 kit and sequenced on an Illumina HiSEq 25000 device. The in-house sample shows substantially better overall coverage.

```
abline(v = 100, col = "gray60")
abline(h = 0.50, col = "gray60")
abline(h = 0.90, col = "gray60")
axis(1, at=c(20,50,80), labels=c(20,50,80))
axis(2, at=c(0.90), labels=c(0.90))
axis(2, at=c(0.50), labels=c(0.50))
```

The coverage plot shows which percentage of the data is covered with at least 20, 50, 80, and 100 nucleotides (Figure 11.1a).

We note that the coverage for this exome is not especially good, which is not surprising because we intentionally used only one of the four sequencing runs available at the Genome in a Bottle FTP site in order to simplify the examples in this book (see page 49). Figure 11.1b shows a coverage plot for an in-house exome with substantially better coverage.

In practice, there is no absolute threshold below which one cannot perform the downstream analysis. Coverage plots are particularly useful to compare multiple exomes from the same sequencing unit. If a new batch of exomes displays substantially worse coverage, this may indicate a technical problem with the library preparation or other parts of the sequencing procedure. If one exome in a batch shows substantially worse coverage, then again, one should work with the staff of the sequencing facility to troubleshoot. Depending on many circumstances, it may be advisable to repeat the experiment or simply to sequence more material in order to attain better coverage.

FURTHER READING

Quinlan AR, Hall IM (2010) BEDTools: a flexible suite of utilities for comparing genomic features. *Bioinformatics* **26**:841–842.

11.3 EXERCISES

Exercise 1

There is a complex relationship between the mean coverage of an exome and the number of true positive, false positive, true negative, and false negative variant calls that will be made.

To get intuition about the relationship between coverage and the statistical power to make a variant call, this exercise asks readers to develop a simulation. We will define a simple rule for calling a heterozygous variant that demands that there are at least two ALT reads and that between 20% and 80% of the reads are ALT and the rest are REF.

Write a program in R which simulates the read counts obtained at various coverage levels for a true, heterozygous variant and determines if the variant would be called with the criteria defined above. For all possible coverage levels between 5 and 100, decide randomly which of the reads show the ALT allele or the REF allele according to a binomial distribution with $p = 0.5$ (using the function **rbinom**). Doing this 100 times for each coverage level will allow you to roughly estimate the proportion of times the heterozygous variant would be called at that depth of coverage in true sequencing data. Plot this proportion as a function of the coverage.

Describe your results. How many reads should cover a base to capture at least 90% of the true heterozygous variants?

IV

Variant Calling

CHAPTER 12

Variant Calling and Quality-Based Filtering

$\mathbf{V}$ ARIANT calling refers to the identification of probable variants (deviations from the reference sequence) in an alignment.[1] Current software practice performs a separate analysis for small variants (less than about 50 nucleotides) and larger structural variants (we will cover structural variants in Chapter 19). There are numerous tools for variant calling. In this chapter, we will concentrate on GATK, which was introduced in Chapter 10. We refer to that chapter for information on installing GATK and basic usage.

12.1 CALLING VARIANTS WITH GATK

Calling variants aims to identify differences with respect to the genome reference sequence in an aligned and sorted BAM file. There are many options for variant calling with GATK. The `HaplotypeCaller` module of GATK is suitable for single-sample or multiple-sample analysis. The basic command for calling variants from the aligned and sorted BAM file of a single sample[2] is as follows:[3]

```
java -jar GenomeAnalysisTK.jar \
```

---

[1]For germline variants, which we discuss in this chapter, the reference sequence is obviously the reference genome; for somatic variants (Chapter 30) the reference typically includes a non-cancerous control sample from the same individual which allows somatic variants to be distinguished from germline variants.

[2]We will discuss joint variant calling in multiple samples in Chapter 18.

[3]For the demonstrations in this chapter, we used the realigned BAM file without subsequent base quality score recalibration (see Sections 10.3 and 10.4).

```
-T HaplotypeCaller \
-R hs38DH.fa \
-I NIST7035_indelrealigner.bam  \
--genotyping_mode DISCOVERY \
-stand_call_conf 30 \
-o raw_variants.vcf
```

The `HaplotypeCaller` works by first determining regions of the genome for which there is at least some evidence of variation, termed "active regions", and then identifies haplotypes that are consistent with the data and realigns each haplotype with the reference genome using the Smith–Waterman algorithm. Then, the `HaplotypeCaller` does a pairwise alignment of each read against each potential haplotype using a Hidden-Markov Model algorithm, which yields a matrix of probabilities of the haplotypes given the read data. The likelihoods are marginalized in order to estimate the likelihoods of alleles at each site with a potential variant. Finally, the most likely genotype is assigned using Bayesian methods.

The output file is a **Variant Call Format** (VCF) file [95], which contains general information in a header followed by one line for each genomic position in which a variant was called. We will explore the VCF format in detail in Chapter 13.

12.2 HARD FILTERING VERSUS VQSR

Most bioinformatic analysis procedures involve some trade-off between sensitivity and noise — it is nearly inevitable that increasing the sensitivity (i.e., reducing the false negative rate) will also add noise (i.e., increase the false positive rate). Variant calling with GATK is no exception to this rule. The initial variant call set produced by GATK as described above is designed to maximize sensitivity at the cost of a certain amount of false-positive calls. Accordingly, the results obtained from GATK or any other variant caller need to be filtered according to their quality. The goal of hard filtering and variant-quality score recalibration (VQSR) is to reduce the number of false-positive calls without greatly reducing the sensitivity.[4]

---

[4]The quality-based filtering of variant calls is always necessary and should be the first postprocessing step before trying to interpret the called variants. In addition, variants can be filtered according to other criteria, e.g., by predicting their impact on protein function, or by considering their inheritance patterns in a family affected by a genetic disorder. These topics will be addressed in Parts V and VI of the book.

As we will see in Chapter 13, VCF files assign a quality score (`QUAL`), which is a Phred-scaled quality score for the assertion made about the alternate (variant) base or sequence (see Equation 12.1). Each variant caller determines this value using its own algorithms, and one cannot directly compare the `QUAL` values returned by different variant callers. Many additional quality metrics are reported for each variant, which can be used to filter out variants that fail to pass a threshold. The application of such fixed thresholds can be referred to as "hard filtering". On the other hand, VQSR is a more sophisticated machine learning procedure that attempts to learn the most appropriate thresholds from the data, using a set of "gold-standard" trusted calls. We will review both procedures in this chapter. In order to understand how hard filtering is performed, we will present BCFtools, and show how it can be used to analyze various quality metrics in the next section. Be aware that much of what we describe here is specific to GATK. Other variant callers may provide different quality metrics or specify the same metrics in different ways. How to filter the called variants therefore depends very much on the variant caller used.[5]

12.3 BCFTOOLS

BCFtools is a program for manipulating VCF files as well as binary variant call format (BCF) files.

To install BCFtools, follow the instructions on the samtools/bcftools website.[6]

```
$ git clone git://github.com/samtools/htslib.git
$ git clone git://github.com/samtools/bcftools.git
$ cd bcftools
$ make
```

This will create an executable named `bcftools` in the `bcftools` sub-directory. We will use BCFtools to extract information from the VCF file. The following command will show a list of samples represented in the VCF file (in our case, there is only one).

```
$ bcftools query -l raw_variants.vcf
NA12878
```

---

[5]In Section 30.3.4, we show how to filter somatic variants called with VarScan?
[6]http://samtools.github.io/bcftools/howtos/install.html

To extract information about selected fields of VCF lines, we use a percent sign (%) followed by the name of the field. For instance, to extract the chromosome (CHROM), the position (POS), the reference base(s) in the genome (REF), and the alternate (variant) sequence (ALT), we use the following command:

```
$ bcftools query -f '%CHROM %POS %REF %ALT\n' \
    raw_variants.vcf | head -5
chr1 14397 CTGT C
chr1 14574 A G
chr1 14590 G A
chr1 14599 T A
chr1 14604 A G
```

FORMAT tags in the sample columns of the VCF file can be extracted using the square brackets [] operator, which loops over all samples in the VCF file. For example, the AD field shows the allelic depths for the REF and ALT alleles in the order listed.[7]

To extract the allelic depth (AD) information, use the following command

```
$ bcftools query -f '[%AD]\n' raw_variants.vcf | head -5
15,5
1,3
0,3
0,3
0,3
```

In this example, there were 15 REF reads and 5 ALT reads in the first called variant, and so on.

We will now use BCFtools to examine the distribution of selected quality attributes we will need to perform hard filtering of the raw variant file.

12.3.1 QualByDepth

The QUAL field of the VCF file is defined as a Phred score that reflects the variant quality

$$-10 \times \log_{10}(1 - p) \tag{12.1}$$

---

[7]See Chapter 13 for more details on FORMAT tags.

where p is the probability of the variant being present given the read data. This probability is calculated by GATK (or by other variant callers). However, each read contributes a little to the QUAL score, and thus variants in regions with deep coverage can have artificially inflated QUAL scores. The QualByDepth (QD) score is the QUAL score divided by the allele depth of the variant (i.e., the ALT allele depth). There is no "normal" range for this value, but a QD under 2 is considered poor quality.

In the following, we extract the QUAL and the allele depth of the variant (AD{1}) from the VCF file using BCFtools, and use awk to calculate QD=QUAL/AD{1} if the allele depth is greater than zero. In this raw VCF file, one of the genotypes is listed as "./.", which has no associated, meaningful allelic depth and leads to a division by zero error unless we test that the AD{1} is not zero. The results are redirected to a file called QD.tab.

```
$ bcftools query -f '%QUAL [%AD{1}]\n' raw_variants.vcf | \
    awk '{ if($2>0) {printf "%f\n",$1/$2}}' > QD.tab
```

Density plots for the QualByDepth values are shown for the Corpas and the NA12878 exomes (Figure 12.1). Both samples show good values for QD.

12.3.2   FisherStrand

This parameter is an estimate of strand bias, a kind of sequencing bias in which one strand is favored over the other. Strand bias can result in inflation or deflation of the evidence for a heterozygous variant. This test uses Fisher's exact test to evaluate whether there is strand bias between forward and reverse strands for the reference or alternate allele. The result of the test is shown as a Phred-scaled p-value; that is, the higher the value for FS, the more likely there is to be bias, which in turn can be a sign of false-positive calls (Figure 12.2). Values of FS over 60 are taken to be strong evidence for strand bias.

12.3.3   RMSMappingQuality

The root mean square of the mapping quality (RMSMappingQuality, MQ) provides an estimation of the overall mapping quality of reads supporting a variant call. The RMS is based on the mapping qualities $q_1, \ldots, q_n$ of the n reads that support the variant call, and is defined

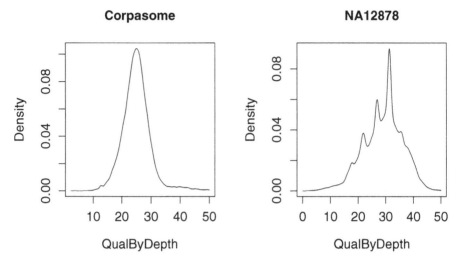

Figure 12.1. Distribution of QualByDepth (QD) values for the Corpas and the NA12878 exomes. QD values higher than 50 were considered outliers and removed. The mean QD for the Corpas exome was 27.2 and that of the NA12878 exome was 25.0.

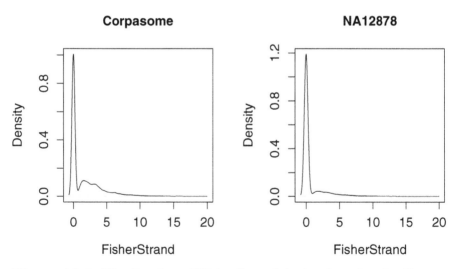

Figure 12.2. Distribution of FisherStrand (FS) values for the Corpas and the NA12878 exomes. There was only one FS value higher than 60 in the Corpas exome (values of FS over 60.0 are taken to be strong evidence for strand bias). The NA12878 exome displayed 46 variants with an FS value above 60.0.

as

$$RMS = \sqrt{\frac{1}{n}\sum_{i=1}^{n} q_i^2} \tag{12.2}$$

The threshold suggested by GATK for `MQ` is 40. A total of 7115, or about 4.9% of all variant calls of the NA12878 dataset, fail to pass this criterion.[8]

12.3.4 MappingQualityRankSumTest

The rank sum test for mapping qualities of `REF` reads versus `ALT` reads (MappingQualityRankSumTest, `MQRankSum`) compares the mapping qualities of the reads supporting the reference allele with those supporting the alternate allele.[9] If there is a statistically significant difference in quality in either direction, then it suggests that sequencing and/or the mapping process may have been biased or affected by an artifact. However, for variant calling, we are interested in whether there is evidence that the quality of the data supporting the alternate allele is comparatively low. If this is the case, then `MQRankSum` takes on highly negative values. GATK suggests filtering if `MQRankSum` is less than –12.5.

We can count these variants using BCFtools and `awk` as follows.

```
$ bcftools query -f '[%MQRankSum]\n' raw_variants.vcf | \
    awk '{if ($1<-12.5){MQRankSum++}} END {print MQRankSum}'
```

In practice, we only filter out highly negative values when evaluating variant quality because the idea is to filter out variants for which the quality of the data supporting the alternate allele is comparatively low. In the NA12878 dataset, there were only 5 such variants.

The `MQRankSum` metrics is not defined for all variants. For example, it requires at least one `REF` read to be present. But this is not a problem because filtering according to the quality of `REF` reads makes little sense for homozygous variants.

---

[8]Compare with the distribution of mapping qualities in Figure 9.5.

[9]The rank sum test (also known as the Mann-Whitney test) ranks the combined mapping qualities of `REF` and `ALT` reads and tests whether there is a deviation from the null hypothesis that it is equally likely for the mapping quality of a `REF` read to be greater or less than that of an `ALT` read.

Table 12.1. Suggested thresholds for hard filtering of GATK variant calls.

Item	Threshold
QD	< 2.0
MQ	< 40
FS	> 60.0
MQRankSum	< −12.5
ReadPosRankSum	< −8.0

12.3.5   ReadPosRankSumTest

The rank sum test for relative positioning of REF versus ALT alleles within reads (ReadPosRankSumTest, ReadPosRankSum) tests whether there is evidence of bias in the genomic position of reference and alternate alleles within the reads that support them. Such a bias, especially if a variant is called only near the ends of reads, can be an indication of error (cf. Figure 10.2 and 10.3). Analogously to the MappingQualityRankSumTest, we are only interested in negative values, which indicate that the alternate allele is found at the ends of reads more often than the reference allele. GATK suggests filtering if ReadPosRankSum is less than –8. Only two such variants were found in the NA12878 exome.

As with the MQRankSum metrics, ReadPosRankSum is not defined for variants without REF reads.

12.4   HARD FILTERING

With this information in hand, it is relatively easy to perform hard filtering with GATK. It should be obvious however, that it is impossible to define one threshold that will perfectly separate true calls from false-positive calls. Users may want to explore their data using the methods described above to avoid using filter criteria that are too stringent and wind up removing large numbers of true variants. Table 12.1 displays recommended values for the hard filtering criteria. Users should consider whether these values are appropriate for individual datasets they are analyzing, and adjust them if necessary.

The following GATK command will perform hard filtering. If a variant fails to pass any one of the criteria, it will be marked with snpFilter in the FILTER column of the VCF file to indicate this. If all criteria are met, the variant will be marked with PASS in the FILTER column.

To use GATK to perform hard filtering, we need to supply a filter expression. The following expression will cause a variant to be filtered if any one of the criteria is not met.

```
QD < 2.0 || FS > 60.0 || MQ < 40.0 || \
  MQRankSum < -12.5 || ReadPosRankSum < -8.0
```

This expression is abbreviated for legibility as <filt-expr> in the GATK command that follows:

```
$ java -jar GenomeAnalysisTK.jar \
    -T VariantFiltration \
    -R hs38DH.fa \
    -V raw_variants.vcf \
    --filterExpression "<filt-expr>" \
    --filterName "snpFilter" \
    -o NA12878_filtered.vcf
```

Using the following BCFtools command, we can count the number of lines that have passed filtering.

```
$ bcftools query -f '%FILTER\n' NA12878_filtered.vcf | \
    wc -l
146000
$ bcftools query -f '%FILTER\n' NA12878_filtered.vcf | \
    grep PASS | wc -l
138083
```

Thus, about 94.6% of the 146,000 called variants pass our filter criteria.

12.5   VARIANT QUALITY SCORE RECALIBRATION

Variant Quality Score Recalibration (VQSR) calculates a new score for each variant that is termed VQSLOD (variant quality score log-odds). VQSR uses machine learning to learn profiles of attributes that separate "good" from "bad" variants. The key assumption of VQSR is that if we find a well-known variant in our data then we can assume that this variant is truly present in the sample. VQSR uses validated variants from resources such as the 1000 Genomes Project or HapMap data and searches for them in the VCF data. VQSR develops an estimate of the relationship between SNP call annotations (such as those described in the section on hard filtering) and the probability that a called variant is a true positive. In contrast to hard filtering, VQSR is an integrated model that exploits the complex correlation structure of the covariates [104].

VQSR calculates a `VQSLOD` score for each variant and adds this score to the `INFO` section of the VCF file (c.f., Chapter 13). The `VQSLOD` score is defined as the log odds ratio of being a true variant versus being false under the error model. The `VQSLOD` score can then be used to filter out low-quality variants. In contrast to the hard filtering approach, VQSR allows users to choose a threshold `VQSLOD` score that maintains a certain sensitivity for the true variants (the recommended threshold is 99%). We refer to the original publication for details on the VQSR algorithm [104].

It would not be a great idea to demonstrate VQSR on our example exome, which is from a sample that was used in the 1000 Genomes Project. This is because some of the resources used for the recalibration rely on 1000 Genomes data which will already largely contain at least the correctly called variants in the sample. Also, VQSR works best with larger amounts of data. It is said to work best for WES with a minimum of 30 samples [60], and the method was not designed to be used on single-sample exome datasets. To give our readers a feeling for how VQSR performs using an easily available dataset, we therefore downloaded a VCF file containing SNPs called in a genome of an individual from Mongolia [24].[10]

The file `Mongolian_genome.snp.vcf` had been called against the hg19 genome and the file required some doctoring before it could be used.[11]

To perform VQSR, we need to download truth and training sets.

---

[10]We downloaded the file `Mongolian_genome.snp.vcf` from `http://gigadb.org/dataset/view/id/100104`

[11]The variants were sorted in lexical order (alphabetically) rather than according to chromosomal position with the chromosomes ordered as chr1, chr2, ..., chrX, chrY, chrM as demanded by GATK. We therefore sorted the VCF correctly with a Perl script. We additionally had to add `INFO` lines to the header for the fields `VDB`, `FQ`, `AC1`, `AF1`, `PV4`, and `DP4` because the VCF format requires that all `INFO` fields used in the data lines have been defined in the header.

These can be downloaded from the FTP server of the Broad Institute.[12] We will need the following four resources for the hg19 genome.

```
hapmap_3.3.hg19.sites.vcf.gz
1000G_omni2.5.hg19.sites.vcf.gz
1000G_phase1.snps.high_confidence.hg19.sites.vcf.gz
dbsnp_138.hg19.vcf.gz
```

These files will also need some doctoring, because the index files offered at the Broad FTP site are no longer recognized by GATK. For each of the four files, we need to unpack them with **gunzip** and recompress them with **bgzip**.[13] For instance,

```
$ gunzip hapmap_3.3.hg19.sites.vcf.gz
$ bgzip -c hapmap_3.3.hg19.sites.vcf > \
    hapmap_3.3.hg19.sites.vcf.gz
$ tabix -p vcf hapmap_3.3.hg19.sites.vcf.gz
```

The GATK command will still need the sequence dictionary created with Picard as well as the FASTA index file for the hg19 reference genome.

Create the sequence dictionary with Picard.

```
$ java -jar picard.jar CreateSequenceDictionary \
  R=hs37d5.fa \
  O=hs37d5.dict
```

Then create a FASTA index file using SAMtools.

```
$ samtools faidx hs37d5.fa
```

We now can perform the VQSR for this file as follows.

```
java -Xmx4g -jar GenomeAnalysisTK.jar \
   -T VariantRecalibrator \
   -R hs37d5.fa \
   -input mongolian.vcf  \
   -recalFile mongolian.recal \
```

---

[12]ftp://gsapubftp-anonymous@ftp.broadinstitute.org/bundle

[13]**bgzip** is a block compression/decompression utility that is designed to work with **tabix** to produce indices of compressed VCF and other files.

```
-tranchesFile mongolian.tranches \
-rscriptFile mongolian_plots.R \
-nt 4 \
[resources] \
-an QD -an MQ -an MQRankSum \
-an ReadPosRankSum -an FS -an DP \
-tranche 90.0 -tranche 99.0 \
-tranche 99.9 -tranche 100.0 \
-mode SNP
```

The option -R specifies the reference genome. The -input option passes GATK the mongolian.vcf file we repaired as described above. The arguments to -recalFile, -tranchesFile, and -rscriptFile provide the names of output files that we will explain below. The option -nt 4 indicates the number of threads to be used.

The [resources] block (which would not have fit on the page) contains four arguments (see below) that specify the truth and training sets. The first argument names the resource as HapMap, which is used as a very high-confidence variant call set [179]. The command causes VariantRecalibrator to use the HapMap data for training the recalibration model (training=true), to assume HapMap variants to be true (truth=true) and to use a prior likelihood of 96.84% (Q15 on a Phred scale; prior=15.0). The HapMap, Omni, and 1000G variants are all marked as known=false, meaning they will not be used to stratify output metrics such as the Ti/Tv ratio (see Section 16.1).

We will also use these sites later on to choose a threshold for filtering variants based on sensitivity to truth sites.

```
-resource:hapmap,\
  known=false,\
  training=true,\
  truth=true,\
  prior=15.0 \
  hapmap_3.3.hg19.sites.vcf.gz \
```

The second argument describes the Omni resource, a set of polymorphic SNP sites produced by the Omni genotyping array.

```
-resource:omni,
  known=false,\
  training=true,\
```

```
truth=true,\
prior=12.0 \
1000G_omni2.5.hg19.sites.vcf.gz \
```

The argument is analogous to the HapMap argument except that the prior likelihood assigned to the variants is Q12 (93.69%).

The third argument describes high-confidence SNP sites produced by the 1000 Genomes Project.

```
-resource:1000G,\
  known=false,\
  training=true,\
  truth=false,\
  prior=10.0 \
  1000G_phase1.snps.high_confidence.hg19.sites.vcf.gz \
```

Again the argument is similar except that the prior likelihood assigned to the variants is Q10 (90.0%).

The fourth argument describes dbSNP [393] version 138. This dataset is substantially noisier than the others, and correspondingly the truth value is set to false to indicate that it has not been validated to a high degree of confidence. The variants are not used for training. However, the `VariantRecalibrator` will use them to stratify output metrics by whether variants are present in dbSNP or not (`known=true`). The prior likelihood assigned to the dbSNP variants is Q2 (36.90%).

```
-resource:dbsnp,\
  known=true,\
  training=false,\
  truth=false,\
  prior=2.0 \
  dbsnp_138.hg19.vcf.gz \
```

The `VariantRecalibrator` constructs its model based on the variant metrics (annotations) indicated in the annotation arguments (`-an QD`, etc.).

The `VariantRecalibrator` tool produces several output files. `mongolian.recal` contains the recalibration data that will be used in the next step to actually perform the recalibration. The VQSR procedure uses gold-standard variants that have been validated in large-scale projects (Omni, 1000 Genomes, HapMap). It finds variants in the VCF file that is being recalibrated that overlap with these gold-standard variants, and looks at their variant call annotations (as specified by the -an flag; in our example, these were QD, MQ, MQRankSum, ReadPosRankSum, FS, and DP). VQSR clusters all variants in the VCF file according to these annotations and uses the clustering to assign a VQSLOD score to each variant. The resulting VQSLOD score (log odds ratio of being a true variant versus being false under the VQSR model) can then be used as a threshold. To do so, the data are divided into tranches[14] that are defined using multiple -tranche arguments (-tranche 100.0 ... -tranche 90.0). We can find the VQSLOD threshold for recalibration as the VQSLOD score that is strict enough to include only the best 90% of the truth data (90% tranche).[15] The VQSLOD score that corresponds to the 90% tranche is 7.0137, as can be seen in the mongolian.tranche file. The VariantRecalibrator outputs a graphic that shows the proportion of truth and other variants at tranches defined by the VQSLOD thresholds that include 90, 99, 99.9, and 100% of the truth data (Figure 12.3).

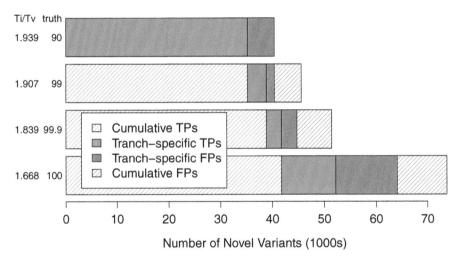

Figure 12.3. Mongolian genome tranches. The bars at each tranche show the number of variants from the truth set (labeled as TPs) included at the given VQSLOD threshold together with the numbers of other variants (labeled as FPs). The number of variants that are included by each new tranche are shown as tranche-specific TPs and FPs.

---

[14]In this usage, the word tranche essentially means a "slice" of the data.

[15]and so on for the tranche values 99.0, 99.9, and 100.

We can finally perform the actual recalibration. For the sake of demonstration, we will choose to recalibrate the variants at the strictest tranche level (90%).

```
$ java -Xmx4g -jar GenomeAnalysisTK.jar \
   -T ApplyRecalibration \
   -R hs37d5.fa \
   -input mongolian.vcf  \
   -recalFile mongolian.recal \
   -tranchesFile mongolian.tranches \
   -o recalibrated90.vcf \
   --ts_filter_level 90 \
   -mode SNP
```

This will add the entry VQSRTrancheSNP90.00to99.00 to the FILTER field of the VCF file if the VQSLOD score for the variant in question falls below the threshold (7.0137 in this case), i.e., fails the filter. Variants with better VQSLOD scores are marked with PASS in the FILTER field. Of the total of 3,731,155 data lines of the original file, 3,040,757 passed according to this filter (82%). If we take a less stringent tranche (99.9%), then 3,567,693 variants pass (96%). In this way, the tranches can be used to define VQSLOD thresholds with a given effect on the truth set. In practice, it is important to tune and validate the VQSR procedure to obtain a good trade-off between sensitivity and specificity.

12.6   CONCORDANCE (OR NOT) BETWEEN VARIANT CALLERS

We have concentrated on GATK in this book because it is educational to consider how individual steps of the GATK pipeline work and what influence they have on the data. However, there are many other variant callers and variant calling pipelines, and there is no clear "winner" in terms of overall accuracy. It should be apparent now to readers who have made it this far that variant calling in NGS data remains a difficult endeavor, and it probably will not come as a surprise that different variant calling pipelines do not identify the same set of variants. For instance, one study sequenced 15 exomes from four families and compared the results of variant calling with five different alignment and variant-calling pipelines. The overall agreement between the five pipelines for single-nucleotide variants was only 57.4%, with up

Table 12.2. A selective list of variant callers for germline variants in WES/WGS data.

Variant Caller	Reference
GATK	[104, 438]
FreeBayes	[129]
SAMtools mpileup	[254]
VarScan2	[204, 205]
SNVer	[456]

to 5.1% of variant calls being unique to each pipeline. Indel concordance was even lower [317]. A number of other studies documented low concordance rates between callers and between technical WES replicates [163, 264, 320, 339, 478]. A more recent study made use of Genome-in-a-Bottle consortium data to assess the performance of thirteen variant calling pipelines consisting of combinations of one of three read aligners and one of four variant callers. This study documented a concordance of 92% for Illumina datasets for three variant callers. Concordance varied not only between individual pipelines but also between datasets, and variant callers displayed tendencies towards different categories of error. In this study, the best performing pipeline for single-nucleotide variants in the germline was BWA-MEM with SAMtools, and the best pipeline for indels was BWA-MEM and GATK [175]. We note however, that different comparison studies have come to different conclusions as to which pipelines are "best", and it is not currently possible to make a general recommendation. One must also distinguish between calling variants in the germline or somatic variants in tumor samples, which we will address in Chapter 30. Bioinformaticians should test various pipelines on data from their own center and should strive to deeply understand the advantages and limitations of the set of tools that they decide to work with. Table 12.2 provides an overview of some currently available (germline) variant callers.

12.7 A WORD ABOUT TERMINOLOGY: SNV VERSUS SNP

A single-nucleotide polymorphism (SNP, pronounced "snip") is a variation at a single nucleotide position in a DNA sequence among individuals. For instance, we might speak about a certain SNP having a minor allele frequency of 0.27 in some population. The SNP thus describes the fact that two (or more) alleles (different nucleotides) can be found at a given position of the genome in different individuals in a population. Some definitions restrict the term "polymorphism" to variation with a minor allele frequency of at least 1% in a population. On the other hand, the phrase "single-nucleotide variant" (SNV) refers to a difference identified in an individual sample with respect to the reference sequence. One occasionally hears people use "SNP" to describe variants found in a VCF file, but strictly speaking this is incorrect and "SNV" should be used.

FURTHER READING

Van der Auwera GA, Carneiro MO, Hartl C, Poplin R, Del Angel G, Levy-Moonshine A, Jordan T, et al. (2013) From FastQ data to high confidence variant calls: the Genome Analysis Toolkit best practices pipeline. *Curr Protoc Bioinformatics* **43**:11.10.1–11.10.33.

12.8 EXERCISES

Exercise 1

Perform hard filtering using a series of progressively more stringent filters and examine their effect on the number of variants that survive filtering.

Exercise 2

Early approaches towards variant calling made use of the SAMtools pileup function. The pileup format is explained in Chapter 30, and essentially provides a summary of the base calls at all positions covered by at least one read. Pileup files can be used to call the numbers of REF and ALT base calls at each position in order to call variants. For this exercise, we will restrict ourselves to calls on the smallest chromosome (chr21). Use samtools to extract the alignment data for chromosome 21 from the BAM file we generated in Chapter 10.

```
$ samtools view \
    -b NIST7035_indelrealigner.bam \
    chr21 > chr21.bam
```

A FASTA file for chromosome 21 can be downloaded from the

UCSC Genome browser (be sure to download the file from the correct genome build, i.e., hg38/GRCh38). Use samtools to generate a FASTA index.

```
$ samtools faidx chr21.fa
```

This creates an index file called `chr21.fa.fai`. Let us now create the pileup file.

```
$ samtools mpileup -f chr21.fa chr21.bam > chr21.pileup
```

Use the `tview` program of SAMtools to compare the read alignments in the BAM file with the position-specific base information in the pileup file.

```
$ samtools tview chr21.bam chr21.fa
```

You will need to tell SAMtools what part of the alignment you would like to see. Since this is an exome file, most of the chromosome has little or no reads aligned to it. You can peak into the pileup file to find a place with enough reads to make the alignment interesting. Then, press the "g" key, and you can enter the position as (for instance) `21:31691775`. Look at individual columns and compare the alignment with the pileup and make sure you understand what the format means. Write a script to perform simple variant calling using the pileup file. Your program should implement the following simple rules.

- Determine if the sample meets the minimum coverage requirement of at least three reads with base quality ≥ 20.

- Determine possible variants for each position based upon the read bases observed. A variant allele must be supported by at least two independent reads and at least 8% of all reads.

- Determine the genotypes for your called variants. For this exercise, variants are called homozygous if supported by 75% or more of all reads at a position; otherwise they are called heterozygous.

How many heterozygous and homozygous variants did you observe?

Exercise 3

FreeBayes [129] is a Bayesian haplotype-aware variant caller for SNVs and indels that performs realignment. For this exercise, download and build FreeBayes.[16] Compare the results of variant calling by GATK and FreeBayes on the chromosome 21 BAM file from the previous exercise. Consult the online documentation for FreeBayes, but the basic command used for variant calling is simple:

```
$ freebayes -f chr21.fa  chr21.bam > chr21fb.vcf
```

You can use grep or BCFTools to count the overall number of variants called by each caller. If you would like to explore the data further, find variants called by both callers or by just one of the callers and inspect them using IGV (Chapter 17). GATK offers a module called **CombineVariants** that allows the union or the intersection of the variants to be calculated (see the online GATK documentation). Is the overall quality of the variants found by both callers higher than those found by just one caller?

---

[16]https://github.com/ekg/freebayes

CHAPTER 13

Variant Call Format (VCF)

Peter Robinson and Peter Krawitz

S EQUENCE variants called by variant callers are stored in Variant Call Format (VCF) files [95]. VCF files are text files that are often stored in compressed form. They contain a header with meta-information and a line that names the eight fixed, mandatory columns and in many cases provides the names of the samples that were sequenced, and finally data lines each containing information about a position in the genome. VCF files can represent genotype information on one or more samples for each position. Together with FASTQ files and SAM/BAM files, VCF files make up one of the three essential formats for computational analysis of WES/WGS data. As with the chapters on FASTQ and SAM format, we will not attempt to comprehensively cover every aspect of the VCF format, and will assume that readers will consult the online specifications as needed.[1] Instead, we will explain the most important features of the VCF format and show how to filter and analyze the files.

VCF files have two main sections, a meta-information section and a data section, that are separated by the column header line.

---

[1]https://samtools.github.io/hts-specs/

13.1 META-INFORMATION SECTION

The lines of the meta-information section all start with two hash marks (`##`) followed by a string with key=value pairs. The purpose of the meta-information section is to provide background information about the analysis results provided in the data section.

The first line in a VCF file is the only mandatory line in the meta-information section; it specifies the version of the VCF format used. There can be an arbitrary number of additional meta-information lines, which are intended to provide a standardized description of tags and annotations used in the data section. This is necessary, because individual VCF files can vary greatly with respect to the amount and type of information they contain. Meta-information sections vary according to the type of analysis performed and the programs used to carry out the analysis. For instance, VCF files used to record variants in WES/WGS studies contain definitions for the `FORMAT` field in the meta-information section describing genotypes and genotype qualities.

In the NA12878 exome VCF file, the meta-information section has the following fields.[2]

`##fileformat` – The `fileformat` field is the only metadata item that is required for a well formed VCF file. It specifies the VCF format version used. We have used version 4.2,[3] and the first line of our VCF file is thus

```
##fileformat=VCFv4.2
```

`##FILTER` – The `FILTER` lines describe filters that have been applied to the data. If a variant fails to pass one of the filters, then the name of that filter will appear in the `FILTER` column of the corresponding line in the data section.

`##INFO` – Each line of the data section contains an `INFO` field with

---

[2]We recommend reading this chapter while examining a concrete VCF file produced by GATK's `HaplotypeCaller`. Although we provide some examples throughout the chapter, browsing a complete VCF file will help with understanding how the single sections and lines are formated and especially how the meta-information is related to the data section.

[3]Recently, version 4.3 has become available, but none of the aspects of the VCF format discussed in this book have changed.

information about the variant that is independent of the genotypes called for each sample. The INFO fields of the meta-information section explain the abbreviations used in the INFO field of the data section.

##FORMAT – Each line of the data section contains a single FORMAT field and one genotype field for each of the samples represented in the VCF file. The FORMAT field of the data lines specifies the format of the genotype fields. The FORMAT lines of the meta-information section explain the abbreviations used in the FORMAT and genotype fields of the data section and are thus similar to the INFO lines.

##GATKCommandLine – Variants callers generally add information about the parameters used to run the program that produced the VCF file. For instance, GATK adds a line to the VCF file that records the parameters used to run the program, ##GATKCommandLine.HaplotypeCaller.

##contig – The contig lines report the names and lengths of the contigs (or reference sequences) represented by the alignment from which variants are called. For instance, the first contig line in our VCF file represents chromosome 1.

```
##contig=<ID=chr1,length=248956422>
```

##reference – The line records the file containing the reference contigs, i.e., the reference genome file.

```
##reference=file:///path.../hs38DH.fa
```

We will explain the meta-information fields in more detail below.

13.2   THE COLUMN HEADER LINE

The meta-information section, whose lines beginning with two hash-marks (##), is separated from the data section by the header line, which is the only line to begin with just one hash mark.

```
#CHROM POS ID REF ALT QUAL FILTER INFO FORMAT NA12878
```

Table 13.1. Fields of the VCF file for sample NA12878.

Field	Example	Explanation
CHROM	chr5	Name of the contig
POS	130076433	One-based position
ID	.	No reference ID available
REF	G	Wildtype base is G
ALT	A	Variant base is A
QUAL	62.74	Phred score for variant
FILTER	PASS	This variant passed filters
INFO	(see below)	(see below)
FORMAT	GT:AD:DP:GQ:PL	Format of genotype field(s)
NA12878	1/1:0,2:2:6:90,6,0	Genotype call for sample NA12878

Note: A multisample VCF file will have one column for the genotype calls of each sample.

The first 8 fields (CHROM to INFO) are mandatory in any VCF file, and if the VCF file reports genotypes then it must contain a ninth field that specifies the format in which the genotypes are reported (FORMAT),[4] and one field for each sample reported in the VCF file. (In this VCF file, only one sample is contained, called NA12878.)

13.3 DATA LINES

The data lines contain information about a position in the genome. In most cases, one data line will refer to one variant at one position of the genome, but there are exceptions that we will explain below. Table 13.1 presents an overview of the 10 fields of the data lines of the VCF file we generated in chapter 12 (NA12878_filtered.vcf).

The fields CHROM, POS, REF, and ALT together specify the location as well as the reference and alternate nucleotides of the called variant. With the example shown in Table 13.1, the variant is a transition from G to A at position 130,076,433 of chromosome 5:

```
chr5:130076433G>A
```

---

[4]VCF files are not just used to record variants and genotypes found in WES/WGS studies. For instance, ClinVar uses the VCF format to record all of the variants in the database, without reporting the genotypes of individual patients. This VCF file does not have a FORMAT field.

The numbering system used in VCF files is one-based rather than zero-based, and is straightforward for single-nucleotide variants (SNV). The scheme used to specify the position of insertions, deletions, and block variants is slightly more complicated and will be explained in Section 13.7.

The ID field can be used to specify the accession number of a variant in databases such as dbSNP, and might contain an entry such as rs25458. This field is not populated by default by GATK, and in our case, a period (.) is used to indicate that the field is empty.

The QUAL field is a Phred-scaled quality score for the assertion made in ALT. If the ALT field contains a variant call, then QUAL reflects the estimated probability that the variant call is wrong:

$$QUAL = -10 \log_{10} p(\text{no variant}) \tag{13.1}$$

where $p(\text{no variant})$ is the probability that there is no variant at this site. If on the other hand, the ALT field is "." (i.e., no variant call), then QUAL is the estimated probability that the call of no variant is wrong:

$$QUAL = -10 \log_{10} p(\text{variant}) \tag{13.2}$$

The FILTER field specifies whether the position has passed the filters indicated in the meta-information section. In our VCF file, two filters were applied, which are described as follows in the meta-information section

```
##FILTER=<ID=LowQual,Description="Low quality">
##FILTER=<ID=snpFilter,Description="QD < 2.0 || \
  FS > 60.0 || MQ < 40.0 || MQRankSum < -12.5 || \
  ReadPosRankSum < -8.0">
```

We can use BCFtools, which was introduced in Chapter 12, to count the number of positions passing all filters.

```
$ bcftools query -f '%FILTER\n' NA12878_filtered.vcf | \
    grep -c PASS
138083
```

Thus, ~94.6% of all 146,000 positions represented in the VCF file passed all filters. On the other hand, 7917 positions did not pass the `snpFilter`. This item represents the result of the hard filter we applied in Section 12.4. Recall that the hard filter we applied would not let a variant pass if it failed to fulfill any one of the criteria used in the filter. For instance, our VCF file reports a `G>A` transition at position 22,646,441 of chromosome 22. This variant is marked with `snpFilter` in the `FILTER` field, because the value of the RMS Mapping Quality (MQ) was less than 40:

```
MQ=39.81
```

13.4   INFO

The `INFO` field provides additional information about the variants called at the genomic position of the line. The exact format of each INFO sub–field should be specified in the meta-information. We will explain how this works for one of the variants called in our VCF file. The first portion of the line is shown below (the `INFO`, `FORMAT` and genotype fields will be explained further on).

```
chr7 1936796 . C T 274.77 PASS  (...)
```

The `INFO` field for this variant contains 15 fields that are separated by semicolons (`;`). Each of the fields has a corresponding explanation in the meta-information section.

Allele count (`AC`). The meta-information contains the following line that explains this field.

```
##INFO=<ID=AC,Number=A,Type=Integer, \
  Description="Allele count in genotypes, \
  for each ALT allele, in the same order as listed">
```

ID specifies the name of the item, which is often a two-letter abbreviation. `Number` specifies the number of values that the field should contain. In this case, the value "A" indicated that the field contains one value per alternate allele. `Type` specifies the data type that is expected for the field, and `Description` is a string that must be surrounded by double quotes. Other possible fields in the `##INFO` are `Source` and `Version`.

The value for `chr7:1936796C>T` is

`AC=1`

This reflects the allele count in the genotype for this variant (`0/1`, i.e., heterozygous). Homozygous variants (`1/1`) have `AC=2`.

Allele frequency (`AF`) and Allele Number (`AN`). The `AF` field shows the Allele Frequency,[5] for each `ALT` allele, in the same order as listed, and the `AN` field shows the total number of alleles in called genotypes. `AF` is calculated as `AC/AN`. For the heterozygous variants (`0/1`), `AC=1`, `AN=2`, and `AF=0.500`. For the homozygous variants (`1/1`), `AC=2`, `AN=2`, and `AF=1.00`. For a slightly more complicated example, consider a VCF file that describes three samples. If a data line indicates a variant that was called homozygous (`1/1`) in sample 1, heterozygous (`0/1`) in sample 2, and heterozygous (`0/1`) in sample 3, then `AF=4`, `AN=6` and, the Allele Frequency is calculated as $4/6$ or `AF=0.667`.

`BaseQRankSum`. This field refers to the Z-score from a Wilcoxon rank sum test of `ALT` vs. `REF` base qualities. Intuitively, we expect that this will take the base qualities for the position of a heterozygous variant from all of the contributing reads and rank them in order, then the reference bases and the alternate bases should appear at random in the list (this would be associated with a Z-score close to zero). If the qualities of the `ALT` base were ranked significantly lower than the qualities of the reference base, this could be a sign of an artifact and the Z-score would have a large absolute value. For `chr7:1936796C>T` , the Z-score is not significant (`BaseQRankSum=-0.386`), meaning there is no clear bias in the rank of the `REF` and the `ALT` reads.

`ClippingRankSum`. This contains the Z-score from a Wilcoxon rank sum test of `ALT` vs. `REF` number of hard clipped bases. The field is analogous to the `BaseQRankSum`, but with respect to the number of hard clipped bases—there should not be a significant difference between `ALT` and `REF` reads.

Approximate Read Depth (`DP`). Approximate read depth (some reads may have been filtered). For multisample VCF files, `DP` refers

---

[5]Note that the phrase allele frequency has a different meaning in the context of population genetics. In VCF files, allele frequency refers to the allele balance of the alleles observed in the samples being sequenced, not the frequency of the variants in the population.

to the combined depth across samples. For this line, DP=23, meaning that 23 reads covered this position. We will see further below in the genotype field that there were 11 REF reads and 12 ALT reads.

Excess heterozygosity (ExcessHet). This field contains the Phred-scaled p-value for an exact test of excess heterozygosity in a population of samples according to the Hardy-Weinberg equilibrium. ExcessHet is not meant to apply to a VCF file with a single or just a few samples.

Strand bias (FS). This field contains the Phred-scaled p-value using Fisher's exact test to detect strand bias. For our variant the value was FS=0.000.

Maximum likelihood expectation for the allele counts (MLEAC). This field shows the Maximum likelihood expectation (MLE) for the allele counts, for each ALT allele, in the same order as listed. This is GATK's Maximum likelihood (ML) estimation of the number of alternate alleles for each individual sample at a site, and is usually equivalent to the AC field, which in turn is equivalent to the sum of "1"s in a genotype (for a biallelic site). The MLEAF field (Maximum likelihood expectation (MLE) for the allele frequency) is analogous and is usually equivalent to the AF field.

RMS Mapping Quality (MQ). The Root Mean Square of the Mapping quality (RMSMappingQuality, MQ) provides an estimation of the overall mapping quality of reads supporting a variant call (see Section 12.3.3).

Mapping Quality Rank Sum (MQRankSum). This field is similar to the BaseQRankSum and ClippingRankSum fields, and indicates the Z-score from the Wilcoxon rank sum test of ALT versus REF read mapping qualities. There was no problem with chr7:1936796C>T, and the value was MQRankSum=0.000.

Variant Confidence/Quality by Depth (QD). The QualByDepth (QD) score is the QUAL score divided by the allele depth for the variant, and was explained in Section 12.3.1. chr7:1936796C>T has QD=11.95, which is good.

Read position rank sum (`ReadPosRankSum`). This field specifies the Z-score from a Wilcoxon rank sum test of `ALT` vs. `REF` read position bias in a way that is analogous to the other rank sum fields. The value `ReadPosRankSum=0.326` does not indicate an issue.

Symmetric Odds Ratio (`SOR`). This field presents the symmetric odds ratio of a 2x2 contingency table to detect strand bias. Strand bias is a type of sequencing bias in which one DNA strand is favored over the other. The `SOR` is related to the Fisher Strand Test (`FS`) but is better at taking into account large amounts of data if there is high coverage at the position of a variant. There is no absolute cutoff between good and bad values for this field, but values of `SOR>4.0` may be considered suspicious. `chr7:1936796C>T` had an unremarkable value of `SOR=0.527`.

We have made the effort to go through each of these fields to demonstrate the type of information that is contained in typical VCF files. Clearly, one will not go through this information by hand for each of the variants in the file, but it is useful to understand the information in the `INFO` (and `FORMAT`) fields especially in situations where a potentially relevant variant has poor or borderline quality metrics.

13.5   FORMAT

The `FORMAT` field specifies the items that will be used for the genotype fields for each sample as well as their order as a colon-separated alphanumeric string. For `NA12878_filtered.vcf`, the `FORMAT` field has the following items

`GT:AD:DP:GQ:PL`

The `FORMAT` field is followed by one genotype field per sample with colon-separated data that corresponds to the items specified in `FORMAT`. Each of the items in the `FORMAT` field should have a corresponding explanation in the meta-information section. The format is the same as for the `INFO` lines in the header. For instance, the following entry states that the `GT` field refers to the Genotype, has one entry per genotype field, and has the data type `String`.

`##FORMAT=<ID=GT,Number=1,Type=String,Description="Genotype">`

The genotype field for the `chr7:1936796C>T` line we examined above is:

```
0/1:11,12:23:99:303,0,290
```

Genotype (GT). The first entry refers to the genotype. The two most commonly encountered genotypes are simple to understand: `0/1` refers to a heterozygous variant, and `1/1` refers to a homozygous variant. In general, the genotype is encoded as allele values separated by one of `/` or `|`. The separator `/` is used if the variants are unphased, and the separator `|` is used for phased variants (see Section 13.6).

The allele values are 0 for the reference allele (i.e., the allele that is shown in the `REF` field), 1 for the first allele listed in `ALT`, 2 for the second allele list in `ALT` and so on. For calls on the Y chromosome, or on the male chromosome X in the non-pseudoautosomal regions, the genome sequence is haploid (meaning there is only one copy as opposed to the diploid sequence of the autosomal chromosomes and the female X chromosomes). For haploid calls, only one allele value should be given, e.g., `1`. If a call cannot be made at a given locus, then '.' should be specified for each missing allele (e.g., `./.` for a diploid genotype).

Consider the genotype call for the following genomic position.

```
chr15 35933537 . A C,T 631.77 PASS  (...)
```

Here, the `GT` field (not shown above) has the value `1/2`. This variant corresponds to the triallelic polymorphism rs11073131. The reference allele is `A`, but there are two variants that are known to occur in the population, `C` and `T`. In the NA12878 sample, the reference base does not occur, but rather both variants occur in heterozygous form.

Allelic Depth (AD). This field indicates the allelic depths (i.e., read counts) for the `REF` and `ALT` alleles in the order listed. As mentioned already in the discussion of the `DP` field of `INFO`, there were 11 `REF` reads and 12 `ALT` reads, and thus the `AD` field is `11,12`.

Read Depth (DP). This field indicates the read depth at the indicated position for the current sample.

Genotype Quality (GQ). The `GQ` field represents a Phred-scaled quality value that represents the probability of the genotype call being wrong conditioned on the site being variant.

$$GQ = -10 \log_{10} p(\text{genotype call wrong}|\text{site is variant}) \qquad (13.3)$$

In the case of `chr7:1936796C>T`, the `GQ` has the value of 99, corresponding to an estimated probability of $10^{-99/10} = 10^{-9.9}$ of the genotype call being wrong.

Normalized, Phred-scaled likelihoods for genotypes (`PL`). The `PL` field contains normalized, Phred-scaled likelihoods for each of the three genotypes `0/0`, `0/1`, and `1/1`, without priors (rounded to the closest integer). For example, in the heterozygous case, this is

$$\mathcal{L}(\text{alignment data} \mid \text{true genotype is } 0/1)$$

The most likely genotype (given in the GT field) is scaled so that its probability is $P = 1.0$ (0 when Phred-scaled), and the other likelihoods reflect their Phred-scaled likelihoods relative to this most likely genotype.

For `chr7:1936796C>T`, the values of the PL field are `303,0,290` Thus, the most likely genotype is `0/1`, and the other two possible genotypes are substantially less likely ($10^{-30.3}$ and 10^{-29}, respectively).

13.6 PHASING

The pipe (`|`) symbol can be used for each sample to indicate that each of the alleles of the genotype in question derive from the same haplotype as each of the alleles of the genotype of the same sample in the previous NON-FILTERED variant record.[6]

```
#CHROM  POS ID  REF ALT QUAL FILTER INFO FORMAT sample
chr1    90  .   A   T   99   PASS   .    GT     0/1
chr1    100 .   A   T   99   PASS   .    GT:PQ  0|1:1427.45
chr1    110 .   T   C   99   PASS   .    GT:PQ  1|0:1427.45
```

The sample thus has the haplotypes TTT and AAC. The first heterozygous genotype (at position 90) in the haplotype has / and not |, because otherwise, it would be a continuation of preceding haplotypes. The second genotype (at position 100) is `0|1` and indicates that the first allele goes with the first allele in the previous genotype `0/1` (thus, the A at position 100 is on the same haplotype as the A at position 90). The third genotype (at position 110) also indicates that the first allele goes with the first allele in the previous genotype. Note here that the order of the "1" and the "0" is reversed in the third genotype, which indicates that the C goes with the A of the previous genotype. The

---

[6]For phasing, variant positions without FILTER=PASS are ignored.

`PQ` value (Phred-scaled phasing quality score) is an estimation of the accuracy of the phase assertion.

In practice, variant callers may use other schemes to record phase information. GATK does not output phase information by default but has a `ReadBackedPhasing` module that identifies haplotypes based on colocalization of variants on individual reads. This module does not use the | notation, but instead indicates haplotype information with the `HP` tag. FreeBayes can also call haplotypes based on read information but represents the entire haplotype using the `REF` and `ALT` fields (Figure 13.1). We refer the reader to the documentation of GATK [285] and FreeBayes [129] for details.

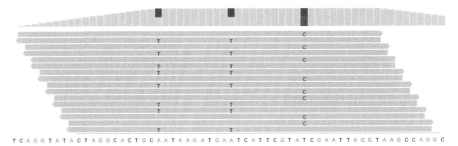

Figure 13.1. Phasing based on physical information. It can be seen that some reads have A→T variants at two positions, and other reads have a single T→C variant. Clearly, the reads originate from different haplotypes (chromosomes). Both GATK and FreeBayes called the corresponding haplotypes. GATK used the `HP` tag 90-1,90-2 for both A→T variants and the `HP` tag 90-2,90-1 for the T→C variant. FreeBayes reported the haplotype of the A→T variants as REF=AATAAGATGAA ALT=TATAAGATGAT.

13.7 REPRESENTING VARIANTS IN VCF FORMAT

A single nucleotide variant is the simplest category of variants, and the VCF representation is straightforward.

For instance, the following line specifies a variant at position 112,275,379 of chromosome 5, whereby the reference base is an `A` and the alternate base is a `G`.

```
chr5    112275379    .    A    G    322.77    PASS    (...)
```

Consider now the following deletion of five nucleotides on chromosome 5

```
Ref: AGTATAGTTTAG
Alt: AGTA-----TAG
```

The deleted nucleotides are located on chromosome at positions 113,024,753 to 113,024,757. Naively, one might expect the VCF file to list TAGTT as REF and "-" as the ALT sequence, but instead, the VCF format specifies that the base preceding the deletion is shown together with the deleted sequence in REF, and only the preceding base is shown as ALT. The position (POS) is corresponding that of the preceding base. For the deletion mentioned above, the correct VCF format is as follows.

```
chr5   113024752   .   ATAGTT   A   489.73   PASS (...)
```

Now consider the insertion of a single nucleotide between the T at position 113,040,194 and the A at position 113,040,195.

```
Ref: GCATGT-AGCC
Alt: GCATGTAAGCC
```

This is represented by the reference base immediately 5' to the insertion being specified in the REF field (and the position of this base in the POS field), and the same base followed by the inserted base(s) being shown in the ALT field.

```
chr5   113040194   .   T   TA   53.70   PASS   (...)
```

The VCF format has additional specifications for mixed records and structural variants that can be found in the online documentation.[7]

13.8   VARIANT NORMALIZATION: VCF

Transforming variant coordinates from the genome level (VCF) to the transcript is not a trivial task, but clinical genomic testing and genetic research are critically dependent on the robust identification and reporting of variant-level information in relation to disease [476].

---

[7]http://samtools.github.io/hts-specs/VCFv4.3.pdf for the most recent VCF version (4.3) at the time of this writing in early 2017.

Compared to SNV calling, indel calling is more difficult and error prone, being susceptible to errors not only in PCR and sequencing procedures, but also in computational mapping and calling. The Single Nucleotide Polymorphism database (dbSNP) of NCBI, now contains not only SNPs but also insertion and deletion (indel) polymorphisms. In 2014 it was demonstrated that up to 10% of the indels polymorphisms in dbSNP were represented by multiple, redundant entries that used different notation to refer to the identical sequence change [260]. The variant `chr22:g.16537622_16537623insCTTT` is listed as rs200449532 in dbSNP. The minor allele frequency is indicated as 0.0423. On the other hand, the variant `chr22:g.16537624_16537625insTTCT` is listed as rs4010175; much less information is available for this dbSNP entry, but we see that the homozygous `TTCT/TTCT` genotype was identified in the genome of Craig Venter, the founder of Celera genomics and proponent of shotgun sequencing for the determination of the human genome sequence. Closer inspection, however, reveals that these two variants are actually identical (Figure 13.2). If different databases use different representations of the same variant, it becomes difficult or impossible to query the databases to get information on the numerous variants found in WES/WGS datasets. For instance, if we were trying to interpret the insertion `chr22:g.16537624_16537625insTTCT` and tried to find out whether others had reported the same variant in dbSNP, we would fail to find the correct variant because of inconsistent variant numbering.

A similar problem pertains to VCF files. Without further specification, the rules that we summarized above are not sufficient to uniquely proscribe how to write a variant. Consider the deletion in Figure 13.3. Above, we indicated that the rule for denoting a deletion in VCF is for the `REF` to show the nucleotide directly preceding the deletion together with the deleted base, and `ALT` to show only the nucleotide preceding the deletion. Both Figure 13.3A and 13.3D satisfy this condition. How do we choose among these options? VCF normalization demands that VCF entries are *left-aligned*.

> A VCF entry is left aligned if and only if its base position is smallest among all potential VCF entries having the same allele length and representing the same variant [419].

The variant representation in Figure 13.3B is left aligned (`POS` is identical to the correct variant representation in Figure 13.3D). However, the change `AGTGT`→`AGT` is clearly identical to the change `AGT`→`A`.

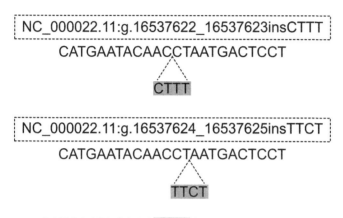

Figure 13.2. The need for variant normalization. The dbSNP entries rs200449532 and rs4010175 result in an identical alternate sequence. The two dbSNP entries denote distinct mutational events: (i) the insertion of the four nucleotides CTTT between positions 16,537,622 and 16,537,623 of chromosome 22 (rs200449532) and the insertion of the four nucleotides TTCT between positions 16,537,624 and 16,537,625 (rs4010175). However, it is impossible to observe whether the actual mutagenesis event was an insertion of CTTT or an insertion of TTCT. The biological effect would be the same for both mutations since the resulting sequence is identical (cf. the alignment at bottom of the panel). It is conventional to use only one denomination of variants such as this, so that all possible mutations that result in the same final sequence are given the same name. This process is referred to as variant normalization.

Therefore, VCF normalization applies the additional requirement that variant representations are *parsimonious*.

> A VCF entry is parsimonious if and only if the entry has the shortest allele length among all VCF entries representing the same variant [419].

Figure 13.3C is not parsimonious, even though only one GT repeat is shown, because the two nucleotides immediately preceding the deletion are shown in REF and ALT.

A number of algorithms and tools for VCF normalization

```
Reference Sequence: ACAGTGTGTACC
Alternate Sequence: ACAGTGTACC
```

		Genome Reference	**Variant Call Format**		
		ACAGTGTGTACC	**POS**	**REF**	**ALT**
(A)	REF	GTG	6	GTG	G
	ALT	G			
(B)	REF	AGTGT	3	AGTGT	AGT
	ALT	AGT			
(C)	REF	CAGT	2	CAGT	CA
	ALT	CA			
(D)	REF	AGT	3	AGT	A
	ALT	A			

Figure 13.3. VCF variant normalization. The reference sequence consists of 12 bases including three GT repeats. The alternate sequence has ten bases with two GT repeats. Panels (A)–(D) show four potential ways of representing the sequence change, but only one way is correct (See text; figure adapted from [419]).

have been presented in the literature, including GATK's [104] LeftAlignAndTrimVariants (LAATV) tool, vt normalize [419], [417], and the Best Alignment Normalisation (BAN) tool, which uses additional alignment steps to identify equivalent variant representations [33].

13.9 VCF VERSUS HGVS NORMALIZATION

Variant reporting formats differ in some ways between VCF and the Human Gene Variation Society (HGVS) format for describing transcript and protein variants (which we will introduce in Chapter 15). For instance, VCF shifts nucleotide repeats left with respect to the genome, while HGVS shifts right with respect to the transcript. While this may appear unnecessarily complicated, it is important to realize that transcript normalization is performed in the direction of the transcript, and VCF normalization is performed in the direction of the chromosome. Since approximately half of all genes are transcribed in the "forward" and half in the "reverse" strand orientation, even if

HGVS normalization involved left-shifting instead of right-shifting, it would "disagree" with VCF normalization about half of the time.

13.10   GENOMIC VCF

If GATK is used in standard VCF mode to call variants in a single sample, it does not emit a genotype call of 0/0 (homozygous REF) for positions with no evidence for a variant. Thus, if a variant at this position is found in another sample, it is impossible to know whether the first sample was called confidently homozygous reference or if there simple was not enough coverage to make a call. Genomic VCF (gVCF) files in contrast store information about both variant and non-variant positions. GATK can be made to produce gVCF files by passing the option --emitRefConfidence GVCF (alternatively, -ERC GVCF). The resulting gVCF file will contain lines that describe blocks of sequence that were not found to have variants. These blocks are interspersed between the variant calls as with standard VCF files. Thus, a gVCF file has information about all sites in the genome, whether or not a variant was called.

One of the main use cases of gVCF files is to be able to combine multiple gVCF files to perform joint analysis of a cohort in later steps. For instance, this can be useful for *de novo* variant calling (Chapter 18). The main disadvantage of gVCF files is simply that they take up more space. We will briefly demonstrate how to use GATK to create a gVCF file. For simplicity, we will restrict the analysis to a 500,000 nucleotide interval on chromosome 2.

```
$ java -jar GenomeAnalysisTK.jar \
    -T HaplotypeCaller \
    -R hs38DH.fa \
    -L chr2:10000000-10500000 \
    -ERC GVCF \
    -I NA12878.bam \
    -o subset.g.vcf
```

This command adds additional lines to the VCF file that represent blocks of sequences with no called variants. For instance, the following four lines from the gVCF file we created describe one called variant (a heterozygous C→T change at position 10,394,526 of chromosome 2), followed by three non-variant blocks.

```
chr2    10394526    .   C   T,<NON_REF>    23.79   .   \
    BaseQRankSum=1.501;(...)    GT:AD:DP:GQ:PL:SB    \
    0/1:4,2,0:6:52:52,0,90,64,96,159:1,3,1,1
chr2    10394527    .   G   <NON_REF>    .   .   END=10394529\
    GT:DP:GQ:MIN_DP:PL    0/0:7:18:7:0,18,270
chr2    10394530    .   C   <NON_REF>    .   .   END=10394544\
    GT:DP:GQ:MIN_DP:PL    0/0:6:15:5:0,15,163
chr2    10394545    .   G   <NON_REF>    .   .   END=10394550\
    GT:DP:GQ:MIN_DP:PL    0/0:5:12:5:0,12,180
...
```

The non-variant blocks have a start position (in the usual POS column) as well as an end position indicated by the corresponding key value pair (END) in the INFO field. For instance, the first non-variant block comprises three nucleotides at positions 10,394,527–10,394,529 of chromosome 2. It would be possible to get one line for each non-variant position in the genome by substituting -ERC BP_RESOLUTION for -ERC GVCF. In order to save space, the gVCF groups non-variant positions into blocks of consistent genotype quality (GQ). Note that the three blocks shown here have GQ values of 18, 15, and 12. The gVCF header defines the ranges of GQ values with lines such as the following

```
##GVCFBlock12-13=minGQ=12(inclusive),maxGQ=13(exclusive)
```

Together with the other lines, the GQ blocks are defined to be integers from 0, 1, ..., 60, and then blocks of 10 values starting at 70,80,90, up to 100. This leads to a file with 49,473 non-variant blocks and 419 variant calls for a region of 500,000 nucleotides. It is possible to adjust the block sizes with the -GQB option to reduce the file size at the cost of losing information about the exact genotype qualities of the non-variant positions. The following command defines just three GQ blocks.

```
$ java -jar GenomeAnalysisTK.jar \
    -T HaplotypeCaller \
    -R hs38DH.fa \
    -L chr2:10000000-10500000 \
    -ERC GVCF \
    --GVCFGQBands 10 \
    --GVCFGQBands 50 \
    --GVCFGQBands 100 \
    -I NA12878.bam \
    -o subset2.g.vcf
```

Correspondingly, the header of the output gVCF file has the following three lines.

```
##GVCFBlock0-10=minGQ=0(inclusive),maxGQ=10(exclusive)
##GVCFBlock10-50=minGQ=10(inclusive),maxGQ=50(exclusive)
##GVCFBlock50-100=minGQ=50(inclusive),maxGQ=100(exclusive)
```

This gVCF file has only 10,711 non-variant block lines.

There are a number of different conventions for gVCF files that we will not further discuss. Interested readers are referred to the web page of gvcftools.[8] Multiple gVCF files can be used for joint variant calling with GATK's `GenotypeGVCFs` module.[9]

FURTHER READING

Tan A, Abecasis GR, Kang HM (2015) Unified representation of genetic variants. *Bioinformatics* **31**:2202–2204.

13.11 EXERCISES

Exercise 1

Assume that a VCF file has the following **FORMAT** field.

```
GT:AD:DP:GQ:PL
```

Assume a particular variant has the following genotype field:

```
0/1:15,10:25:99:338,0,485
```

Determine the genotype, the number of **REF** and **ALT** reads, the overall depth, the genotype quality and the likelihoods of the three possible genotypes.

Exercise 2

Review the VCF documentation about how insertions and deletions are represented in VCF format. Use the VCF file described in Chapter 12 or a VCF file of your choice, and find several deletions and insertions in the VCF file. Determine the exact location of the variant in the genome. Check your answer by examining the location of the indels in the UCSC Genome Browser [379].

---

[8]`https://sites.google.com/site/gvcftools/home/about-gvcf/gvcf-conventions`

[9]See the online GATK documentation for details.

Exercise 3

Using the VCF file from Exercise 2, determine the number of heterozygous and homozygous variants using Unix/Linux command line tools.

Exercise 4

How many variants in the VCF file from Exercise 2 display a variant Phred score less than 30?

Exercise 5

In this exercise, you will examine the relationship between depth and variant quality. Write a script that will extract the DP (depth) and GQ (genotype quality) fields from the VCF file from Exercise 2.

Write the results to a file that you can read into R in order to plot the relationship between read depth and genotype quality. What do you notice?

Exercise 6

Assume the values for the PL field are 417,0,12. Write a function in R to calculate the genotype likelihoods.

Exercise 7

Plot the distribution of values for the allele frequency (AF) using R. What do you notice?

Exercise 8

Use the Unix tool **grep** to convert a gVCF file to a standard VCF file.

See hints and answers on page 500.

CHAPTER 14

Jannovar

J ANNOVAR is a stand-alone Java application as well as a Java library designed to be used in larger software frameworks for exome and genome analysis [184] using the standardized Variant Call Format (VCF). The current version of Jannovar uses a compact variant of an interval tree to swiftly identify all transcripts affected by a given variant. The interval tree is implemented as an array using the idea of augmented trees [89, Section 14.3]. Further, Jannovar provides Human Genome Variation Society (HGVS)–compliant annotations both for variants affecting coding sequences and splice junctions as well as untranslated regions and noncoding RNA transcripts. Jannovar can also perform family-based pedigree analysis with data from members of a family segregating a Mendelian disorder, and integrate data on variant frequency.

We will use Jannovar to demonstrate these analyses in Chapters 15,20, and 22. The current chapter will explain how to download and install Jannovar and prepare for the analysis.

14.1 JANNOVAR

Jannovar solves the basic problem of translating chromosomal coordinates to transcript variants. Variant calling programs usually generate chromosomal coordinates as reads are aligned to the reference while medical and biological interpretation usually require the prediction of the variant effects on individual transcripts and the corresponding proteins.

VCF is the *de facto* standard for variant calling output. VCF files denote variants according to their chromosomal position (Chapter 13).

For example, the following variant on chromosome 15 alters a thymine nucleotide at position[1] 48,463,207 to a cytosine nucleotide.

```
chr15  48463207 . T C (...)
```

However, in order to provide a biological or medical interpretation of `chr15:48463207T>C`, we need to know if an how it affects genes, transcripts, and other important elements of the genome. In this case, the variant affects transcript `NM_000138.4` of the *FBN1* gene, resulting in a missense mutation that is associated with the disease acromelic dysplasia: `c.5099A>G:p.(Y1700C)` [185].

The core functionality of Jannovar is to translate chromosomal variant coordinates to gene-based variant annotations, and to provide several variant analysis functionalities including frequency and pedigree filtering.

14.1.1 Downloading Jannovar

The complete source code of Jannovar should be downloaded as a Maven[2] repository with additional JUnit code for unit testing from GitHub at `https://github.com/charite/jannovar`. To perform the following steps, you will need to have `git` and `maven` installed on your computer.

To download the repository and checkout the version of the book, enter:

```
$ git clone https://github.com/charite/jannovar
```

In this book we used version 0.20 of Jannovar.[3]

This will download the Jannovar maven package. In the directory, you will see a number of directories and files, including the **pom.xml** file, which is the file that **maven** uses to coordinate the build process (roughly similar to a Makefile). To build the package, enter

---

[1]With reference to the GRCh38 genome assembly.

[2]Maven is a very powerful and popular software build and management tool for Java and will be discussed during this section. Maven allows all phases of the build process to be managed, including unit testing and the packaging of the executable jar file.

[3]If readers want to use the same version of Jannovar to ensure the results in the book can be reproduced exactly, it is possible to obtain the v0.20 by changing to the jannovar directory and entering the following command: `$ git checkout tags/v0.20` before continuing with the rest of the chapter.

```
$ cd jannovar
$ mvn package
```

This will download a number of archives needed for the build process, perform unit testing, and build various jar files if unit testing was successful. We will be using the file
jannovar-cli/target/jannovar-cli-0.20.jar.

If you prefer, you can download the jannovar-cli-0.20.jar directly from Maven Central (the easiest way to find it is to go to http://search.maven.org/ and search for "jannovar-cli-0.20.jar").
for the exercises in this book. To avoid having to enter the full path in the exercises, readers may want to create a symbolic link in the directory in which they will use Jannovar (on Linux and Mac OS X).

```
ln -s /path/to/jannovar-cli/target/jannovar-cli-0.20.jar \
    Jannovar.jar
```

14.1.2 Checking the Jannovar installation

To ensure that you have correctly installed and built Jannovar, enter the following command, and you should see a help message showing the usage and arguments for Jannovar.

```
$ java -jar Jannovar.jar -h
usage: jannovar-cli [-h] [--version] (...)
```

If you do not see this message, then check the online documentation (https://jannovar.readthedocs.io) for detailed installation instructions.

14.1.3 Reserving memory for Jannovar

Jar files are Java software packages that contain the compiled code that can then be executed with the java -jar command. This command starts the Java Virtual Machine (JVM) that allows the execution of Java code in a platform-independent manner. The JVM runs with fixed available memory. Once this memory is exceeded, a java.lang.OutOfMemoryError will occur. The JVM tries to make an intelligent choice about the available memory at startup, but if the program runs out of memory, it will crash.[4]

---

[4]In contrast, C programs may be able to obtain more memory from the operating system at runtime.

There are two important settings to adjust the amount of memory used by the JVM, -Xms and -Xmx. For instance the option -Xms1G would set the amount of initial available memory available to the JVM to 1 gigabyte, and the option -Xmx4G would set the maximal amount of memory available during a program run to 4 gigabytes.

These parameters can be set from the command line or as system-wide settings. On Windows, for instance, one would go to the Control Panel, click on **Programs**, go to Java settings, click on "Java Control Panel", and then go to "Runtime Parameters" and change the value, or if it is blank decide for the new value, of the Java memory (the two parameters shown above).

On Linux systems, this can be done by adding the following line to your $HOME/.bashrc file.

```
export JAVA_TOOL_OPTIONS="-Xms2G -Xmx2G"
```

It may be preferable just to pass these arguments on the command line as follows (here we have reserved a minimum of 2 and a maximum of 4 GB memory for the JVM):

```
$ java -Xms2G -Xmx4G -jar Jannovar.jar [options]
```

14.1.4   Understanding and creating the transcript definition file

Jannovar is designed to use transcript data from sources including the UCSC Genome Browser database [379], Ensembl [10], or NCBI Ref-Seq [344]. The data comprises information about the chromosomal locations of transcripts, their exons, strands, the corresponding mRNA sequences, and several cross-references to other databases.

To show the full range of supported databases, enter the following command.

```
$ java -jar Jannovar.jar db-list
(...)
Available data sources:

    hg18/ucsc
    hg18/ensembl
    hg18/refseq
    hg18/refseq_curated
    hg19/ucsc
```

```
hg19/ensembl
hg19/refseq
hg19/refseq_curated
hg38/ucsc
hg38/ensembl
hg38/refseq
hg38/refseq_curated
mm9/ucsc
mm9/ensembl
mm9/refseq
mm9/refseq_curated
mm10/ucsc
mm10/ensembl
mm10/refseq
mm10/refseq_curated
```

Users can automatically download transcript definitions for human (genome assemblies hg19 [GRCh37] and hg38 [GRCh38]) and mouse (assemblies mm9 and mm10) as defined by the UCSC Genome Browser database [379], Ensembl [10], or NCBI's RefSeq resource [344]. Jannovar will automatically download the corresponding files.

```
$ java -jar Jannovar.jar download -d hg38/refseq
```

For the hg38/refseq resource, Jannovar will create a file called hg38_refseq.ser in the data directory. This is a serialized file that allows the transcript data to be quickly loaded rather than parsing the original database files each time the program is run.[5] The internal format of the serialized database may change with each Jannovar version, and thus one should not use a serialized database file created with a different version of Jannovar.

14.1.5 Setting the proxy

If you live behind a firewall, you may need to set the proxy and the proxy port accordingly. This is done by adding the --proxy and the --proxy-port flags to the download command

```
$ java -jar Jannovar.jar download -d hg38/refseq \
        --proxy proxy.example.edu --proxy-port 123
```

---

[5]A serialized file contains a direct binary representation of Java objects that can be quickly loaded into memory.

We will use Jannovar in Chapter 15 to annotate VCF files, in Chapter 20 to perform pedigree analysis, and in Chapter 22 to analyze and filter VCF files according to variant frequency.

14.2   WRITING JAVA CODE WITH THE JANNOVAR LIBRARY

In this book, we present Jannovar as a standalone application. However, it can also be used as a Java programming library. Interested readers should consult the online documentation and examples to get started. Jannovar is used as a software library to perform variant annotation in the Exomiser [373, 399] and Genomiser [403], which we will present in Chapter 27.

The Application Programming Interface (API) is documented at `https://javadoc.io/doc/de.charite.compbio/jannovar-core`.

14.3   OTHER VARIANT ANNOTATION SOFTWARE

Several useful software packages for variant annotation are available, including ANNOVAR [450], SnpEff [77], and VEP [287].

FURTHER READING

Jäger M, Wang K, Bauer S, Smedley D, Krawitz P, Robinson PN (2014) Jannovar: a java library for exome annotation. *Hum Mutat.* **35**:548–55.

`https://jannovar.readthedocs.io`: Online documentation for Jannovar

CHAPTER 15

Variant Annotation

Peter Robinson and Manuel Holtgrewe

V ARIANT nomenclature must be standardized for robust computational analysis of variants and interoperability between databases. Over the last decade or so, the community has widely adopted the standards of the Human Genome Variation Society (HGVS) [19, 103]. Previous to that, there were multiple mutually inconsistent ways of numbering nucleotides and denoting various classes of variant. For example, in an early publication from the year 1997 by one of the authors of this book, a mutation in the fibrillin-1 gene was denoted as G1760A [43]. Hopefully it was clear to the readers of the publication from the context that G1760A referred to a change of the nucleotide sequence, an exchange of a guanine for an adenosine base at position 1760 of the coding sequence. But in principle, G1760A could also refer to an exchange of a glycine residue for an alanine residue at position 1760 of a protein. The HGVS recommendations intend to supply unambiguous, precise descriptions of genetic variants, and are today nearly universally used. This chapter explains the basics of the HGVS nomenclature, and then shows how it is used to describe the various classes of variant that are encountered in WES and WGS data.

15.1 HGVS NOTATION AND MUTATIONAL CATEGORIES ENCOUNTERED IN WES AND WGS ANALYSIS

The analysis of WES and WGS files in translational research contexts focuses on the identification or prediction of sequence variants that cause or in some way influence disease.

An essential step in the interpretation of this data is the annotation of these variants with respect to their potential effects on genes and transcripts. This step is the translation of chromosomal variants from the VCF file, which reflect chromosomal coordinates (e.g., denoted as chr15:g.48463207T>C in HGVS notation), to the gene-based variant description (e.g., NM_000138.4(FBN1):c.5099A>G for the base change c.5099A>G in transcript NM_000138.4 of the *FBN1* gene). Often, a prediction of the molecular impact on the corresponding protein is also indicated (e.g., here the missense mutation p.(Y1700C)), as the effect on the gene product is required for the biological and medical interpretation.

The Human Genome Variation Society (HGVS) has published standards for reporting sequence variants that are in near universal use [103]. The HGVS Website about the mutation nomenclature (http://varnomen.hgvs.org) is very useful for learning about the HGVS standard. Thus, we will only provide a high-level review in this chapter by introducing the HGVS standard for describing each of the variant types that can be annotated by Jannovar.

15.2 HGVS NUMBERING CONVENTIONS

For genomic coordinates, the nucleotide numbering is g.1, g.2, g.3, ..., etc. from the first to the last nucleotide of the reference sequence of a chromosome or scaffold. For coding sequences, numbering starts with c.1 for the first nucleotide of the ATG start codon and ends with the last nucleotide of the stop codon (TAG, TAA, or TGA). For non-coding transcripts, the positions are referred to by n.1, n.2, n.3, ... Nucleotides on the transcript upstream of the start codon are numbered with a minus sign and positions downstream of the stop codon are prefixed with an asterisk (∗). Nucleotides in introns are numbered by offsets from the closest nucleotide in an exon (Figure 15.1). In the 5' half of an intron, the nucleotides are denoted with positive offsets from the last base of the exon on the 5' side of the exon (e.g, c.42+1, c+42+2, ...). Conversely, in the 3' half of an intron, the nucleotides are denoted with negative offsets from the first base of the exon on the 3' side (e.g., `c.43-1`, `c.43-2`, ...). Nucleotides in introns are numbered in the direction of transcription with respect to the last or first exonic position of the exon that is directly up- or downstream. Nucleotides upstream and downstream of the transcript (i.e., even outside the UTRs) are denoted similarly to intronic nucleotides with offsets from the first/last nucleotide in the transcript (e.g., `c.-30-10` or `c.*30+10`).

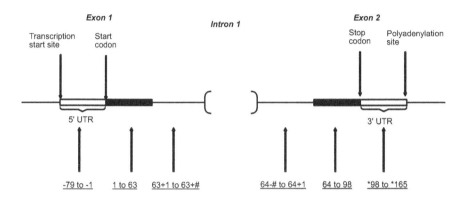

Figure 15.1. HGVS numbering scheme. The numbering scheme used by HGVS notation. # stands for any integer.

15.3 ANNOTATING VCF FILES

Jannovar can be used to annotate the variants in a VCF file. It is important to use a transcript description file that matches the genome build being analyzed: the coordinates of genes change on major genome build version changes (e.g., from GRCh37 to GRCh38) as single bases or whole segments are added to or removed from a chromosome. In this chapter, we will use the VCF file we created in Chapter 12 starting from a BAM file representing an alignment to the GRCh38 genome build. We will use the RefSeq transcript definitions created by Jannovar (`hg38_refseq.ser`).

To annotate the VCF file, enter the following command (if necessary, create the serialized file as described in Chapter 14).

```
$ java -jar Jannovar.jar annotate-vcf \
    -d data/hg38_refseq.ser \
    -i NA12878_filtered.vcf \
```

```
-o NA12878_annotated.vcf
```

Jannovar will add the variant type and a representative annotation to each line of the VCF file (review Chapter 13 for details on the VCF format if necessary). The annotated VCF file has a new `INFO` field (as described in the header line `##INFO=<ID=ANN,...`). The variants are each annotated with a corresponding `ANN` field in their `INFO` column (Table 15.1).

Table 15.1. The items contained in the Jannovar `ANN` field with an example for `chr4:g.74750631A>T`.

Item	Example
Allele	T
Annotation	missense_variant
Annotation_Impact	MODERATE
Gene_Name	BTC
Gene_ID	685
Feature_Type	transcript
Feature_ID	NM_001729.3
Transcript_BioType	Coding
Rank	4/6
HGVS.c	c.370T>A
HGVS.p	p.(Leu124Met)
cDNA.pos / cDNA.length	731/49928
CDS.pos / CDS.length	370/537
AA.pos / AA.length	124/179
Distance	
ERRORS / WARNINGS / INFO	

Note: The items are separated by pipe ("|") symbols.

15.4   VARIANT CATEGORIES

Jannovar can annotate a total of 30 variant categories, each of which are mapped to terms of the Sequence Ontology [113] (Table 15.2).

We can use BCFtools to count the number of variants in each category. For instance, to count the number of missense variants in our file, we use BCFtools to extract the `ANN` item from the `INFO` field. Then, we use the UNIX tool `cut` to extract the second field delimited

Table 15.2. Jannovar variant descriptions.

Term	Annotation	Putative impact
FRAMESHIFT_ELONGATION	Primary	High
FRAMESHIFT_TRUNCATION	Primary	High
FRAMESHIFT_VARIANT	Additional	High
COMPLEX_SUBSTITUTION	Primary	High
MNV	Primary	High
INTERNAL_FEATURE_ELONGATION	Additional	High
FEATURE_TRUNCATION	Additional	High
STOP_GAINED	Primary	High
STOP_LOST	Primary	High
START_LOST	Primary	High
SPLICE_ACCEPTOR_VARIANT	Additional	High
SPLICE_DONOR_VARIANT	Additional	High
MISSENSE_VARIANT	Primary	Moderate
INFRAME_INSERTION	Primary	Moderate
DISRUPTIVE_INFRAME_DELETION	Primary	Moderate
INFRAME_INSERTION	Primary	Moderate
DISRUPTIVE_INFRAME_DELETION	Primary	Moderate
STOP_RETAINED_VARIANT	Additional	Low
INITIATOR_CODON_VARIANT	Additional	Low
SYNONYMOUS_VARIANT	Primary	Low
SPLICE_REGION_VARIANT	Additional	Low
NON_CODING_TRANSCRIPT_EXON_VARIANT	Primary	Low
NON_CODING_TRANSCRIPT_INTRON_VARIANT	Primary	Low
FIVE_PRIME_UTR_EXON_VARIANT	Primary	Low
THREE_PRIME_UTR_EXON_VARIANT	Primary	Low
FIVE_PRIME_UTR_INTRON_VARIANT	Primary	Low
THREE_PRIME_UTR_INTRON_VARIANT	Primary	Low
DIRECT_TANDEM_DUPLICATION	Additional	Modifier
UPSTREAM_GENE_VARIANT	Primary	Modifier
DOWNSTREAM_GENE_VARIANT	Primary	Modifier

Note: Jannovar combines primary and possibly multiple additional variant descriptions for richer annotation. For example, a synonymous or missense variant (primary description) might have the additional annotation of being located in the splice region. Additionally, each variant description is assigned a "putative" impact.

by pipes ("|"), use `grep` to filter out only those variants annotated as `missense_variant`, and use word count (`wc`) to count the lines.

```
$ bcftools query -f'[%ANN]\n' NA12878_annotated.vcf.gz | \
    cut -d '|' -f 2 | \
    grep missense_variant | \
    wc -l
9512
```

15.4.1 Creating comprehensive annotations

Sometimes it is desirable to know all of the potential annotations associated with some variant. While the VCF annotation functionality produces one representative annotation, it is also possible to generate a file with a comprehensive annotation list (one for each affected transcript). To do so, just add a `--show-all` flag to the Jannovar command.

```
$ java -jar Jannovar.jar annotate-vcf \
    --show-all \
    -d data/hg38_refseq.ser \
    -i NA12878_filtered.vcf \
    -o NA12878_annotated.vcf
```

The `--show-all` flag indicates that Jannovar should create an entry in the `ANN` field for all transcripts affected by the current variant. By default, Jannovar will only display the transcript with the highest predictive impact.

For example, the following shows all predicted variant effects for the chromosome 7 variant rs3218660 (`g.44073274G>T`) on all transcripts of the *POLM* gene. Only a subset of the `ANN` fields is shown for clarity.

```
missense_variant    XM_005249708.1   c.1405C>A    p.(His469Asn)
missense_variant    XM_005249709.1   c.1315C>A    p.(His439Asn)
missense_variant    XM_005249711.1   c.1135C>A    p.(His379Asn)
missense_variant    XM_005249712.1   c.763C>A     p.(His255Asn)
missense_variant    XM_005249713.1   c.763C>A     p.(His255Asn)
non_coding_transcript_exon_variant   POLM  XR_242073.1 n.1472C>A
non_coding_transcript_exon_variant   POLM  XR_242074.1 n.1525C>A
3_prime_UTR_exon_variant   NM_013284.2 c.*17C>A    p.(=)
3_prime_UTR_exon_variant   XM_005249710.1  c.*17C>A    p.(=)
```

15.4.2 Missense variant

A `missense_variant` is defined as a sequence variant that changes one or more bases, resulting in a different amino acid sequence but where the length is preserved (Sequence Ontology term SO:0001583). Most commonly, a single nucleotide is changed; for instance, the single nucleotide substitution `chrX:154465991G>C` alters a codon for glutamic acid (GAG) to a codon for aspartic acid (GAC) in transcript `NM_001163257.1` of the *PLXNB3* gene. The corresponding change is `c.3537G>C`, which is predicted to result in a missense variant in the protein: `p.(Glu1179Asp)`. The NA12878 exome contained a total of 9512 missense variants with respect to GRCh38.

15.4.3 Stop gained (nonsense variant)

A `stop_gained` is defined as a sequence variant whereby at least one base of a codon is changed, resulting in a premature stop codon, leading to a shortened polypeptide (SO:0001587). This class of variant is often referred to as a nonsense variant. The NA12878 exome contained a total of 88 `stop_gained` variants, including one in the `NM_001076780.1` of the *PKD1L2* gene caused by a single nucleotide substitution: `c.658C>T`. The consequence is that the protein is truncated at position 220, where a glutamine residue (Gln) is transformed into a premature stop codon (denoted by an asterisk [*]), i.e., `p.(Gln220*)`.

15.4.4 Frameshift variants

Frameshift variants cause a disruption of the translational reading frame, because the number of nucleotides inserted or deleted is not a multiple of three (SO:0001589).

For instance, the duplication of a single nucleotide at position 196 of the NM_053003.3 transcript of the *SIGLEC12* gene (c.196dup) is predicted to alter an alanine residue at position 66 of the protein to a glycine residue, with the frameshifted protein continuing until a stop codon 50 codons downstream: p.(Ala66Glyfs*50). Fifty-eight such variants were found by Jannovar in the NA12878 exome. Jannovar can additionally recognize several more specific types of frameshift variant.

15.4.5 Frameshift elongation (frameshift insertion)

A frameshift_elongation is a frameshift variant that causes the translational reading frame to be extended relative to the reference feature (SO:0001909). The variant c.28_29insGGCGGCGGCGGCGC describes the insertion of the bases GGCGGCGGCGGCGC between nucleotides 28 and 29 of the XM_005259580.3 transcript of the *POLRMT* gene and is predicted to cause a stop codon 11 amino acids downstream: p.(Pro10Argfs*11). There were 46 frameshift_elongation variants in the NA12878 exome.

15.4.6 Frameshift truncation (frameshift deletion)

A frameshift_truncation is a frameshift variant that causes the translational reading frame to be shortened relative to the reference feature (SO:0001910). For instance, the deletion of the two nucleotides at position 9 and 10 of the NM_001297624.1 transcript of the *ZNF480* gene (c.9_10del) leads to a frameshift at the position of the deletion at codon 3 of the protein: p.(Cys3fs). There were 67 frameshift_truncation variants in the NA12878 exome.

15.4.7 Start lost

A start_lost variant (SO:0002012) is defined as a variant that changes at least one base of the canonical start codon. Start lost variants probably prevent translation of a functional protein product, and their effect on the protein is indicated as p.0?. Several types of variant fall into this class, including those that delete all or part of the start codon and missense changes of the methionine residue that is encoded by the start codon. Some examples are:

- `c.-3_3del`: Deletion of three bases upstream of the start codon as well as of all three bases of the start codon itself

- `c.1A>G`: Substitution of the first nucleotide of the invariant ATG of the start codon

- `c.2T>C`: Substitution of the second nucleotide of the invariant ATG of the start codon

- `c.2_16del`: Deletion of nucleotides 2 to 16 of the transcript, including the second and third nucleotides of the start codon.

15.4.8   Inframe deletion

An `inframe_deletion` is defined as an inframe non-synonymous variant that deletes bases from the coding sequence (SO:0001822). For instance, the deletion of three nucleotides at positions 415, 416, and 417 from the `NM_021026.2` transcript of the *RFPL1* gene, denoted `c.415_417del`, leads to the deletion of a single amino acid from the protein without otherwise affecting the reading frame: `p.(Leu139del)`. There were 50 such variants.

A `disruptive_inframe_deletion` is an `inframe_deletion` with an inframe decrease in the length of the coding sequence (CDS) that deletes bases from the CDS starting within an existing codon (SO:0001826). For instance, the deletion of the six nucleotides at position 234 to 239 of the `NM_000044.3` transcript of the *AR* gene, leading to the deletion of two amino acids: `p.(Gln79_Gln80del)`. 97 such variants were called in the NA12878 exome.

15.4.9   Inframe insertion

An `inframe_insertion` is defined as an inframe nonsynonymous variant that inserts bases into in the coding sequence (SO:0001821). For example, the insertion of the three nucleotides AGC between positions 642 and 643 of the `NM_007367.3` transcript of the *RALY* gene (`c.642_643insAGC`) corresponds to the insertion of a serine residue between the alanine residue at position 214 and the glycine residue at position 215: `p.(Ala214_Gly215insSer)`.

Some `inframe_insertion` variants can be classified with the term disruptive_inframe_insertion, which is defined as an inframe increase in CDS length that inserts one or more codons into the coding sequence within an existing codon (SO:0001824). For instance, the insertion of three nucleotides ACA between positions 607 and 608 of the transcript NM_004711.4 of the *SYNGR1* gene (c.607_608insACA) leads to the insertion of the asparagine residue between positions 202 and 203 of the protein. There are 69 such variants in the NA12878 exome.

15.4.10 Splice variant

Pre-mRNA splicing consists of the recognition of intron/exon boundaries and the subsequent excision of the introns. The vast majority of human genes undergo splicing, and many genes have multiple alternate isoforms. Alternative splicing is controlled both spatially and temporally and results in the expression of different splice variants in different tissues, cells, or stages of development. Alternative splicing is also observed in response to pathological processes [56, 452].

Splicing is mediated by the spliceosome, a multicomponent molecular machine consisting of small nucleolar RNAs and proteins (Figure 15.2). The exonic and intronic sequences flanking the 5' end of an intron are termed the donor site, and the exonic and intronic sequences flanking the 3' end of an intron are termed the acceptor site. The first two nucleotides of an intron at the donor site as well as the last two nucleotides of the intron at the acceptor site are nearly invariant (GT----AG), and it can generally be assumed that a variant at these sites will affect splicing deleteriously. Mutations that affect pre-mRNA splicing account for at least 15% of disease-causing mutations [222], and the actual proportion may be much higher.

Jannovar does not perform a prediction of whether a given variant deleteriously affects splicing. Instead, it classifies variants according to the SO classes splice_acceptor_variant (SO:0001574), a splice variant that changes the 2 bp region at the 3' end of an intron, splice_donor_variant (SO:0001575), a splice variant that changes the 2 bp region at the 5' end of an intron, and splice_region_variant (SO:0001630), a sequence variant in which a change has occurred within the region of the splice site, either within 1-3 bases of the exon or 3-8 bases of the intron.

Here are two examples of splice acceptor variants.

c.68-1G>C

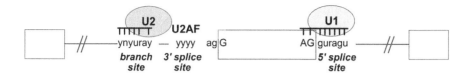

Figure 15.2. Pre-mRNA Splicing. The three critical sequence components for exon recognition and splicing: the 5' and 3' splice sites at the 5' and 3' ends of the intron and the branch point sequence. Intronic nucleotides are shown in lower case, exonic in upper case. The pre-mRNA is shown beneath elements of the spliceosome, and exons are shown as gray boxes. Introns are typically 10 times longer than exons (symbolized by the slanted lines). Ambiguity codes: Y is C/U and R is A/G.

```
c.2223-2A>G
```

Jannovar can add multiple annotations to fully describe a variant. For example, the above two variants were annotated as

```
splice_acceptor_variant&coding_transcript_intron_variant
```

Non-coding acceptor variants such as `n.1479-1G>A` are annotated correspondingly.

```
splice_acceptor_variant&non_coding_transcript_intron_variant
```

More complicated variants can affect the splice site. The following mutation is annotated as

```
frameshift_truncation&splice_acceptor_variant
```

because it involves a deletion that begins in the intron near the acceptor site and extends into the exon

```
c.593-8_621del   p.(Val198Alafs*42)
```

Here are two examples of splice donor variants.

```
c.1704+1G>T
c.207+2T>G
```

Here is an example of a splice donor variant in a 5' UTR intron

```
splice_donor_variant&5_prime_UTR_intron_variant
c.-136+1G>A    p.(=)
```

In Jannovar, all of these variants are subsumed under the class `splice_region_variant`, but variants are also annotated to this class if they are located in the splice region but are not acceptor or donor variants. Here are some examples.

```
c.2566-3T>C
c.308-4C>T
c.1033-5del
c.26-6T>G
c.5488-7del
c.737-8C>T
c.2673+3G>A
c.436+4_436+5insGT
c.262+5_262+6insTAG
c.923+6C>T
c.3299+7A>G
c.538+8T>A
```

The large majority of splice site mutations reported to date involve the invariant GT dinucleotide at the 5' donor splice site or the invariant AG dinucleotide at the 3' acceptor splice site [452]. Mutations at other positions of the 5' or 3' splice sites can also lead to missplicing, although bioinformatic tools currently are not able to reliably predict the consequences of these variants [56, 466]. The most common consequences of a mutation at the 5' or 3' splice site are exon skipping or intron retention (Figure 15.3).

While the most common consequence of a mutation of a splice donor or splice acceptor sequence is exon skipping [30], another common consequence is the activation of "cryptic" 5' or 3' splice sites. A cryptic[1] splice site is defined as a sequence that matches the consensus splicing motif that is not detectably used in wild-type pre-mRNA, but rather is "exposed" by a mutation in the gene. Most commonly, a mutation at the authentic splice site will reduce the strength of this splice site, causing the spliceosome to choose the cryptic splice site instead. Alternatively, a mutation at another position of an exon or intron of a gene may create a novel splice site that is "stronger" than the wild-type splice site (Figure 15.4).

---

[1]The word "cryptic" derives from a Greek word meaning "hidden".

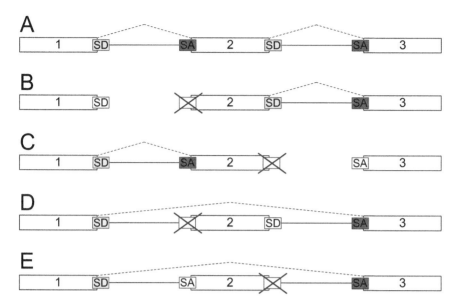

Figure 15.3. Exon skipping and intron retention. (A) A hypothetical gene with three exons and two introns is shown. Splicing joins exons at the splice donor (SD) and splice acceptor (SA) sites. **(B)** A mutation of the splice acceptor site of exon 2 can cause intron 1 (shown as a blue bar) to be retained in the mature mRNA transcript. **(C)** A mutation of the splice donor site of exon 2 can cause intron 2 (shown as a blue bar) to be retained in the mature mRNA transcript. **(D)** A mutation of the splice acceptor site of exon 2 can cause exon 2 to be "skipped", with the mature mRNA transcript consisting only of exons 1 and 3. **(E)** A mutation of the splice donor site of exon 2 can also cause skipping of exon 2.

Note that variants in the exonic portion of the splice region can also cause splice defects. Variants at the first and last nucleotides of exons have been most commonly implicated. This type of variant can be screened for using Jannovar by searching for combined annotations such as the following:

```
splice_region_variant&synonymous_variant
missense_variant&splice_region_variant
```

For instance, NM_001242307.1:c.3024A>G affects the last nucleotide of an exon of the *PITRM1*. It is a synonymous variant, i.e.,

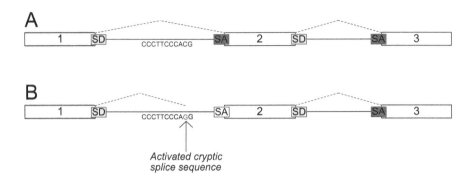

Figure 15.4. Cryptic splicing (A) wildtype and **(B)** mutant sequence of a hypothetical gene with three exons. A C→G mutation in intron 1 creates a novel splice acceptor site that is stronger than the wildtype splice acceptor sequence, causing missplicing with a portion of intron 1 (blue bar) being included in the mature mRNA transcript.

`p.(=)`, but could conceivably interfere with splicing because of its location just upstream to the splice acceptor.

15.4.11 Stop lost variant

A `stop_lost` variant (SO:0001578) is defined as a sequence variant where at least one base of the terminator codon (stop) is changed, resulting in an elongated transcript. For instance, the SNP `rs5065` is present in the Corpasome and located at position `chr1:g.11846011A>G`. The transcript-level change is `NM_006172.3:c.454T>C` and the predicted impact on the protein is `p.(*152Argext*2)`. Thus, the protein is extended by two residues (here by two arginine residues).

15.4.12 Coding transcript intron variant

A `coding_transcript_intron_variant` is a transcript variant occurring within an intron of a coding transcript (SO:0001969). These variants are numbered with respect to location to the nearest exon. For instance, the variant `c.1252-99G>T` is located 99 nucleotides 5' of position `c.1252` of transcript `NM_001161683.1` of the *OTOA* gene. By assumption, there is no effect on the protein, which is symbolized as `p.(=)`. Note that in the VCF file, this is printed as `p.(%3D)` because the "equals" symbol (=) is used in the key value pairs in the `INFO` field.

Writing the equals symbol as %3D, a hexadecimal code for =, avoids potential parsing errors. There were 55063 coding transcript intron variants in the NA12878 exome, meaning that roughly half of the variants called were off the exome target, which is a standard result for deeply sequenced exomes.

15.4.13   3′ UTR exon variant

A 3_prime_UTR_exon_variant (SO:0002089) occurs in the exonic sequence of the 3' untranslated region (UTR) of a protein-coding transcript. Nucleotides downstream of the stop codon are marked with an asterisk (*), and so the variant chrX:155774974G>T, which affects position 236 downstream of the stop codon of the transcript NM_001304990.1 of the *SPRY3* gene, is denoted c.*236G>T. The corresponding protein notation is simply p.(=), because the variant is not predicted to change the protein sequence. Again, in the VCF file generated by Jannovar, this is written p.(%3D).

15.4.14   5′ UTR exon variant

A 5_prime_UTR_exon_variant (SO:0002092) occurs in the exonic sequence of the 5' untranslated region (UTR) of a protein-coding transcript. Nucleotides upstream of the start codon are numbered with a minus sign, and so the variant chr17:54968937C>T, which affects the position 291 nucleotides upstream of the start codon of transcript XM_011524342.2 of the *COX11* gene, is denoted c.-291G>A. Again there is no effect on the predicted protein sequence.

15.4.15   UTR intron variants

Some untranslated regions are spliced together from more than one exon. In this case, there are one or more introns within the UTR. Variants in the intron are numbered with respect to the nearest UTR exonic nucleotide. A C→A transversion located 52 nucleotides 3' to a UTR exon that extends up to 45 nucleotides from the start codon is written c.-45+52C>A, and a C→A transition that is located 129 nucleotides 5' to a UTR exon that begins 210 nucleotides 5' to the start codon is written as c.-210-129C>T. Both of these variants are 5_prime_UTR_intron_variant (SO:0002091). This class also encompasses insertions and deletions in 5' UTR introns such as c.-194+16222_-194+16223del.

The category `3_prime_UTR_intron_variant` (SO:0002090) is analogous and includes variants such as `c.*7+796C>T` and `c.*5-71A>G` as well as more complicated variants such as `c.*11+26_*11+28del`.

15.4.16 Synonymous variant (silent variant)

A `synonymous_variant` (SO:0001819) is defined as a sequence variant in a coding region where there is no resulting change to the encoded amino acid. For instance, the variant `c.4980G>A` of transcript `NM_001163257.1` of the *PLXNB3* does not alter the encoded amino acid sequence. Its effect on the protein is therefore written as `p.(=)`.

15.4.17 Downstream, upstream, and intergenic variants

A `downstream_gene_variant` (SO:0001632) is a sequence variant located 3' of a gene (up to 1000 bp in Jannovar's definition). An `upstream_gene_variant` (SO:0001631) is a sequence variant located 5' of a gene (up to 1000 bp). An `intergenic_variant` (SO:0001628) comprises upstream and downstream variants and is defined as any sequence variant located between genes with a distance larger than 1000 bp to the closest gene. These variants do not have HGVS notation, but Jannovar indicates the distance of the variants to the nearest gene.

15.4.18 Noncoding transcripts

Numbering of non-coding transcripts begins with the first transcribed nucleotide and the letter "n" is used instead of "c". The following are examples of exonic and intronic variants in a non-coding transcript.

```
n.1391G>A
n.87-598C>G
```

Variants in non-coding transcripts are even more difficult to interpret than candidate splice variants, but there are non-coding RNA genes that are associated with disease, such as *RNU4ATAC* [290].

15.5 VARIANT NORMALIZATION (HGVS)

Insertion and deletion variants can give rise to ambiguities. Such ambiguities are the most common reason for discrepant variant representations. For example, the following sequence contains a guanine homopolymer tract (four G's) from position 1535 to 1538. In principle, if any one of these four G bases were deleted, the result would be the same: a variant allele with only three G bases from positions 1535 to 1537, followed by a T base at position 1538 of the alternate allele.

```
Ref:           GCCTCTGGGGTGAAC
c.1535delG:    GCCTCT-GGGTGAAC
c.1536delG:    GCCTCTG-GGTGAAC
c.1537delG:    GCCTCTGG-GTGAAC
c.1538delG:    GCCTCTGGG-TGAAC


Alt allele:    GCCTCTGGGTGAAC
```

It is not possible to observe which of the four bases was actually deleted.[2] There are also more complex cases that also lead to ambiguities in variant representation.

The HGVS introduced a normalization rule for such cases to disambiguate different variant representations (e.g., c.1535delG and c.1538delG). The rule is to use the variant representation where the variant is closest to the 3' end of the transcript. In the example above, the description c.1538delG should thus be selected. For proteins, the HGVS enforces shifting the variant towards the terminal residue. The aim is to improve database interoperability and to facilitate re-identification of recurrent variants. In the worst case, candidate disease-associated variants in WES/WGS data would be misinterpreted because the database entries could not be found [51].

If instead an additional 'G' nucleotide had been inserted into this polyguanine tract, then the correct notation would not be c.1538_1539insG but instead c.1538dupG. According to the HGVS definition, a duplication is a sequence change where an exact copy of one or more nucleotides are inserted directly 3' of the original copy of that sequence. Although such variants could be described as insertions, the dup notation is preferred.

Note that normalization according to HGVS differs from VCF normalization (see Section 13.8). The HGVS enforces shifting the variant

---

[2]It is also of little importance for the functional effect of the variant.

towards the 3' end *with respect to the transcript* while the VCF standard enforces shifting *towards the 5' end (left aligned) of the genome reference.*

15.6 MUTALYZER AND VARIANT VALIDATOR

Mutalyzer [459] is a website that is designed to check the correctness of mutation nomenclature and convert between different coordinate systems: `https://mutalyzer.nl/`. Variant Validator is new tool for the variant lookup using different coordinate systems and querying for the variant in different databases: `https://variantvalidator.org`. These programs are both extremely useful for checking the correctness of HGVS representations of variants.

15.7 EXERCISES

Exercise 1

What is the matter with the old-fashioned mutation nomenclature "Prothrombin G20210A" (which is still commonly seen in publications)?

Exercise 2

What are the basic components of the HGVS description of the following mutation? What do they refer to?

`NM_000130.3(F5):c.1601G>A   p.Arg534Gln`

Exercise 3

Why do we need to indicate the transcript (e.g., `NM_000130.3`) in addition to the gene symbol? Isn't it enough to know what gene is affected by the mutation?

Exercise 4

What's the difference between

`[c.515C>T;c.2598delT]`

and

`[c.515C>T];[c.2598delT]`

If necessary, consult the HGVS documentation[3] to understand the notations. Why might this be medically important if the two variants in question are both pathogenic and are located in a gene associated with an autosomal recessive disease?

Exercise 5

Consider the following variant:
Wildtype:

```
... 506 507 508 509 510 ...
... Ile Ile Phe Gly Val ...
... ATC ATC TTT GGT GTT ...
```

Variant (deletion of CTT):

```
... Ile Ile Gly Val ...
... ATC ATT GGT GTT ...
```

If the gene in question has the symbol *CFTR* and the accession number NM_000492 with the version number 3, how do we denote the mutation on the DNA level? How do we denote the mutation on the protein level?

Exercise 6

Enter the following variant in the Mutalyzer (`https://mutalyzer.nl/`) and describe the consequences on the nucleotide level and on the protein level.

NM_003002.3:c.274delGinsAAT

See hints and answers on page 501.

---

[3]`http://varnomen.hgvs.org`

CHAPTER 16

Variant Calling: Quality Control

QUALITY CONTROL of the number and distribution of called variants is the third important quality control step in WES/WGS analysis.

There are a number of QC packages available that can be used to calculate the following metrics, including QC3 [150] and `bcftools stats`. However, in the chapter, we will illustrate how to calculate the metrics using command-line tools in the exercises so that readers can gain some intuition.

16.1 THE TRANSITION-TRANSVERSION RATIO

Transitions (Tis) are defined as single-nucleotide substitutions from one pyrimidine base to another (C↔T) or one purine base to another (A↔G). Transversions (Tvs) are defined as nucleotide changes that substitute a pyrimidine for a purine base or vice versa (C→A, C→G, T→A, T→G, A→C, A→T, G→C, G→T). There are twice as many possible transversions as transitions (see Figure 16.1). Although one might expect to observe more transversions than transitions, the reverse is true. Comparisons of the DNA sequences of metazoa show an excess of transitional over transversional substitutions. Part of this bias is due to the relatively high rate of C→T transitions in the context of CpG dinucleotides [39, 193].

The Ti/Tv ratio is a commonly used quality control parameter for the distribution of variants in exome and genome datasets. It is

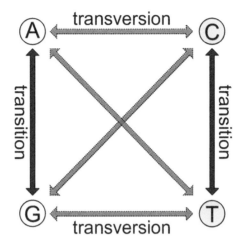

Figure 16.1. A transition is defined as a single-nucleotide variant that changes one pyrimidine base to another or one purine base to another. A transversion is defined as an SNV from a purine to a pyrimidine base or from a pyrimidine to a purine base.

computed as the number of transition single-nucleotide variants (SNVs) divided by the number of transversion SNVs. The following empirical observations can be used for quality control:

- The observed Ti/Tv ratio for SNVs inside of protein coding exons is around 3.0.

- The Ti/Tv ratio genome wide is about 2.0.

- Because exome data includes not only exonic regions but also flanking intronic and intergenic sequences, the Ti/Tv ratio observed in WES data is often between 2.0 and 3.0.

For WES data, a Ti/Tv ratio below 2.0 may indicate poor quality [149, 449].

16.2 THE PROPORTION OF NOVEL VARIANTS

As the effort of cataloging the human genetic variation has been ongoing on a large scale for years, most variants called in any exome or genome will not be novel. Instead, the majority will have been observed before and be present in databases such as dbSNP or ExAC (see Chapter 22). A higher than expected number of novel variants is an indicator that many of the "novel variants" are sequencing or variant calling errors.

The total number of variants, as well as the number of novel variants, will vary from sample to sample and may be related to the populations background of the individual being sequenced. There are many populations not currently well represented in variation databases such as dbSNP and ExAC. Thus, many of the variants common to these populations may not be in the databases and be interpreted as novel.

We will see in Chapter 22 that Jannovar can be used to annotate VCF files with population frequency data from the ExAC database. ExAC currently comprises data from over 60,000 exomes, and thus a majority of common variants will be represented in ExAC.[1] This gives us one quick way to estimate the proportion of novel variants of specified categories present in an exome sample.

```
$ zgrep missense_variant sample_exac.vcf.gz | grep -c EXAC
8776
$ zgrep -c missense_variant sample_exac.vcf.gz
9279
```

Thus, $9,279 - 8,776 = 503$ (or, a little over 5%) of the missense variants present in this VCF file are not represented in ExAC. For the Corpas exome, the figures are a total of 5,206 missense variants and only 132 (2.5%) of these are not in ExAC. Higher than expected counts of "private variants" may indicate quality problems.[2]

In general, production pipelines for WES/WGS analysis will compare called variants with data from dbSNP, 1000 Genomes, and in-house data (at least) in order to characterize the proportion of novel variants in each sample. Samples whose values differ starkly from expected values will be inspected closely to determine if there are quality problems.

---

[1] Recently, an extended version of ExAC called the Genome Aggregation Database (gnomAD) was released at http://gnomad.broadinstitute.org/about.

[2] We are using the term "private variant" to mean a variant that is not found in databases such as ExAC. It is hard to give an exact threshold for what a higher than expected amount of private variants is, especially since some populations are better represented than others in current variant databases. However, if a sample differs substantially in the number of private variants as compared to other similar samples, it should be examined carefully.

16.3 THE RATIO OF HETEROZYGOUS TO HOMOZYGOUS VARIANTS

The heterozygosity to non-reference homozygosity (het-hom) ratio is another good quality control parameter for DNA sequencing data. The het-hom ratio tends to correlate with the population being sequenced, varying between 1.4 and 2.0 in one study [449]. It is recommended to check this parameter and to flag samples for closer inspection if the het-hom ratio deviates strongly from the ratio observed in other samples in a cohort.

16.4 EXERCISES

Exercise 1

Calculate the Ti/Tv ratio for the NA12878. To do so, consider using bcftools (see Section 12.3) to extract the REF and ALT sequences and using a script to count transitions and transversions in the single nucleotide variants. Now calculate the same for the Corpas exome. Do you find a difference?

Exercise 2

Calculate the proportion of novel stop_gained (nonsense) variants in the NA12878 genome and the Corpas exome.

CHAPTER 17

Integrative Genomics Viewer (IGV): Visualizing Alignments and Variants

Peter Robinson and Tomasz Zemojtel

T HE Integrative Genomics Viewer (IGV) is a visualization tool for interactive exploration of large, integrated genomic datasets [429]. IGV is a Java application that can be started directly from the Website of the Broad Institute[1] as a Java Web Start application, or downloaded from there and started from the command line:

```
$ java -jar igv.jar
```

As of this writing, the current version is 2.3.90. In this chapter, we will demonstrate how to use IGV to view alignments and small variants in the NA12878 exome. We will assume that you have started the program from the Web or locally.[2]

IGV is a really nice piece of software with lots of bells and whistles, and we will only attempt to present a small subset of these capabilities

---

[1]http://www.broadinstitute.org/igv/

[2]If necessary, the IGV Website provides detailed information on how to start IGV.

that are relevant to the analysis of small variants in exome sequencing. In Chapter 19, we will show how to use IGV to visualize structural variants.

17.1 GETTING STARTED: LOADING THE DATA INTO IGV

IGV can be used to view alignments in SAM or BAM format. In both cases, the alignment file needs to be indexed. We will use the BAM file from the NA12878 exome together with its index file that we created in Chapter 8.

```
NIST7035_indelrealigner.bam
NIST7035_indelrealigner.bai
```

IGV offers many reference genomes by default, so that users can choose a genome by clicking on it in the Genomes drop-down menu (Figure 17.1).

Figure 17.1. IGV: genomes. Choosing Genomes|Load Genome From Server shows a list of available genomes. The GRCh38 genome assembly (hg38) is highlighted in the figure.

Following this, the selected genome will be displayed in the Genomes menu at the top left of the IGV window (Figure 17.2).

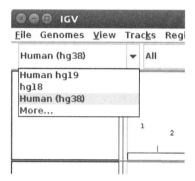

Figure 17.2. The GRCh38 genome assembly can now be chosen from the Genomes drop-down menu.

Now the sequencing data can be loaded into IGV. Use

`File|Load from File...`

and choose the file `NIST7035_indelrealigner.bam`.

You should now see tracks for visualizing the chromosomes 1 to 22, X, and Y, the coverage, the alignments, and genes (Figure 17.3).

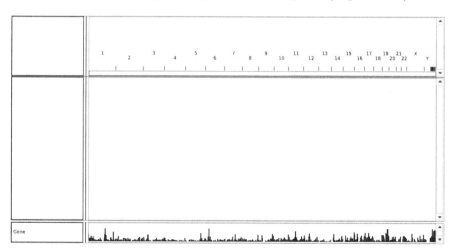

Figure 17.3. IGV summary view of the aligned genome.

IGV only begins to show the actual alignment after it has been sufficiently zoomed in. Choose one of the chromosomes and use the zoom tool at the top right corner to zoom in. Alternatively, a chromosomal location or a gene symbol can be entered. For instance, entering the gene symbol for the hemoglobin subunit beta gene (*HBB*) shows the alignment in the region of this gene. The alignment track will display the three exons of the *HBB* gene and the typical enrichment pattern for exomes — the reads are concentrated over the targeted exons, and very few reads are mapped to the intervening intronic sequences (Figure 17.4).

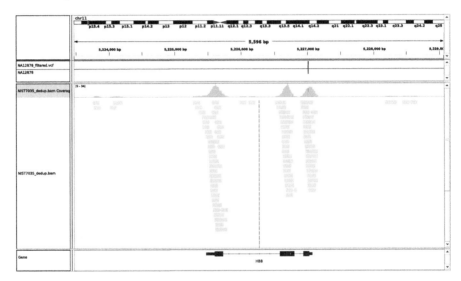

Figure 17.4. IGV view of the alignment in the region of the *HBB* gene.

17.2 VISUALIZING VARIANTS WITH IGV

In order to inspect the called variants, it is extremely useful to load the VCF file NA12878_filtered.vcf together with the alignment. Note that the VCF file must be indexed for viewing in IGV (the file NA12878_filtered.vcf.idx must be present in the same directory).[3]

Choose File|Load from file... from the IGV menu and load NA12878_filtered.vcf. This will add an additional track for the variant calls. One can navigate to the positions of the variants in the VCF

---

[3]GATK should have generated the index file together with the VCF file as a part of its processing. If not, IGVtools can be run from the IGV menu to generate a VCF index file.

file, and examine the data supporting the call in IGV. Mismatched bases are shown in a different color. The base counts at the variant position can be seen in a pop-up text box if the user hovers over the coverage track at the variant position with the mouse. Figure 17.5 shows a homozygous missense variant in the *NECTIN1* gene, whereby IGV displays the variant C bases in blue.

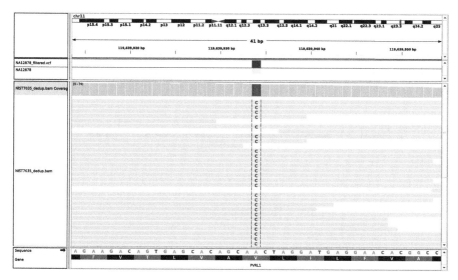

Figure 17.5. Missense variant. IGV view of a homozygous missense variant in the *NECTIN1* gene (also known as *PVRL1*). The variant c.1082T>G, is predicted to lead to a missense variant, p.(Val361Gly). Note that the *NECTIN1* gene is located on the minus strand, but IGV shows the variant on the plus strand (A>C).

Insertions are shown as a purple bar. One can hover over the bar to view the inserted nucleotides. The inserted bases are symbolized with a purple bar (Figure 17.6).

Deleted bases are shown as a blank space with a black line to represent the "gap" caused by the deletion. Figure 17.7 shows an 18 bp deletion in the *ERICH6B* gene.

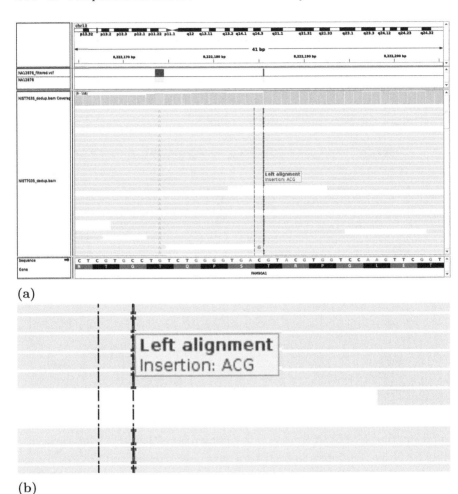

(a)

(b)

Figure 17.6. Insertion. IGV view of a homozygous insertion of three nucleotides in the *FAM90A1* gene (`c.1031_1032insCGT`). The position of the insertion is shown as a purple bar, and hovering over the bar shows the inserted nucleotides.

17.3   HOW TO RECOGNIZE POOR-QUALITY ALIGNMENTS AND VARIANTS IN IGV

For heterozygous variants it can be useful to examine the allele frequency[4] for both read orientations. For instance, the position of a het-

---

[4]Although the term is often used in this context, it would be best to speak of "allele fraction" because "allele frequency" is also used to indicate the commonness of an allele in the population.

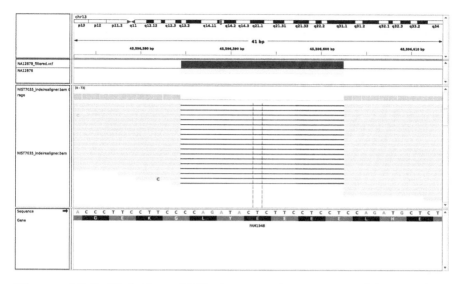

Figure 17.7. Deletion. IGV view of a homozygous deletion of 18 bases in the *ERICH6B* gene. The deletion is indicated by a blank space crossed by a black line.

erozygous variant in the 5' UTR of the esterase D gene (*ESD*) is covered by a total of 14 reads, all of which map to the negative strand, including 6 variant reads. This variant is thus of relatively low quality and one would want to validate it by an orthogonal method such as Sanger sequencing if it was found to be relevant to the experiment or case at hand (Figure 17.8).

On the other hand, a heterozygous missense variant in the *ATP7B* gene is covered by a total of 51 reads. There is a good balance of alleles and read orientations. 27 reads have the reference sequence (15 plus strand, 12 minus strand), and 24 reads have the alternate sequence (11 plus strand, 13 minus strand). This variant is of much higher quality than the variant in the *ESD* gene (Figure 17.9).

The mucins are a family of large heavily glycosylated proteins that form a major part of the mucous barrier in tissues such as the lung. There are over 20 paralogous human mucin genes with a high degree of sequence identity because of relatively recent gene duplication events in evolution. Reads that are colored white instead of gray indicate extremely poor mapping quality (MQ=0), which is commonly seen if short reads can be mapped to multiple repetitive sequences in the genome equally well. The homozygous variant that was called in the

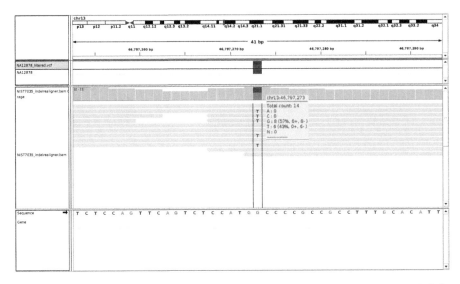

Figure 17.8. IGV view of a heterozygous substitution in the 5' UTR *ESD* gene. By hovering over the coverage plot of this variant, one can inspect the number of reads supporting the REF and the ALT base in each orientation. All of the reads map to the negative strand including 8 that support the reference base G and 6 that support the alternate base T. The fact that all reads map to only one of the two strands can be a sign of an artifact, and thus this variant should be regarded with caution unless it can be validated by an orthogonal method such as Sanger sequencing.

MUC12 gene should be regarded with suspicion because it could be the result of misaligning closely related but not identical paralogous sequences from another mucin gene (Figure 17.10).

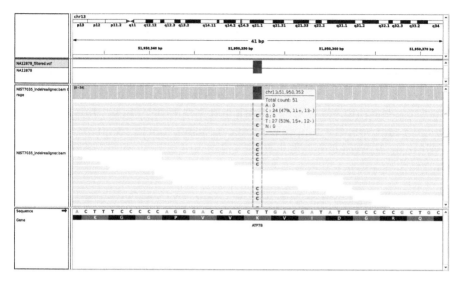

Figure 17.9. IGV view of a relatively high quality heterozygous missense variant in the *ATP7B* gene.

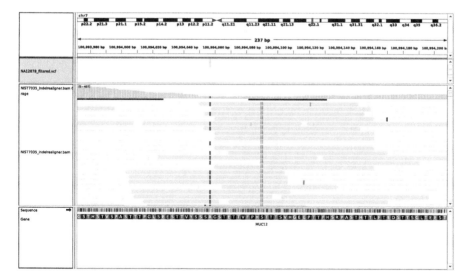

Figure 17.10. IGV view of the alignment in the region of the *MUC12* gene. Reads with very poor mapping quality (MQ=0) are shown in white instead of gray.

Another potential sign of paralogous alignments is the observation that the allele frequency of a variant differs substantially from what one would expect for homozygous (100%) or heterozygous (50%) variants.[5] In Figure 17.11, one of the variant calls displays an allele frequency of 24% (22 of 91 reads were ALT, the rest were REF). It is likely that the 22 ALT reads are misaligned reads from a related mucin gene whose sequence differs at this particular position. These three observations: many poorly mapped reads, unusual allele frequencies, and a high overall frequency of variant calls, are classic signs of artifacts owing to paralogous alignments.

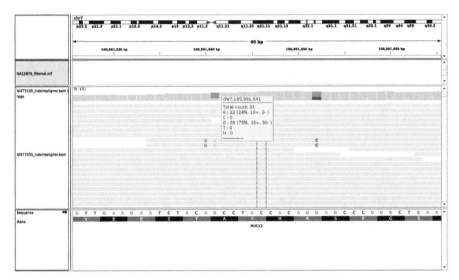

Figure 17.11. IGV view of a called variant in the region of the *MUC12* gene. 24% of the 91 reads support an alternate allele. This allelic frequency differs from the expected value for heterozygous (~50%) and homozygous (~100%) variants, which is a sign that this variant call may be artefactual.

Figure 17.12 shows a region of the alignment covering the *MUC16* gene. There are numerous called variants in a short stretch. The allele frequency is often far from 0%, 50% or 100%, but the variant at the very right is probably truly homozygous ALT. Most of the other "variants" are probably mapping artifacts from paralogous sequences.

---

[5]It is important to keep in mind that this is expected only for germline variants. The expectation does not hold for somatic variants in heterogeneous tumor tissues as we shall discuss later in the book.

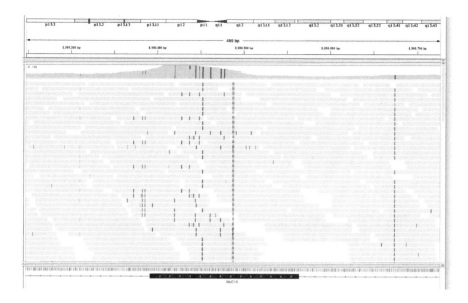

Figure 17.12. IGV view of the *MUC16* gene. There are numerous positions with ALT bases whose frequency does not correspond to the frequencies expected for heterozygous or homozygous variants.

Finally, we have previously shown examples of erroneous alignments around indel variants in Chapter 10 in Figures 10.2 and 10.3. Such problems can be recognized in IGV if there is an accumulation of mismatched bases close to an indel, but the substitutions are preferentially called at the ends of reads (Figure 17.13).

In summary, it is a good idea to inspect any variants considered to be candidates for further experimental or medical workup in IGV to identify potential false-positive (poor-quality) variants early. Depending on the situation, it may be worthwhile to validate such candidates with an orthogonal method such as Sanger sequencing before proceeding with the workup. The following are the most important signs of potential trouble when evaluating variants in IGV.

1. Multiple reads with poor mapping quality in the vicinity of a candidate variant

2. Deviation from allele ratios expected for heterozygous or homozygous germline variants

3. Unusually high overall frequency of variants

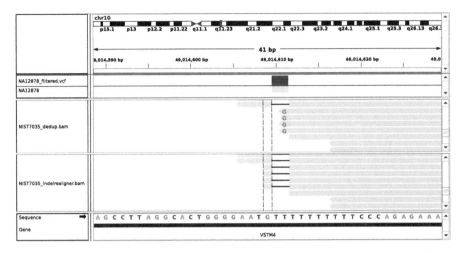

Figure 17.13. IGV view of an alignment before (top) and after (bottom) realignment. The mismatched bases in the top alignment are erroneous, and local realignment by GATK indicates that there is a deletion of two thymine bases from a poly-T tract of ten T bases.

4. Missense variants concentrated at the very beginning or ends of reads (may indicate the presence of an indel rather than a missense variant).

We will show how to use IGV to assess structural variants in Chapter 19.

FURTHER READING

Thorvaldsdóttir H, Robinson JT, Mesirov JP (2013) Integrative Genomics Viewer (IGV): high-performance genomics data visualization and exploration. *Brief Bioinform* **14**:178–192.

17.4 EXERCISES

Exercise 1

Load the GRCh38 genome and the NA12878 exome into IGV. Enter `chr2:20624090` into the search window and click on GO. Where is the variant located? Does it affect a transcript? What is the total number of reads in this position? How many reads support the ALT nucleotide. Are they balanced between both orientations? What is your call, heterozygous or homozygous?

Exercise 2

Take a VCF file of your choice and use a script to extract the five variants with the lowest and five variants with the highest Phred variant quality (QUAL column). Inspect the variants in IGV and try to explain the reasons for the low (high) quality of the variants.

Exercise 3

We encourage readers to use IGV to visualize alignments and variants for Exercises in later chapters. It can be very useful to view more than one alignment file at once, which can be done by choosing the files one after another with File|Load from file... from the IGV menu. For this exercise, readers are asked to do Exercise 1 of Chapter 10 in this way. It will also be useful to examine *de novo* variant candidates in Chapter 18, as well as somatic variants in Chapter 30 (to compare a tumor sample to a matched control from the same individual).

De Novo Variants

C HILDREN inherit half of the genome of each parent, but a small number of additional changes of the nucleotide arise *de novo* (spontaneously) during gametogenesis (production of sperm or oocytes), in the fertilized oocyte, or postzygotically. As we have seen, a typical human genome differs in up to 5 million positions compared with the human reference genome; all of these variants must have occurred as *de novo* mutations at least once during the course of human evolution. Current estimates place the *de novo* mutation rate at 1.0–1.8×10^{-8} mutations per nucleotide per generation, corresponding to about 44–82 *de novo* single nucleotide mutations in any individual [5]. About 1–2 of these variants are located in coding sequences [441]. Most *de novo* mutations are paternal in origin, and offspring of older fathers tend to have a higher number of *de novo* mutations [123, 216]. *De novo* mutations continue to occur throughout life in somatic and germ cells. We will discuss the topic of somatic mutations in cancer samples in Chapter 30.

Although *de novo* chromosomal anomalies had been recognized for decades as the cause of disease (e.g., trisomy 21, the basis of Down syndrome), it was only with the advent of WES that it became possible to fully characterize the role of *de novo* point mutations in human disease. A key insight was the understanding that the size of the mutational target[1] is related to the population frequency of genetic diseases that

---

[1]The mutational target can be defined as the combined length of all the genes that are associated with a certain clinical syndrome. In the example to follow in the text, Schinzel-Giedion syndrome corresponds to a target of just one gene, and Mendelian forms of mental retardation correspond to a target of many hundreds of genes.

are largely caused by *de novo* mutations. For instance, *de novo* mutations in the SET-binding protein 1 (*SETBP1*) cause Schinzel–Giedion syndrome, a rare disorder characterized by severe mental retardation, distinctive facial features and multiple congenital malformations [167]. On the other hand, spontaneous cases of intellectual disability are relatively common in the population, because there are hundreds of genes which, if mutated, can cause this disorder [441, 443].

18.1 COMPUTATIONAL ANALYSIS OF *DE NOVO* MUTATIONS

Conceptually, the process of characterizing *de novo* mutations is simple. One performs exome sequencing of an affected child suspected of having a *de novo* mutation together with his or her parents. Then, one looks for variants that are present only in the affected child. However, a major problem with a naive approach to *de novo* calling is that false-negative (FN) calls in one of the parents can make an inherited variant called in the child appear as a (false positive [FP]) *de novo* call in the child (Figure 18.1). One simple approach to reduce such FP calls is to remove known or common variants using a database such as dbSNP since it is unlikely that such variants represent true *de novo* variants and are causative for the disease in question [443].

Early computational approaches to *de novo* mutation (DNM) calling called variants separately for each of the persons in a trio (two parents and affected child), and then inferred DNMs by comparing the three genotypes. A number of computational approaches have now been developed to further improve the accuracy of *de novo* variant calling. Joint (simultaneous) variant calling in all three samples can reduce FP DNM calls. Additionally, algorithms exploit data about the family structure and transmission of variants, sequence-context specific mutability rates and other sequence characteristics, as well as characteristics of variants as assessed by GATK [256, 266, 326, 352, 453]. In this chapter, we will show how to perform *de novo* variant calling using a simple heuristic and will then review a more sophisticated approach. To start with, we will need to generate a joint VCF file from all three samples from a trio.

18.2 JOINT VARIANT CALLING WITH GATK

The variant calling procedure outlined in Chapter 12 was designed for a single sample. In order to call DNMs, we need to combine the variant

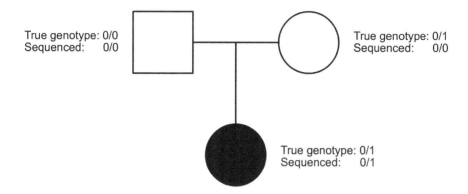

Figure 18.1. False positive *de novo* **variant call.** In this example, the mother is heterozygous for a variant at some genomic position but the variant is not called (false-negative). The affected daughter has inherited the variant, which is called in her WES data. Therefore, the call of a heterozygous variant at this position in the daughter's sequence is true positive, but the inference that the variant represents a *de novo* mutation is incorrect (false positive).

calls from three samples. In principle, there are several ways of doing this.

Single-sample calling: Each of the samples is analyzed separately and the three VCF files are subsequently combined.

gVCF single sample calling: A genome VCF (gVCF) file is generated for each of the samples and they are subsequently combined.

Joint calling: Variants are called simultaneously for all three samples.

Joint calling is the most straightforward way of doing it, and we will discuss it in this chapter. Chapter 13 discussed the gVCF format, and DNM calling can be performed exactly as will be described below with a joint VCF file that has been produced by combining the gVCF files from a trio.

With joint calling, GATK generates a single call set for all analyzed samples. A VCF line is emitted for each position at which one or more of the samples is called as having a variant. We will investigate a trio (NA12981, father; NA12892, mother; NA12878 daughter) from the 1000 Genomes Project. Chapter 4 explains how to download and prepare the data. We will prepare a joint VCF file representing these three samples.

The following command assumes that we have three alignment files as well as the FASTA index file and the sequence dictionary file for the reference genome in the path of GATK. See Chapter 8 to review these topics if necessary.

```
$ java -jar GenomeAnalysisTK.jar \
   -T HaplotypeCaller \
   -R hs38DH.fa \
   -I NA12981.bam -I NA12892.bam -I NA12878.bam \
   --genotyping_mode DISCOVERY \
   -stand_call_conf 30 \
   -o trio.vcf
```

The resulting VCF file has genotypes for each of the three persons in the trio. The column-header line of the VCF file shows the names of the three samples

```
#CHROM  POS     ID      REF     ALT     QUAL    FILTER \
  INFO     FORMAT    NA12878    NA12891    NA12892
```

The data lines allow us to compare the genotypes of each of the three samples. For instance, the following line shows the genotypes of the daughter NA12878, the father NA12891, and the mother NA12892 (the INFO field has been shorted for legibility). All three samples were found to be homozygous for the ALT base.

```
chr1    926250      .   G   A   839.04    . \
  AC=6;AF=1.00;AN=6;DP=26;(...) \
  GT:AD:DP:GQ:PL  \
  1/1:0,5:5:15:189,15,0 \
  1/1:0,4:4:12:145,12,0 \
  1/1:0,17:17:51:531,51,0
```

The following line provides a good example of the problems that could occur if we had performed separate variant calling on each of the three samples and then attempted to search for *de novo* mutations by comparing the resultant VCF files. The daughter (first sample) is heterozygous for a G→A transition at position 113,662,256 of chromosome 1. The father is homozygous for the reference sequence. No call was made for the mother. Thus, we do not know whether the mother is homozygous reference at this position and the daughter has a *de novo* mutation, or whether the mother is heterozygous or homozygous for the G→A transition and transmitted this allele to her daughter. In both cases, neither the father nor the mother would have an entry for this position if VCF files had been generated separately for all three samples.

```
chr1    113662256    .   G   A    83.13   . \
  AC=1;AF=0.250;AN=4;(...) \
  GT:AD:DP:GQ:PL \
  0/1:1,4:5:23:110,0,23 \
  0/0:1,0:1:3:0,3,38
  ./.
```

18.3   A SIMPLE FILTER FOR *DE NOVO* CALLING

We do not recommend performing *de novo* calling with a simple filter (especially if this is the only method used), but it is instructive to consider how one might implement such a filter. In practice, one can identify *de novo* variants in this way, but it is impossible to define a measure of statistical confidence in the calls. Although the calls made with this filter can be inspected visually in IGV (Chapter 17) prior to deciding on further validation, it is impossible to estimate the overall false negative rate.

We will present a simple filter derived from recommendations made in the course of a project that sequenced 10 Danish trios at high depth in order to estimate the *de novo* mutation rate [39]. The authors of that paper first examined each site across the 30 individuals from the 10 trios and removed sites with pronounced strand bias (**FS**; see page 190) and with a bias in the position of the variant in the reads (**ReadPosRankSum**; see page 191). They termed these criteria that were applied across the cohort "site filter". We will examine only one trio (see above: NA12981, father; NA12892, mother; NA12878 daugher), and for this heuristic we will use only the "individual filter".

The purpose of the filter is to exclude *de novo* calls that are likely to be false-positive because one or more of the variant calls for the three persons in the trio exhibits low quality with respect to one or more attributes. For a high-quality *de novo* call, we expect the parents to be homozygous for the reference sequence. We would feel confident about the homozygous reference calls in the parents if the following criteria are fulfilled.

1. There is a good genotype quality (GQ; see page 192);

2. The read depth is not too low, because if we only see a few reads, we cannot be sure that we have not missed a read with the alternate allele by pure chance;

3. The read depth is not too high (which can indicate an artifact). Thus, we demand that the read depth (DP; see page 192) is between 10 and 120 (or perhaps higher for a deeply covered exome);

4. No ALT reads at all are observed in either parent (variant callers may call a homozygous reference genotype even if a small number of reads have the alternate allele because they interpret this as a sequencing error. However, if the child is heterozygous for this allele, it is much more likely that the parent is also heterozygous for the allele).

For the child, we are searching for truly heterozygous variants. Thus, we only consider variants called homozygous reference in the parents (0/0) and heterozygous in the child (0/1). The criteria for genotype quality (GQ) and read depth (DP) are the same as for the parents. To evaluate heterozygosity, we apply an additional criterion for the allelic balance. For a true *de novo* heterozygous mutation, we expect the allelic balance to be approximately 50%. In practice, one also finds many variants called heterozygous in the child with much lower allelic balances. These could be artifacts or postzygotic *de novo* mutations. To reduce false-positive calls, we therefore apply a threshold of $0.3 \leq$ AlleleBalance ≤ 0.7.

When we applied this filter to the joint VCF file described above, we found 14 candidate *de novo* variants. Seven of these were located on alternate scaffolds (see Section 8.7), and were discarded from further analysis because variant calling on these sequences is prone to artefacts [183]. Of the remaining seven candidates, two were located in exonic sequences, including one synonymous substitution and one substitution in a 3' UTR (Figure 18.2). These candidates appear at least not implausible because all three persons are well covered, neither of the parents has even one read with the alternate allele, and the child has a heterozygous call with a well balanced allelic ratio.

(a)

(b)

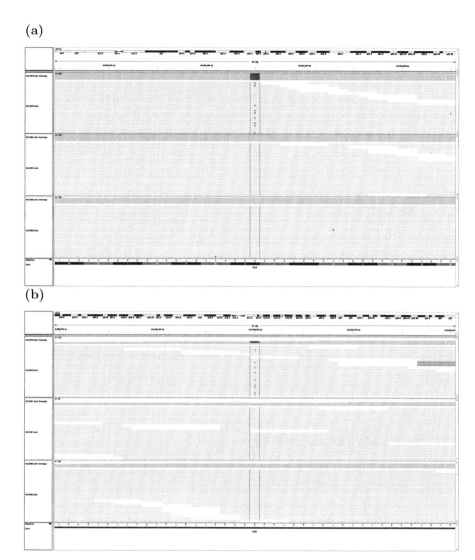

Figure 18.2. **(a)** A synonymous variant in the *PELI2* gene was found in 56 of 138 reads (41%) in the daughter: `c.732C>T:p.(=)`; None of the 101 reads of the father or the 106 reads of the mother showed the change. **(b)** A nucleotide substitution in the 3' UTR of the *ATG3* gene was found in 18 of 47 reads (38%) in the daughter: `c.*1606G>T:p.(=)`. None of the 21 reads of the father or the 76 reads of the mother had the variant.

It should be noted, however, that all such candidates require validation by an orthogonal method such as Sanger sequencing before any diagnostic conclusions are made. The trio we investigated is from the 1000 Genomes project, and to the best of our knowledge, the child is not affected by Mendelian disease; correspondingly, the *de novo* mutations identified here are not predicted to affect the functions of the genes they are located in. To a first approximation, it seems to be a matter of chance whether any of the roughly ~70–100 *de novo* mutations affecting each child (on average) "hits" a medically relevant gene in a way that will produce dysfunction and lead to Mendelian disease.

18.4   TRIODENOVO

TrioDeNovo [455] is Bayesian algorithm that compares the likelihood of two models: M_1, that the offspring harbors at least one DNM in the two alleles, and M_0, that the offspring's genotype follows Mendelian transmission. TrioDeNovo extracts the genotype likelihood values (PL; see page 193) as calculated by variant calling programs such as GATK, from the input VCF files. PL is defined as the Phred-scaled genotype likelihood, and is calculated for each of the genotypes homozygous REF, heterozygous, and homozygous ALT.

We will demonstrate TrioDeNovo with the same VCF file as used above.

18.5   RUNNING TRIODENOVO

TrioDeNovo requires a PED file (see Chapter 20) and a VCF file as input. Note that because TrioDeNovo is searching for *de novo* mutations and not performing pedigree filtering or linkage analysis of the type described in Chapter 20, the pedigree file only requires the first five columns. For our example, the pedigree file can be written as follows.

```
trio1 NA12891 0       0       1
trio1 NA12892 0       0       2
trio1 NA12878 NA12891 NA12892 2
```

Download TrioDeNovo from its website.[2] For this book, we used version v0.05. The program can be compiled by the following operations.

---

[2]http://genome.sph.umich.edu/wiki/Triodenovo

```
$ tar xvfz Triodenovo.0.05.tar.gz
$ cd triodenovo.0.05/
$ make
```

This will create an executable program file called `triodenovo` in the `bin` subdirectory. We can now run `triodenovo` using default settings.

```
$ triodenovo --ped trio1.ped \
  --in_vcf trio.vcf \
  --out denovo.vcf
```

The output file is itself a VCF file (`denovo.vcf`) that identifies 246 candidate *de novo* variants. TrioDeNovo has added two new lines to the header to define the new fields:

DQ: Denovo Quality, or the $\log_{10}(\text{BF})$ where `BF` is the Bayes Factor calculated as $L(M1)/L(M0)$ for M1 and M0 representing models with and without *de novo* mutations.

DGQ: Denovo Genotype Quality: $-10*\log_{10}(\text{post})$ where `post` is the posterior probability of the called trio genotypes among all trios with *de novo* mutations.

The `FORMAT` fields for the candidate *de novo* mutations now look like this:

```
GT:DQ:DGQ:DP:PL \
  A/A:6.91:100:38:0,41,1371 \
  A/A:6.91:100:90:0,160,3222 \
  A/C:6.91:100:69:94,0,1506
```

Here, the parents were found to be homozygous for the A allele, while the daughter has a heterozygous C allele that might be a *de novo* mutation.

If desired, the candidates can be filtered additionally by applying minimum Denovo Quality or depth filters. For instance, the following command applies a threshold `DQ` value of 10 and a minimum depth of 20, and flags only 74 candidates.

```
$ triodenovo --ped trio1.ped \
  --in_vcf trio.vcf \
  --out denovo.vcf \
  --minDQ 10 \
  --minDepth 20
```

Manual inspection of the alignment files reveals that not all of the candidates are of high quality. For instance, there is a *de novo* call where the daughter has 3 of 15 reads supporting a G→A transition, the mother has only `REF` reads, but the father has 7 `ALT` reads from a total of 69. Thus it seems more likely that either the daughter inherited the G from her father or that there is some sequence artefact that is affecting both samples. There is another *de novo* call whereby the daughter has 52 of 339 reads supporting a G allele (`NM_001289396.1:c.2933A>T` in *SMARCA2*), with none of the 241 or 257 reads of the father and mother supporting this allele. However, closer inspection reveals that this variant corresponds to a known polymorphism (rs78915420), and thus it can be regarded as relatively unlikely that a *de novo* mutation recreates a common known polymorphism. A number of candidates are found in which the variant allele is present at a level substantially less than 50% (Figure 18.3). These variants could potentially represent post-zygotic *de novo* mutations that only affect a subset of cells, and thus are said to be present in a mosaic state.

18.6 OUTLOOK

De novo mutations are an important cause of sporadic disease. Although *de novo* mutations often may occur during gametogenesis (production of sperm or oocytes), *de novo* mutations may also occur in postzygotic tissues and then affect only a subset of tissues and cells in the affected person. In this chapter, several approaches and tools for the identification of *de novo* variants have been presented. Bioinformaticians should be aware that the raw predictions of *de novo* mutations tend to be very noisy. A full pipeline to identify *de novo* mutations in medical settings would involve additional filtering of the candidate *de novo* list, and depending on the goals of the experiment might involve removal of low confidence predictions, removal of variants common in the population (that are likely to represent inherited rather than *de novo* variants), and prioritization according to clinical relevance. Readers are referred to the methods sections of publications on the topic for further information [4, 167, 306, 367, 443].

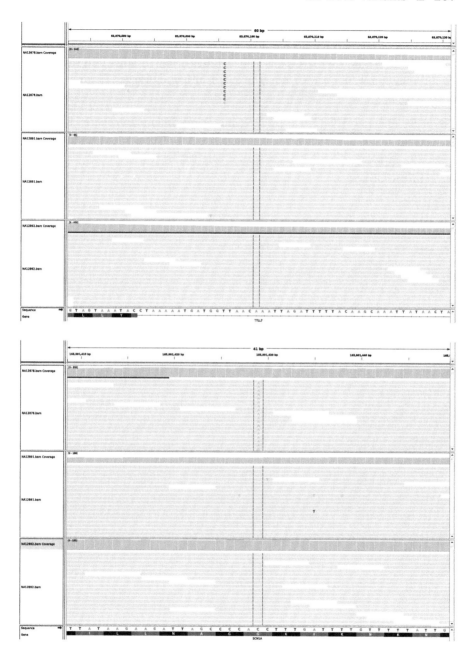

Figure 18.3. The upper panel shows a variant (c.2544-14A>G) in *TTLL7* transcript NM_024686.4 which was found in 11 of 92 reads (12%) in the daughter. In the lower panel, a missense variant in *SCN1A* (c.5846G>T; p.Gly1949Val in transcript NM_001165963.1) was identified in 24 of 240 reads (10%) in the daughter. For both variants there were no reads in the mother or father. Reads were sorted by base.

FURTHER READING

Acuna-Hidalgo R, Veltman JA, Hoischen A (2016) New insights into the generation and role of de novo mutations in health and disease. *Genome Biol.* **17**:241.

Veltman JA, Brunner HG (2012) De novo mutations in human genetic disease. *Nat Rev Genet* **13**:565–75.

18.7 EXERCISES

Exercise 1

Explain why the accuracy of DNM calling should improve as coverage increases.

Exercise 2

Write a script or program in a language of your choice to implement the heuristic *de novo* filter described in Section 18.3.

CHAPTER 19

Structural Variation

A<small>N</small> average genome has over 4 million single-nucleotide variants (SNVs) and contains up to a few thousand structural variants (SVs), but since SVs can comprise up to a few million nucleotides, SVs are the most important source of genetic variation on a per-nucleotide basis. Previous chapters have concentrated on the identification of SNVs; while this is still a far-from-trivial task, accurate and comprehensive detection of SVs is substantially more difficult, especially with short-read NGS technologies. Advances in genomic technologies including array CGH and WGS have revealed roles for SVs in the etiology of diseases such as autism, schizophrenia, major depressive disease, epilepsy, and many others [338], and so methods for accurate detection of SVs in WGS data would be of high medical relevance.

This chapter will briefly review the genetics of structural variation, provide an overview of the major algorithmic approaches that have been developed to characterize structural variation in WES/WGS data, and will present three algorithms in greater detail.

19.1 CAUSES OF STRUCTURAL VARIATION

Approximately 50% of the human genome consists of repeat sequences including Alu and LINE sequences and low-copy repeats (LCRs) such as segmental duplications (SDs). SDs are defined as segments of DNA at least 1 kilobase (kb) in length with $\geq 90\%$ sequence identity; about 4–5% of the human genome is made up of SDs [25]. Genomic regions with SDs and other LCRs are prone to genomic rearrangements. Non-allelic homologous recombination (NAHR) results from crossing-over in meiosis between homologous sequences that are not in allelic po-

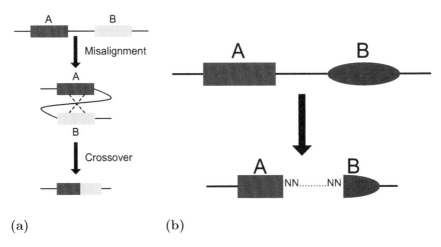

(a)   (b)

Figure 19.1. NAHR and NHEJ. (a) Nonallelic homologous recombination. **(b)** Nonhomologous end joining. Both (a) and (b) show an intrachromosomal event. Interchrosomomal NAHR and NHEJ occur in a analogous manner but involve two different chromosomes.

sitions (i.e., paralogous sequences) [62]. NAHR can result in deletion or duplication (Figure 19.1a). In contrast, nonhomologous end joining (NHEJ) does not require a homologous template for repair of the DNA lesion, and the junctions usually do not show more than short stretches of sequence identity.[1] Small deletions or the insertion of random nucleotides can also be observed at breakpoint junctions (Figure 19.1b). Both NAHR and NHEJ can also be the cause of interchromosomal translocations in case they involve two distinct chromosomes. Numerous other mechanisms have been proposed to explain the origin of various classes of structural variation, including break-induced replication (BIR), microhomology-mediated break-induced replication (MMBIR), serial replication slippage (SRS) and fork stalling and template switching (FoSTeS). Interested readers are referred to the recent review article by Carvalho and Lupski for further details [62].

---

[1]The word microhomology is used to describe the stretches of 2-4 identical nucleotides that are occasionally observed at the junctions of CNVs formed by NHEJ. Although the term "microhomology" is widely used, it is a misnomer because homology implies evolutionary relatedness and is not defined on the basis of mere sequence identity.

19.2 CATEGORIES OF STRUCTURAL VARIATION

An arbitrary but widely used definition is that SVs are genomic rearrangements that affect more than 1 kilobase (kb). SVs can be divided into "balanced" and "unbalanced" forms. With balanced SVs, there is no net loss or gain of genetic material, and with unbalanced SVs, there is (Figure 19.2). All of the structural variant categories described below can occur as germline variants in inherited disease or as somatic mutations in cancer. Cancer-specific structural variation will be discussed in Chapter 29.

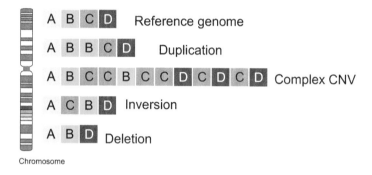

Figure 19.2. Categories of structural variation in the human genome. A, B, C, and D represent segments of DNA. In this figure, only the inversion is a balanced SV, and the remaining SVs are unbalanced.

19.2.1 Copy-number variants

Copy number variants (CNVs) represent a type of SV that results in an alteration of the number of copies of a genomic segment. CNVs can result in gains (duplication), losses (deletion) or complex rearrangements of the genome. On average, each individual has over 1000 CNVs, accounting for about a million bp [339]. A *deletion* is an unbalanced SV that results in the loss of a genomic segment. A *duplication* is an unbalanced SV that results in the gain of a genomic segment.[2] The duplicated segment can be inserted in tandem with the original segment or at another location in the genome. Duplications and deletions are the major classes of Copy Number Variants (CNVs), usually defined as a

---

[2]In some cases, especially in cancer, multiple copies of a segment can be gained resulting in what is often called an *amplification.*

copy number change involving a DNA fragment that is ~1 kilobase (kb) or larger. CNVs are commonly observed near segmental duplications (also known as low-copy repeats), which are segments of duplicated (highly similar) DNA fragments one kb or larger in size [125].

CNVs may be neutral polymorphisms. When CNVs do cause disease, there are two main mechanisms, an alteration of gene dosage or a disruption of genetic regulation. Haploinsufficiency refers to the situation in which a mutation causes a ~50% reduction in protein production (the Greek word *haplos* means single as opposed to *diploos*, double — thus, only the wildtype but not the mutant copy of a gene contributes to protein production, leading to a roughly 50% reduction in the level of the protein encoded by the gene that is affected by a heterozygous deletion). In some, but not all cases, the loss of function of one copy of a gene leads to clinical disease, and this phenomenon is referred to as "haploinsufficiency".

The phenotypic abnormalities seen in some CNV-associated diseases are thought to be related to haploinsufficiency of one or more genes located within the CNV. For instance, Williams syndrome (WS) is a multisystem disorder that results from heterozygous deletion of 1.5 to 1.8 Mb on chromosome 7q11.23, which contains approximately 28 genes. Some of the phenotypic abnormalities of WS have been attributed to hemizygosity of individual genes located within the deleted region. Thus, hemizygosity for the *ELN* gene is thought to cause the supravalvular aortic stenosis, *LIMK1* hemizygosity is implicated in the impaired visuospatial constructive cognition and *GTF2I* hemizygosity is thought to contribute to the mental retardation in WS patients [108]. Alteration of gene dosage by deletion or duplication or by disruption of genes located at the boundaries of CNVs thus represents a plausible pathomechanism for many phenotypic abnormalities seen in CNV disorders. However, structural variations such as CNVs, inversions or translocations can also change the regulatory context of genes, thereby disturbing the delicate balance between enhancers, silencers and insulators by interfering with the complex chromosomal looping and interaction mechanisms of promoters and one or more cis-regulatory elements. These changes in the regulatory environment of genes can result in misexpression and subsequent deregulation of signaling [176].

19.2.2 Inversions

An inversion is a balanced SV in which a genomic segment between two breakpoints is "reversed", so that the reverse complement of the segment replaces the original sequence. Inversions can be neutral (no noticeable effects on phenotype or health) or can cause disease. If an inversion interrupts the coding sequence of a gene, then the gene will no longer be expressed, which in some cases can cause disease. If an enhancer is located on one side of an inversion breakpoint, and a gene targeted by the enhancer is located on the other side of the breakpoint, then the inversion can increase the distance between the enhancer and its target gene, and thereby disrupt proper genetic regulation and cause disease (Figure 19.2).

19.2.3 Translocations

A chromosomal translation is a kind of chromosomal rearrangement whereby a segment of one chromosome is transferred to a nonhomologous chromosome (*inter*chromosomal translocation), or to a new site on the same chromosome(*intra*chromosomal translocation). The exchange of two nonhomologous chromosomal segments is called a reciprocal translocation (Figure 19.3). These events may not have phenotypic consequences, especially if no chromosomal material is gained or lost. However, if the translocation breakpoint results in the truncation of a gene, or otherwise disrupts genetic regulatory interactions, diseases such as infertility, cancer, or malformation syndromes may result. A Robertsonian translocation refers to the fusion of two acrocentric chromosomes at their centromere (acrocentric chromosomes—chr13, chr14, chr15, chr21, and chr22 in humans—have very short p arms). Robertsonian translocations can occur between homologous or between nonhomologous chromosomes. The p arms of acrocentric chromosomes appear to be largely redundant and their loss generally has no medical consequences for the translocation carrier themselves [298].

In routine medical diagnostics, currently the most commonly used methods for searching for translocations are routine cytogenetics, fluorescence *in situ* hybridization or targeted RT-PCR, depending on the clinical situation.

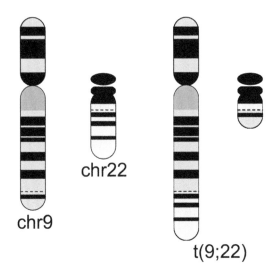

Figure 19.3. Interchromosomal translocation. An interchromosomal translocation exchanges two genomic segments on different chromosomes. The result of the translocation between chromosome 9 and 22 shown in the figure is denoted as t(9;22)(q34;q11), meaning that region 3, band (4) of the long arm (q) of chromosome 9 has been exchanged with region 1, band 1 of the q arm of chromosome 22.

19.3   DETECTING SVS IN HUMAN GENOMES

The earliest methods for genome-wide screening for SVs were array comparative genomic hybridization (CGH) and single-nucleotide polymorphism arrays.[3] Array CGH still remains the gold standard for identifying CNVs in diagnostic settings [240]. Array CGH exploits the fact that if two DNA strands are separated, they still "recognize" their opposite (reverse complementary) strand. Denaturation utilizes heat or other methods to separate DNA strands from one another, and renaturation refers to letting DNA in solution slowly cool, which allows the reverse complementary strands to anneal to one another, a process that is called hybridization. Array CGH involves a "microarray" (glass slide) with oligonucleotide probes that are spaced throughout the genome. Patient and control DNAs are labeled with separate fluorescent dyes and are applied to the microarray. The patient and control DNA compete with each other to hybridize to the probes on the mi-

---

[3]Affectionately known as "SNP chip" (pronounced snip chip).

croarray. A scanner measures the intensity of each signal and the ratio of the two signals is used to determine whether the patient sample has a loss or gain of a genomic segment.

In human genetics, Array-CGH is a commonly used screening investigation to investigate nearly the entire genome for CNVs in an untargeted fashion, and is used in the workup of intellectual disability or developmental delay of unknown cause, congenital malformations or facial dysmorphism, and autism. Array CGH still remains the gold standard for the diagnostic investigation of structural variation in these cases, although it is likely to be displaced by WGS within the next decade, as prices drop and the technology and computational analysis of WGS data matures. SNP arrays measure the intensity of probes at known SNP loci as well as allele ratios at heterozygous sites [289].

In principle, WGS has the potential to improve on the results of array CGH and SNP arrays, but substantial technical and computational challenges remain [2]. WGS detects over 1000 SVs of a size of 2.5 kb or more in human genomes [217]. However, over the last decade, it has become apparent that WGS with the short-read Illumina technology is not capable of detecting all SVs [27].

Four major strategies have emerged for the detection of SVs from NGS data. Paired-end mapping approaches investigate the orientation and inter-read distance of paired-end or mate-pair reads. Split-read approaches search for reads that contain the breakpoints of structural variants and can be "split" into segments from different chromosomal locations. Read-depth approaches search for increases or decreases in the depth of reads mapped to a given chromosomal segment. *De novo* assembly or local reassembly algorithms can perform fine mapping of breakpoints. A number of hybrid approaches have been developed that encompass two or more algorithms from the other classes [484]. We will discuss the signals and approaches used by these algorithms, and present some case studies in more detail to conclude the chapter.

19.4 READ DEPTH

Read-depth (RD) based methods analyze the amount of coverage in genomic regions, whereby deletions are expected to reduce coverage and duplications are expected to increase coverage above normal (Figure 19.4). RD-based methods are suitable for detecting CNVs but not balanced forms of structural variation.

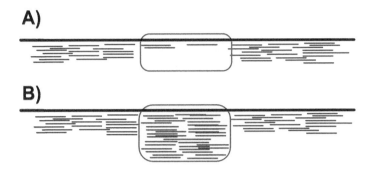

Figure 19.4. Read depth. Alterations in depth of coverage of mapped reads with a deletion and a duplication of a genomic segment.

Most RD-based methods follow a four-step procedure to discover CNVs:

1. Mapping/Data preparation. The first step of most RD-based methods consists in filtering and mapping the reads, and then counting the number of mapped reads in non-overlapping genomic windows ("bins") of a predetermined length (e.g., 100 or 1000 nt; we will denote the window length by W). Filtering comprises removal of duplicate reads and low quality reads as explained in Chapter 8. For quality assessment, one aspect that requires particular care is the handling of reads with low mapping quality (usually $MQ < 30$, see Chapter 12). Low-MQ sequences usually are located in repetitive regions of the reference genome or have low base quality [477]. RD methods then count the number of reads that are located in each bin (for instance, the location of the 5' end of the reads may be counted). If the bin size is too small, then too few reads will map to each bin and downstream statistical tests will lack power. If the bin size is too large, then there would be a loss of precision in defining the boundaries of CNVs. The ideal bin size depends on the overall depth of sequencing, but is often set to 50–1000 bp.

2. Normalization. Read counts are affected by two major sources of bias: the local GC content and the genomic mappability.

The GC content of a bin is the fraction of G and C bases in that bin according to the reference genome. GC abundance is heterogeneous across the genome and there are regions of high GC content and regions of high AT content. A GC pair is bound by three hydrogen bonds, while AT pairs are bound by two hydrogen bonds. DNA with high GC-content is more stable than DNA with low GC-content, and this affects the behavior of GC rich fragments in hybridization and any number of molecular biological experiments. Thus, GC content can confound the results of genomics experiments. In fact, the average measured read depth of diploid genomic segments (copy number 2) correlates with GC content [37]. Figure 19.5 shows an example of how read counts can be correlated with GC content.

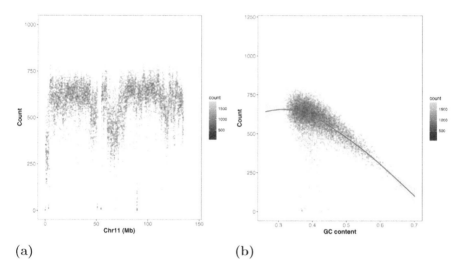

(a)   (b)

Figure 19.5. Read depth and GC content. (a) Read counts in bins of 10,000 bp across chromosome 11 from a genome taken from the 1000 Genomes Project. (b) The read count plotted against the GC content of the bins on chromosome 11. There is an imperfect correlation between read counts and GC content, with higher levels of GC content being correlated with lower overall read counts.

Normalization for GC content can be performed by applying the following correction:

$$\bar{r}_i = r_i \cdot \frac{m}{m_{GC}}, \tag{19.1}$$

where r_i is the raw read count for bin i, m is the overall (genome-wide) median read count per bin, m_{GC} is the median read count for all windows with the same GC content as bin i, and $\bar{r}_i$ is the adjusted read count.

Mappability bias relates to the difficulty in mapping reads to regions of the genome with many repetitive elements, which can lead to artificially low read counts. An analogous normalization can be performed based on the mean mappability of sequences in the genomic bins.

3. Estimation of copy number. A deletion or duplication is evident as a decrease or increase in read depth across multiple consecutive windows. Many current algorithms make use of the Poisson distribution, which is used to find the probability of a rare event randomly occurring in a specific interval.

A Poisson experiment is a statistical experiment that has the following properties:

1. The experiment results in outcomes that can be classified as successes or failures.

2. The average number of successes (λ) that occurs in a specified region is known.

3. The probability that a success will occur is proportional to the size of the region.

4. The probability that a success will occur in an extremely small region is virtually zero.

For read mapping, we have divided up the genome into small bins, say 100 bp, and each read mapping is regarded as an experiment. The read could be mapped to the 100 bp window ("success" in terms of this model) or to somewhere else in the genome ("failure"). The probability that a given read will map to a particular genomic bin is very small, but the probability is proportional to the size of the bin. The expected number of successes, λ, can be calculated based on the coverage that would result from uniformly distributing all N reads of a sequencing experiment across a genome of size G into bins of size W.

$$\boxed{\lambda = \frac{NW}{G}} \quad \text{where} \quad \begin{cases} N & \text{total number of reads} \\ W & \text{size of window} \\ G & \text{size of genome} \end{cases} \quad (19.2)$$

The Poisson distribution is then defined as follows, where k is the number of occurrences actually observed in a certain bin.

$$P(X = k) = \frac{\lambda^k e^{-\lambda}}{k!} \tag{19.3}$$

To calculate the probability of an event, we can determine the tail sum of the Poisson distribution to find the probability of observing at least as many (or as few) reads

$$P(X \geq 25) = 1 - P(X < 25) = 1 - \sum_{k=0}^{24} \frac{\lambda^k e^{-\lambda}}{k!} \tag{19.4}$$

For $X \sim$ Poisson(λ), both the mean and the variance are equal to λ (Figure 19.6). In real datasets, however, it can be observed that the variance is higher than the mean, a phenomenon referred to as overdispersion. For this reason, statistical tests in some read depth approaches are based on the negative binomial distribution [292].

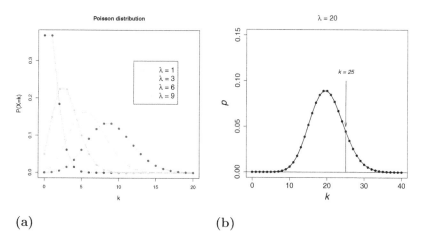

(a)   (b)

Figure 19.6. Poisson distribution. (a) Poisson distribution for various values of λ. **(b)** Tail sum.

4. Segmentation. The windows or bins used for calculating raw p-values are much smaller than typical CNVs, and a deletion or duplication typically manifests as a decrease or increase in coverage across multiple consecutive windows. Therefore, an additional segmentation step is used to estimate the overall size of candidate CNVs. Many different approaches for this have been published. At a basic level, neighboring bins with copy number changes in the same direction are merged [477]. Finally, the copy number within each segment is estimated.

Prominent RD-based algorithms for SV detection include CNVnator [3], cn.MPOS [200], GROM-RD [406], and ReadDepth [292]. Note that for WES data read depth analysis is much less straightforward because the capture efficiency is far from uniform across the targeted regions (see Chapter 11). We will discuss a possible approach to CNV analysis for exomes in Section 19.9.

19.5 READ-PAIR MAPPING ANALYSIS: ORIENTATION AND DISTANCE OF ALIGNED MATES TO ONE ANOTHER

Read-pair mapping approaches exploit the distance and the orientation of reads from paired-end or mate-pair sequencing (see Chapter 3) to search for certain types of SV. A key to understanding this type of analysis is the concept of insert size (Figure 19.7). Read-pair sequences are extremely useful for read mapping in whole genome sequencing because we not only have the information about the DNA sequences but also the distance and orientation of the two mapped reads to one another.

Read-pair methods compare the average insert size of read pairs mapped to a specific location of the genome to the expected size based on the genome-wide average. As was explained in Chapter 3, the library preparation steps involve fragmentation of input DNA to produce fragments that (optimally) have a narrow size distribution. A substantial deviation of the inferred from the expected fragment size of mapped reads at a specific location of the genome can be a sign of a CNV (Figure 19.8). The resolution of the method is limited by the experimental variability of fragment size, and so small insertion and deletion events are difficult to detect by this method.

If a read pair contains an insertion, the paired reads map closer to one another on the reference sequence than one would expect based on the insert size, since the reference doesn't have this insertion. The size of the insertion can be roughly estimated by

$$\text{Insertion size} = \text{insert size} - \text{mapping distance}$$

Let us put aside the issue of fragment size variability for a moment and imagine that all fragments sequenced in some WGS experiment have a length of exactly 500 nucleotides (nt). Imagine we perform

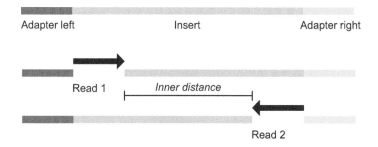

Insert size: Length of read 1 and 2 plus inner distance

Fragment size: Insert size plus length of both adapters

Figure 19.7. Insert size. Imagine that the adapters are 60 nucleotides (nt) in length, the reads are each 50 nucleotides in length, and the not sequenced portion of the insert inner distance is 170 nt long. The insert size is then $2 \times 50 + 170 = 270$ nt, and the fragment size is $270 + 2 \times 60 = 390$ nt. The term "fragment" is sometimes also often used to refer to the fragment of genomic DNA that needs to be sequenced (which is the insert without adapters). The usage should be clear from the context.

paired-end sequencing with 2×100 nt reads. The insert size is then 500 nt (2×100 nt read sequences with an inner distance of 300 nt). If there is no SV in this region of the genome, then the distance between the two 5' ends of the mapped reads is also exactly 500 nt (Figure 19.8A). The calculated insertion size is then $500 - 500 = 0$ nt.

If the sequenced fragment contains a deletion, then the sequence of the fragment does not match with the reference sequence anymore. The physical insert size is still 500 nt (because all fragments have the same actual length), but the mapping distance is greater because the reference sequence comprises the sequence that was deleted in the fragment in addition to the other sequences. For instance, if the deletion is 100 nt long, the mapping distance would be 600 nt, and the insertion size would be calculated as $500 - 600 = -100$ nt (Figure 19.8B).

If the sequenced fragment contains an insertion, then the fragment has a segment of sequence that is not present at this position of the reference genome. The mapping distance is thus reduced. For instance, if the insertion comprises 100 nt, then the mapping distance would be 400 nt and the insertion size is calculated as $500 - 400 = 100$ nt (Figure 19.8C).

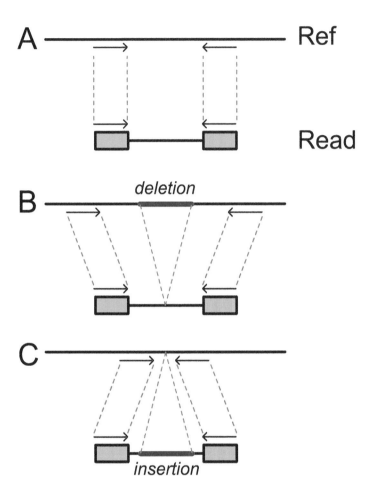

Figure 19.8. Read-pair signatures for insertions and deletions.
The sequenced fragments all have the same size distribution, but if the
patient sample has an insertion or a deletion compared to the genomic
reference, this will disrupt the normal read mapping pattern. (**A**) If a
sequence read contains wildtype reference sequence, then the mapping
distance corresponds to the physical length of the sequenced fragment.
(**B**) If the sample contains a deletion, then the reads map further apart
than the physical fragment size. (**C**) On the other hand, if the sample
contains an insertion compared to the reference genome, the reads map
closer together than the physical fragment size.

In reality, the actual fragment size varies, and there is a distribution of actual sizes that also causes the calculated insertion size to vary. In addition, if the insertion or deletion (indel) is heterozygous, then about half the reads will contain the reference sequence, and the calculated insertion sizes will be composed of a mixture of two distributions. Numerous read-pair algorithms have been developed that employ various clustering and machine-learning approaches that exploit these signals to detect indels. Additionally, inversions and tandem duplications create characteristic read-pair signatures. If a read spans a tandem duplication, the forward read may originate in the 3' region of the duplicated block, and the reverse read may originate in the 5' region of the duplicated block. The aligner attempts to place the reads in the single block of the reference genome. This causes the mapped reads to be oriented away from each other rather than pointing towards each other as is usual with paired-end sequencing (Figure 19.9A). If a fragment spans an inversion breakpoint, then both reads are mapped in the same orientation, and the distance between the reads is increased (Figure 19.9B).

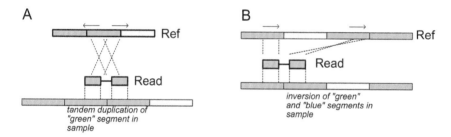

Figure 19.9. Read-pair signatures for (A) tandem duplications and (B) inversions.

Two noteworthy read-pair algorithms are Breakdancer [66] and MoDIL [244].

19.6 SPLIT READS

The read-pair methods exploit the distance and orientation of the two reads of paired-end or mate-pair sequencing. In contrast, split-read (SR) methods attempt to detect deletions and small insertions on the basis of the alignment of the reads themselves to the reference genome. A read is considered to be "split" if the alignment to the genome is interrupted or broken. A stretch of gaps in the read indicates a deletion, and a stretch of gaps in the reference indicates an insertion [274]. The SR approach may examine read pairs for which one read has a reliable mapping and the other fails to map to the genome or has been soft-clipped by the aligner. The unmapped or soft-clipped read may contain a breakpoint which interferes with the alignment of the read, but the other read in the pair allows the fragment to be localized to a certain position in the genome. SR algorithms search for multiple reads in a given location that display comparable break point positions. In some cases it is possible to map the precise start and end positions of the inserted or deleted segments.

SR algorithms use different methods to find the "split". Many divide the incompletely mapped reads into multiple segments which they then attempt to map separately. Often, the first and last segments of the split read can be aligned to the reference genome, and further steps may allow the precise start and end positions of the insertion/deletion to be determined (Figure 19.10). Widely used split-read based methods include CREST [448], SplitRead [191], and Pindel [475].

A related approach can be used to search for translocations; the basic signal is that either read pairs are mapped to two different chromosomes, or that one read of a pair is split between two chromosomes (Figure 19.11). It should be noted that typically one finds very numerous artefactual chimeric reads, and in practice algorithms or scripts need to search for translocations supported by multiple read pairs.

19.7   FINE MAPPING/ALIGNMENT

In contrast to the RD, PEM and SR approaches that first align NGS reads to a known reference genome before the detection of SVs, the assembly-based methods (AS) first reconstruct DNA fragments by assembling overlapping reads (*de novo* assembly). AS methods then compare the resulting assembled contigs with the reference genome in order to detect SVs. In general, *de novo* assembly for SV detection is restricted to a subset of reads. For instance NovelSeq first identifies orphan reads and one-end anchored reads and uses *de novo* assembly as part of a pipeline that is able to identify novel sequence insertions and map their insertion locations [154].

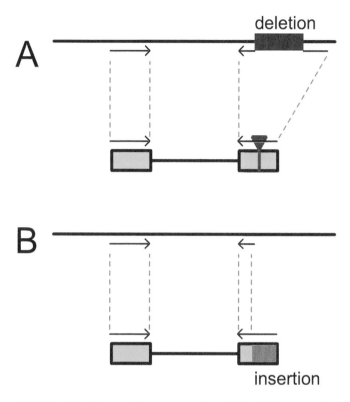

Figure 19.10. Split read approach. (A) If the read overlaps a deletion, then the portions of the read 5' and 3' to the deletion will map to locations on the reference genome that are separated by the deleted segment. **(B)** If the read contains an insertion, than only a portion of the read may map to the genome (as in this figure) or there may be two segments of the read that map to the genome and are separated by a completely contained insertion.

19.8 COMBINED APPROACHES

The four strategies described above each have their own strengths and weaknesses (Table 19.1), and none of them allows accurate and comprehensive detection of a wide variety of structural DNA variants by WES or WGS [16, 420]. This has led to the development of methods that combine the results of multiple independent methods in order to boost sensitivity and specificity [422]. For instance, DELLY integrates paired-end and split-read analysis [358].

We will conclude the chapter with presentations of the CoNIFER

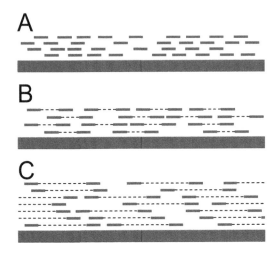

Figure 19.11. Translocations. An exemplary translocation between two chromosomal segments (shown as green and violet). Read mapping results are shown for three sequencing technologies. (A) Single-end sequencing. Two split reads (chimeric between the two chromosomes) are identified. (B) Paired-end sequencing. Two split reads are seen as well as two reads, each of whose mates maps to a different chromosome (termed gap read pairs). (C) Mate-pair (large-insert) library. There is one split read and four gap read pairs.

method for SV detection in WES data as well as CNVnator and DELLY for SV detection in WGS data.

19.9 DETECTING CNVS IN WES DATA: CONIFER

The SV analysis methods presented above were developed in the context of WGS data. WES is still substantially cheaper than WGS and since it is extensively used for diagnostic applications in medical genetics, a number of algorithms have been developed to search for CNVs in WES data. As we have seen in previous chapters, exome capture leads to non-uniform coverage (read-depth) between the targeted regions.[4] The capture procedure often is associated with strong systematic bi-

---

[4]In addition, coverage is obviously mainly in the targeted and not in the non-targeted regions. However, the substantial non-uniformity between different target regions is the main challenge that is addressed by methods such as CoNIFER.

Table 19.1. SV detection approaches.

	RD	RP	SR	AS
Determine copy number	✓			
Detect exact breakpoints			✓	✓
Detect large CNVs	✓	✓	✓[1]	✓[1]
Detect translocations		✓	✓	✓
Can detect SVs other than indels		✓	✓	✓

Note: Computational approaches to SV detection in WES/WGS data. Check marks indicate that the method in question is well suited to detect the indicated category of SV. RD: read depth; RP: read-pair; SR: split read; AS: assembly. [1]Mostly deletions (insertions are more problematic with these methods).

ases between batches of samples. Therefore, raw read-depth approaches are not suitable for WES data, and additional normalization steps are required before read-depth (RD) approaches can be applied to WES data.[5] An advantage of WES data is the higher coverage of most of the targeted regions, which may improve the accuracy of RD analysis. Owing to the discontinuity of the captured exonic regions, there is a much higher chance of missing the actual breakpoints of CNVs, and therefore, SR, RP, and AS approaches have not been extensively applied to WES data.

CoNIFER (copy number inference from exome reads) is an algorithm designed to detect copy number variants (CNVs) in WES data [228]. CoNIFER uses Singular Value Decomposition (SVD) to "denoise" the WES data and enable accurate CNV calling.[6] We will explain the algorithm as presented by the authors and show how to implement a simplified version of the algorithm in R.[7] The CoNIFER method requires WES data from a large cohort of patients (e.g., groups of 8, 366, and 533 probands were analyzed in the original publication). For each patient, a read count is generated for each captured exon.

---

[5]An exception is CopywriteR [231] which uses raw read-depth but only from non-targeted regions to detect CNVs in WES data. We will discuss an example in Section 30.4.

[6]For readers unfamiliar with SVD, we have provided an intuitive introduction in an online Appendix that is available at the book's website, see page xvii.

[7]Readers are referred to the original publication for the full algorithm, which we will not explain here.

The first step of the method is to calculate RPKM values for each exon. RPKM (reads per kilobase per million reads) is a length normalization which is necessary because longer exons are likely to have more reads than shorter exons.

RPKM is defined as

$$RPKM = \frac{10^9 \cdot C}{N \cdot L},\qquad(19.5)$$

where C is the number of reads that are mapped to the exon, N is the total number of mappable reads in the experiment, and L is the length of the exon in base pairs.

The formula and the original definition of the RPKM as reads per kilobase transcript per million reads can be related to one another as follows.

$$
\begin{aligned}
RPKM &= \frac{\text{reads mapped to transcript}}{\frac{\text{total reads}}{1{,}000{,}000} \cdot \text{transcript length in kb}} \\[2mm]
&= \frac{\text{reads mapped to transcript}}{\frac{\text{total reads}}{1{,}000{,}000} \cdot \frac{\text{transcript length in bp}}{1000}} \\[2mm]
&= \frac{10^9 \times \text{reads mapped to transcript}}{\text{total reads} \cdot \text{transcript length in bp}}
\end{aligned}
$$

Here, RPKM is being used for exon counts instead of transcript counts. For example, if an exon is 1 kb in length with 2000 alignments in a sample of 10 million reads (out of which 8 million reads can be mapped), then its RPKM value is

$$\text{RPKM} = \frac{10^9 \cdot 2000}{8 \times 10^6 \cdot 1000} = \frac{2 \times 10^{12}}{8 \times 10^9} = 250$$

The RPKM values are then transformed into standardized z-scores (termed ZRPKM values) based on the mean and standard deviation of the RPKM values across all analyzed exomes. Assuming there are M exons with valid ZRPKM scores and N samples, the data is organized into an $M \times N$ matrix $\mathbf{X}$.

To demonstrate the algorithm we will employ a file with simulated data from a small chromosome with exon capture RPKM values for the genes located on that chromosome (which we have named *YFG1*, *YFG2*, ..., for *Your Favorite Gene-1*, *-2*, ...). This file (`conifer.dat`) is available at the book's website. To input the data and create the

$M \times N$ matrix **X**, download `conifer.dat`, start R from the same directory (or adjust the path), and enter the following commands.

```
> dat <- read.table("conifer.dat",header=T,row.names=1)
> X <- as.matrix(dat)
> dim(X)
[1] 782 144
```

Thus, there are 782 exons and 144 samples in this simulated WES experiment. We go on to filter out exons with a median RPKM of less than one, leaving 778 valid exons.

```
> filtered.rows <-  which(rowMeans(X)>1)
> length(filtered.rows)
[1] 778
> X <- X[filtered.rows,]
```

We now wish to calculate the z-score of the RPKM values, which we will denote ZRPKM. The z-score is defined based on the mean μ and standard deviation σ as

$$z = \frac{x - \mu}{\sigma} \tag{19.6}$$

The z-score of a vector of data W can be calculated in R as

```
> m <- mean(W)
> s <- sd(W)
> W.zscore <- (W-m)/s
```

However, it is easier to use the equivalent built-in function `scale`. We will transform RPKM values, organized into an exon-by-sample matrix **X**, into ZRPKM values based on the mean and standard deviation across all analyzed exomes. Note that `scale` is a generic function whose default method centers and scales the columns of a numeric matrix. In our case, we want to normalize across the values of individual exons, but these are in the *rows* of the matrix. Therefore, we use the transpose operator twice:

```
> ZRPKM <- t(scale(t(X)))
```

By means of the command > `sum(abs(ZRPKM)>3)`, we see that in 1662 cases the value of an exon in a specific sample was more than 3 standard deviations away from the mean. We can now perform the SVD on ZRPKM. As explained in the online Appendix, the SVD decomposes a matrix **A** into three matrices.

$$\mathbf{A} = \mathbf{U\Sigma V}^T \tag{19.7}$$

In R, the SVD can be performed on the ZRPKM matrix with a single command.

```
> s<-svd(ZRPKM)
```

The result of the SVD is stored in the variable **s**, which contains the fields **s$u**, a 778×144 matrix, **s$d**, a 144-component vector that contains the diagonal entries of $\mathbf{\Sigma}$ ($\mathbf{\Sigma}$ is a diagonal matrix with all off-diagonal entries being equal to zero), and **s$v**, a 144×144 matrix. The data structure can be explored using the **str** command:

```
> str(s)
List of 3
 $ d: num [1:144] 300.8 87.2 35.6 27 23.7 ...
 $ u: num [1:778, 1:144] -0.037 -0.0382 -0.0385 -0.0384 ...
 $ v: num [1:144, 1:144] -0.00556 0.00942 -0.14918 ...
```

19.9.1 Scree Plots

Principal component analysis (PCA) is used for dimensionality reduction of high-dimensional datasets in a way that allows projection of items onto a transformed two or three-dimensional space in which it is often easier to discern clusters or groupings of related items. In the case of CoNIFER, the SVD is used with the opposite intention. The authors reasoned that the rare CNVs that they are interested in are actually responsible for only a small part of the overall variance. They therefore remove the primary singular vectors under the assumption that this will remove the major sources of bias in the WES read-depth data (including uneven capture efficiency and other factors such as GC content). What remains will better allow CNVs in individual datasets to be detected, according to this assumption.

The scree plot is a standard visualization (for principle component analysis and SVD) that displays the relative amount of contributed variance from each component. Scree plots are said to resemble the side of a hill.[8] It is a useful way to visualize the contributions of the individual singular vectors to the overall variability and to decide how many singular vectors to remove from the data. Although the decision where to make the cut is subjective, it is recommended to choose the point where there is a clear drop in the magnitudes of the singular values.

The diagonal entries of Σ, known as the singular values, reflect the relative amount of variation contributed by each component. The amplitude of the singular values is related to their importance in explaining the data, whereby the square of a singular value is proportional to the variance explained by its corresponding singular vector (Figure 19.12).[9]

```
> L <- dim(X)[2]   # number of samples
> x <- seq(1:L)
> y <- s$d                        # singular values
> par(las=1, mar=c(4, 4, 2, 4))
> plot(x, y, ylab="singular values", xlab="Number (n)",
             cex.axis=1.25, cex.lab=1.25,
             pch=23, cex=1.5, col="blue")
> box(bty="u")
> var_k <- s$d*s$d / sum(s$d*s$d)          # variances
> y2 <- cumsum(var_k) * max(y)
> lines(x, y2, col="red", lwd=2)
> abline(h=max(y), lty=3, col="red", lwd=2)
> axis(4, at=c(0, 0.5*max(y), max(y)),
          labels=c("0%", "50%", "100%"))
```

It can be seen that the first 4 of the 144 singular values capture over 60% of the overall variance.

```
> y <- s$d
> sum(y[1:4]*y[1:4])/sum(y)
[1] 62.00402
```

---

[8] The word "scree" means a slope of loose rock debris located on the side of a hill or cliff.

[9] The amount of overall variance explained by the ith pair of SVD vectors corresponds to $\frac{\sigma_i^2}{\sum_j \sigma_j^2}$, and this is the usual quantity that is displayed in scree plots.

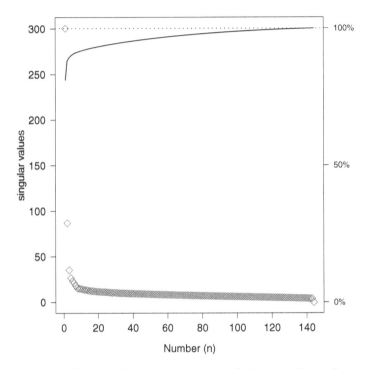

Figure 19.12. Scree plot. A scree plot of the singular values derived from the simulated YFG dataset. It can be see that the first handful of singular values (blue) represent a large proportion of the total variance (red).

It can be expected that the largest part of the variance in these components comes from experimental noise. If these components are eliminated, then the remaining data will better reflect individual (rare) CNVs.

One of the samples has a deletion for the *YFG18* gene. To plot this, we will extract the range of row indices for the gene using `grep`.

```
> idx <- grep('^YFG18\\.exon\\d+',rownames(X))
```

The variable `idx` now contains the indices $182, \ldots, 207$.

Now plot the values from the unnormalized data matrix **X** for the *YFG18* gene

```
> ylim <- range(X)
```

```
> ydim <- dim(X)[2]
> gscale = gray.colors(ydim)
> plot(idx,ylim=ylim,type="n",xlab="exon index",
  ylab="Z-RPKM",main="YFG18")
> for (i in 1:ydim) { points(X[idx,i],col=gscale[i]) }
```

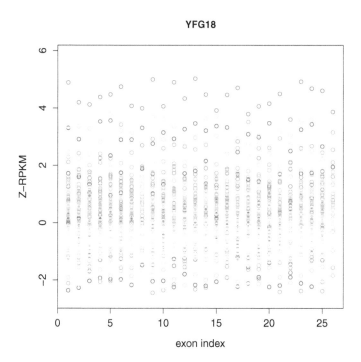

Figure 19.13. Unnormalized ZRPKM values for the exons of the *YFG18* gene. Each data point represents the ZRPKM value for an exon for the *YFG18* gene for one of the 144 samples.

It is difficult to see which sample, if any, has a deletion in these exons (Figure 19.13). This is because the noise in the data is overwhelming the signal. Therefore, we subtract the strongest singular vectors from the dataset. Inspection of the scree plot in Figure 19.12 suggests that removing the first 4 singular vectors (which contain over 60% of the total variance) will be a good start.[10] Algorithmically, in order to

---

[10] The decision of how many singular vectors to choose is arbitrary, in that there is no algorithmic way to determine the "optimal" number.

remove the strongest k components, the singular values $\sigma_1, \ldots, \sigma_k$ are set to zero, following which we recalculate $\mathbf{X}$ by multiplying the three matrices (noting that $\sigma_1, \ldots, \sigma_k$ have been set to zero in $\mathbf{\Sigma}$). Finally, we generate the corresponding z-scores and store them in the matrix X2.

```
k<-4 ## we will set four SVs to zero
d <- s$d
d[1:k] = 0
X2 <- s$u %*% diag(d) %*% t(s$v)
rownames(X2) <- rownames(X)
colnames(X2) <- colnames(X)
```

Each of the values in X2 represents the normalized relative copy number of an exon in a sample, with the "noise" that was contained in the first four singular vectors removed.

We will now show how to plot the result. We first define an R function that will take a vector of values, calculate the standard deviation, and return a vector with only those values that lie more or less than 3 standard deviations from the mean (the other values are replaced by NA).

```
call.exon <- function(data) {
  threshold <- 3
  ret = as.numeric(scale(as.matrix(data)))
  ret[abs(ret) <= threshold] = NA
  return(ret)
}
```

To plot the data, we first set up several variables that define the range and a set of colors to use, and plot points in gray for each of the exons of *YFG18* for each of the samples.

```
ylim <- range(X2)
ydim <- dim(X2)[2]
gscale <- gray.colors(ydim)
hcol <- rainbow(ydim)
plot(idx, ylim=ylim, type="n", xlab="Exon index",
  ylab="SVD corrected coverage",
  main="YFG18 (first 4 singular values = 0)")
for (i in 1:ydim) {
```

```
    points(X2[idx,i], col=gscale[i])
}
```

We add some color to all points that are more than three standard deviations from the mean with the following code.

```
for (i in seq_along(idx)) {
  exon.vars = call.exon(X2[idx[i],])
  for (j in 1:ydim) {
    if (!is.na(exon.vars[j])) {
      points(i, X2[idx[i],j], col=hcol[j])
    }
  }
}
```

And finally, we plot black lines to show the values at three standard deviations from the mean and plot the values for sample 36 in red.

```
hsd = c()
lsd = c()

threshold <- 3
for (i in 1:length(idx)) {
  hsd = c(hsd, threshold*sd(X2[idx[i],]))
  lsd = c(lsd, -threshold*sd(X2[idx[i],]))
}

lines(hsd, col="black")
lines(lsd, col="black")
lines(X2[idx,"sample36"], lwd=2, col="red")
```

We see that the ZRPKM values for sample 36 are more than three standard deviations below the mean for multiple exons of the *YFG18* gene (Figure 19.14). This is a pretty good indication there may be a deletion of these exons in sample 36 and should prompt further diagnostic workup.

We have presented a simplified version of the CoNIFER algorithm with the intention of providing an intuitive explanation of SVD-based (or PCA-based) algorithms for CNV detection in exome data. We did not present the methods that are used to flag candidate CNVs, but

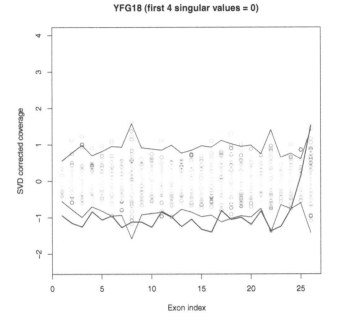

Figure 19.14. Normalized ZRPKM values for the exons of the *YFG18* gene.

assumed that we knew which gene was deleted in which sample. Readers are referred to the original publication for the full algorithm. In addition to CoNIFER, a number of other methods use SVD or PCA for noise reduction prior to CNV calling including cnvOffSeq [36], ExoCNVTest [84], FishingCNV [394], and XHMM [126].

19.10   CNVNATOR AND DELLY ANALYSIS

CoNIFER was designed to analyze cohorts of patients with WES data. In this section, we will present two methods that call SVs from single WGS samples. In practice, the sets of SVs identified by different SV callers never perfectly overlap, and in fact usually the amount of overlap seen is less than 50%. We will present two methods in the section, one of which is based on read-depth analysis and the other on read-pair and split read analysis. The example is characteristic of what one should expect to see with SV analysis by NGS technologies, which remains a difficult undertaking with sensitivities and specificities that are still far from ideal.

If you would like to follow along, download the WGS datasets for NA12878, NA12891, and NA12892 as described in Chapter 4. To perform this analysis, you will need the following alignment files (produced and post-processed as described in previous chapters).[11]

```
NA12878_dedup.bam
NA12891_dedup.bam
NA12892_dedup.bam
```

19.10.1   CNVnator

CNVnator uses approaches including GC correction and binning to process read-depth data. It additionally applies the mean-shift technique to group bins into segments prior to statistical testing [3]. To follow along with the exercises in this section, download and compile CNVnator.[12]

Several steps need to be performed to run CNVnator.

19.10.1.1   Preprocessing

This step produces a list of potential CNVs. The -root argument is the name of the output file that will be produced by CNVnator containing the preprocessed data. The -chrom parameter indicates the names of the chromosomes for which CNVnator should perform the analysis. The chromosome names used for the -chrom parameter (here 1, 2, ..., X, Y) must match those in the reference sequence used for the alignment. The -tree parameter is the BAM file with the alignment of the patient sample. Finally, -unique is used to avoid mapping reads with zero mapping quality (Q=0, i.e., without unique mapping coordinates).

```
$ cnvnator -root NA128787.root \
  -chrom 1 2 3 4 5 6 7 8 9 10 11 12 \
    13 14 15 16 17 18 19 20 21 22 X Y \
  -unique \
  -tree NA12878_dedup.bam
```

---

[11] In the following examples, we used hg19 (hs37d5.fa) as the reference genome.

[12] At the time of this writing, the current version was v0.3.3, available at https://github.com/abyzovlab/CNVnator/releases/. Follow the instructions in the README file in the CNVnator_v0.3.3.zip archive to generate the command-line tool cnvnator.

19.10.1.2 Coverage histogram

The next step generates a histogram of coverage. CNVnator requires single files for each chromosome (`1.fa`, `2.fa`,...). If you only have a single FASTA file (denoted here as `ref.fa`) with all of the chromosome data, you can split the file using the following `awk` command.

```
$ awk '/^>/ {OUT=substr($1,2) ".fa";print " ">OUT}; \
  OUT{print >OUT}' ref.fa
```

The bin size for the histogram (here, 100 nucleotides) is specified as the argument for `-his`. The appropriate value for this parameter depends on factors such as the overall sequencing depth. In this command, `<directory>` stands for the path to a directory containing one FASTA file for each chromosome to be analyzed.

```
$ cnvnator -d <directory> \
  -root NA128787.root \
  -chrom 1 2 ... X Y \
  -his 100
```

19.10.1.3 Calculating the p-value for each bin

We then calculate the p-values for each of the bins. The `-stat` parameter also controls the bin size, which should be in general the same as the bin size used to generate the histogram.

```
$ cnvnator -root NA128787.root \
  -chrom 1 2 ... X Y \
  -stat 100
```

19.10.1.4 Segmentation

The next step performs read-depth signal partitioning (segmentation), meaning that CNVnator attempts to find longer contiguous regions where individual bins have a similar read depth and are therefore inferred to have the same copy number. The `-partition` argument specifies the bin size.

```
$ cnvnator -root NA128787.root \
  -chrom 1 2 ... X Y \
  -partition 100
```

19.10.1.5 Calling CNVs

Finally the CNVs are called. In this case, the `-call` argument specifies the bin size.

```
$ cnvnator -root NA128787.root \
  -chrom 1 2 ... X Y \
  -call 100 > NA12878_calls.cnv
```

This will produce a table with CNV candidates and the following columns:

```
CNV_type  coordinates  CNV_size  normalized_RD \
    e-val1  e-val2  e-val3  e-val4  q0
```

normalized RD: read depth, normalized to 1

e-val1: probability value from the t-test statistic.

e-val2: probability of RD values within the region to be in the tails of a Gaussian distribution describing frequencies of RD values in bins.

e-val3: same as e-val1 but for the middle of the CNV.

e-val4: same as e-val2 but for the middle of the CNV.

q0: fraction of reads mapped with `MAPQ=0`.

It is useful to perform an additional step to filter for only statistically significant CNVs according to `e-val1` and `e-val2`, and simultaneously generate a BED file for use in IGV. The following R script performs these tasks

```
args<- commandArgs(trailingOnly = TRUE)
inFile<- args[1]
outFile<- args[2]
dat<- read.table(inFile, header=F, sep="\t")
dat<-dat[intersect(which(p.adjust(dat$V6) < 0.05), \
    which(p.adjust(dat$V5) < 0.05)),]
write.table(matrix(unlist(strsplit(paste(dat$V2,dat$V1,sep="-"), \
    split="[:-]")),byrow=T, ncol=4),outFile,col.names = F,row.names=F, \
    sep="\t", quote=F)
```

With the current WGS samples, this reduced the number of CNV calls from several thousands per sample (NA12878: 30251; NA12891: 15355; NA12892: 13804) to ∼1200 (NA12878: 1299; NA12891: 1252; NA12892: 1140).

19.10.2 DELLY2

DELLY analyzes paired-end and split-read signals to make CNV calls. In principle, DELLY is able to detect not only CNVs (like CNVnator), but also tandem duplications, inversions, or reciprocal translocations [358].

DELLY2 (currently at version v0.7.6) releases can be downloaded as precompiled packages for Linux from the DELLY GitHub page.[13] In the following, we will only show how to use DELLY to call deletions (DEL), but the procedure is the same for the other structural variant types that DELLY can call.

DELLY, unlike CNVnator, does not need any preprocessing steps and can call structural variants directly by running

```
$ delly call -t DEL \
  -g hs37d5.fa \
  -o DEL_NA12878.bcf \
  -x human.hg19.excl.tsv \
  NA12878_dedup.bam
```

DELLY provides a file (`human.hg19.excl.tsv`) with regions of known high variability in the human genome that we use here to exclude these regions.

```
$ delly call -t DEL \
  -g hs37d5.fa \
  -o DEL_NA12878_NA12891_NA12892.bcf \
  -x human.hg19.excl.tsv \
  NA12878_dedup.bam NA12891_dedup.bam NA12892_dedup.bam
```

Following this, the per-sample calls are regenotyped using also breakpoint spanning read informations from the whole batch.

```
$ delly call -t DEL -g hs37d5.fa -v DEL_NA12878_NA12891_NA12892.bcf \
    -o DEL_NA12878.bcf -x human.hg19.excl.tsv NA12878_dedup.bam
$ delly call -t DEL -g hs37d5.fa -v DEL_NA12878_NA12891_NA12892.bcf \
    -o DEL_NA12891.bcf -x human.hg19.excl.tsv NA12891_dedup.bam
$ delly call -t DEL -g hs37d5.fa -v DEL_NA12878_NA12891_NA12892.bcf \
    -o DEL_NA12892.bcf -x human.hg19.excl.tsv NA12892_dedup.bam
```

We then merge and index the regenotyped BCF files using BCFtools.

---

[13]https://github.com/dellytools/delly/releases/

```
$ bcftools merge -O b -o DEL_NA12878_NA12891_NA12892.geno.merged.bcf \
    DEL_NA12878.bcf DEL_NA12891.bcf DEL_NA12892.bcf
$ bcftools index DEL_NA12878_NA12891_NA12892.geno.merged.bcf
```

DELLY has predefined filters for germline or *de novo* analysis. We will use the germline filter here.

```
$ delly filter -t DEL -f germline \
    -o DEL_NA12878_NA12891_NA12892.geno.filtered.bcf \
    DEL_NA12878_NA12891_NA12892.geno.merged.bcf
```

Finally, we generate a VCF file for downstream analysis and index it.

```
$ bcftools view DEL_NA12878_NA12891_NA12892.geno.filtered.bcf \
    | bgzip -c > DEL_NA12878_NA12891_NA12892.geno.filtered.vcf.gz
$ tabix ${OUTFOLDER}/${TYPE}.geno.merged.filtered.vcf.gz
```

It can be seen in Figure 19.15A that DELLY called 419 deletions for NA12878 and CNVnator 1120.[14] The overlap was only 162. While we might expect DELLY to call more deletions than CNVnator because DELLY can also analyze deletions smaller than the CNVnator bin size, there were 1120-162=958 CNVs that CNVnator called that were not called by DELLY. It is impossible to say which of these calls are false positive or false negative without experimental validation, but inspection of the alignment files suggests that both CNVnator and DELLY fail to detect at least some true SVs. Figure 19.15B shows a deletion that was called by both programs. Figure 19.15C shows a large deletion that was called by CNVnator but missed by DELLY, presumably because there were no supporting read pairs or split reads. Finally, Figure 19.15D shows a small deletion that was called by DELLY but not CNVnator. CNVnator tends to have difficulties calling small SVs because there is not sufficient read-depth evidence. DELLY detected several split-read alignments and an increased insertion size between read pairs.

19.11 LONG-READ SEQUENCING

The approaches we have presented above are based on an analysis of Illumina (short-read) data. As mentioned on page 18, recently two technologies for long-read sequencing have been introduced to market. Longer reads can in principle improve read alignment and CNV

---

[14]CNVnator called a total of 1299 CNVs for NA12878, comprising 1120 deletions and 179 duplications.

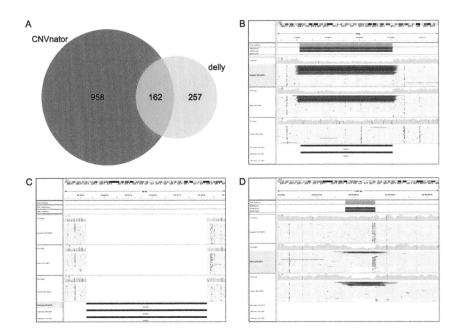

Figure 19.15. CNVnator and DELLY2. Structural variant callers based on coverage (CNVnator) or split read and insert size (DELLY) have advantages and disadvantages. (A) The Venn diagram shows the number of deletions for NA12878 following the filter steps for DELLY and CNVnator. (B) Both DELLY and CNVnator can call deletions of intermediate size (here ∼5,900 bp). (C) A typical deletion (∼50,000 bp), that was called by CNVnator but not by DELLY, due to the size and the lack of supporting read pairs or split reads. (D) The size of this deletion (813 bp) is too small to be called by CNVnator but not for DELLY with support from several split-read alignments and the increased mapping distance of read pairs.

detection, especially in repetitive regions of the genome that pose substantial difficulties for alignment of short reads. Initial experiences have shown that PacBio sequencing could substantially improve discovery of structural variation in a haploid genome sample [174]

The 10x Genomics platform, Chromium, segregates samples into individual reaction compartments (droplets in an emulsion), and labels the DNA in each droplet with a unique barcode sequence. Following this, the libraries are sequenced using the usual Illumina technology. Since the amount of DNA in each droplet contains only a small fraction of the human genome, the barcodes can be used to computationally "link" reads from individual droplets in a way that produces high-quality phased assemblies [299]. Early experiences suggest that this technology may be able to detect some structural variants that cannot be identified by standard Illumina sequencing owing to repetitive sequences that are difficult to map with short reads.

FURTHER READING

Carvalho CM, Lupski JR (2016) Mechanisms underlying structural variant formation in genomic disorders. *Nat Rev Genet.* **17**:224–238.

Krumm N, Sudmant PH, Ko A, O'Roak BJ, Malig M, Coe BP; NHLBI Exome Sequencing Project, Quinlan AR, Nickerson DA, Eichler EE (2012) Copy number variation detection and genotyping from exome sequence data. *Genome Res* **22**:1525–1532. (Original publication on the CoNIFER method)

Abyzov A, Urban AE, Snyder M, Gerstein M (2011) CNVnator: an approach to discover, genotype, and characterize typical and atypical CNVs from family and population genome sequencing. *Genome Res* **21**:974–984.

Rausch T, Zichner T, Schlattl A, Stütz AM, Benes V, Korbel JO (2012) DELLY: structural variant discovery by integrated paired-end and split-read analysis. *Bioinformatics* **28**:i333–i339.

19.12   EXERCISES

Exercise 1

To learn more about structural variants and SV calling in WES/WGS data, readers are advised to read the articles mentioned under Further Reading, and to use CNVnator and DELLY on data of their choice and examine the candidate SVs with IGV (see Chapter 17). Readers should consult the online documentation of IGV about the color schemes for read pairs to show anomalies of orientation or distance (*Interpreting color by insert size*, and *Interpreting color by pair orientation*). IGV is a tool that requires some amount of practice to use well, but it is invaluable in understanding alignment files and anomalies of alignments that point to various classes of structural variation.

V

Variant Filtering

CHAPTER 20

Pedigree and Linkage Analysis

Peter Robinson and Max Schubach

C OSEGREGATION refers to the inheritance of disease-causing mutations together with clinical diseases in a family. The analysis of cosegregation of variants with disease is a powerful tool for gene discovery research, and diagnostics. In the context of exome and genome sequencing, datasets are usually relatively small and encompass nuclear families or parts thereof. Often, variants or genes are filtered to those showing a pattern of inheritance that matches the disease diagnosis. For instance, if we are considering an autosomal dominant disease, then a variant that is not present in a patient or that is present in an unaffected relative can usually be excluded from further analysis. In this chapter, we will show how the analysis of cosegregation in nuclear families (e.g., parents and siblings) can be used to filter genes and variants in WES/WGS studies. The family structure is recorded in PED files, and we will begin the chapter with an explanation of the PED file format and the symbols used for describing a pedigree.

20.1 BASIC PEDIGREE ANALYSIS

Publications on genetic diseases as well as patient records use a standard set of symbols for pedigree analysis. Males are represented by squares, females by circles. Matings are represented by drawing a line between the mother and father, and children are shown (in the order of birth) beneath them. Twins are indicated by the use of diagonal instead of vertical lines, and monozygotic (identical) twins are connected by a horizontal line. In some cases the sex of a person in a pedigree is unknown or irrelevant, and this is indicated by a "diamond" symbol. It is possible to summarize multiple children by indicating their number within the pedigree symbol (Figure 20.1).

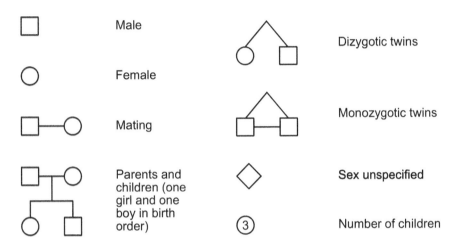

Figure 20.1. Pedigree symbols (1).

Individuals who are affected by the disease (or trait) under investigation are shown as filled symbols. In some occasions, it is useful to show heterozygotes (carriers) for an autosomal recessive (AR) disease by half-filled symbols. Similarly, female carriers of X-linked recessive diseases can be shown with a dot in the middle of the circle. Deceased persons are symbolized with a slash through the pedigree symbol, and consanguineous (blood-related) matings are symbolized by two horizontal lines. The proposita (for females) or propositus (for males), also known as the index patient, is the individual who is the first one to be investigated in a pedigree. Often, this is the patient who first presented with a disease that physicians or researchers are trying to diagnose. The proposita(-us) is indicated with an arrow (Figure 20.2).

20.2   THE PED FILE

A PED file ("pedigree file") describes the family relationships of each sample along with their gender and phenotype. PED files are typically used by software for genetic linkage analysis. The PED file is

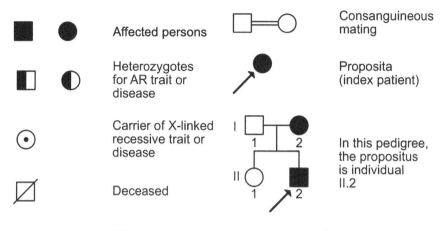

Figure 20.2. Pedigree symbols (2).

white-space (space or tab) delimited and should have no header (Table 20.1). The first six columns are mandatory. Additional columns may be used for linkage analysis but are typically not used in the context of WES/WGS analysis.

The family ID and the individual IDs may be made up of letters and digits, and the combination of family and individual ID should uniquely identify each person represented in the PED file. The parents of a person in the pedigree are shown with the corresponding individual IDs. Individuals whose parents are not represented in the PED file are known as founders; their parents are represented by a zero (0) in the columns for mother and father. Finally, the sex and the phenotype (disease) status are shown in columns 5 and 6.

20.2.1 Autosomal recessive pedigrees

Autosomal recessive inheritance is characterized by the following characteristics.

1. Affected persons have healthy parents who are heterozygous for a disease-causing mutation.[1]

2. Vertical transmission is not observed.[2]

---

[1]Exceptions to this rule such as pseudodominant inheritance may be observed, for instance, if one parent is affected and the other is a carrier for a disease-causing mutation).

[2]With the exception of pseudodominance.

Table 20.1. Columns of the PED file.

	Item	Example	Explanation
1	Family ID	FAM001	An identifier for the family.
2	Sample ID	859_A	A unique identifier for the sample (person).
3	Father's sample ID	922_B	The sample ID for the father of the current individual (or 0 if the father is not represented in the pedigree)
4	Mother's sample ID	923_B	The sample ID for the mother of the current individual (or 0 if the mother is not represented in the pedigree)
5	Gender	2	1 for male, 2 for female, 0 if unknown.
6	Phenotype	1	1 for unaffected, 2 for affected, 0 if unknown.

3. Inheritance is independent of the sex of the child.

4. This recurrence risk for siblings of an affected child or children of heterozygous parents is 25%.

5. Autosomal recessive inheritance is commonly observed in consanguineous matings.

Figure 20.3 shows the pedigree and the corresponding PED file for a family that is segregating a disease with autosomal recessive inheritance.

20.2.2 Homozygous and compound heterozygous variants

With rare exceptions, the parents of children with autosomal recessive diseases are heterozygous mutation carriers. If both parents carry the same mutation, then each affected child is homozygous for the mutation. Alternatively, if each parent carries a distinct mutation, then each affected child is compound heterozygous for these two distinct mutations.

In some situations it may be desirable to search specifically for homozygous or compound heterozygous variants that are compatible with autosomal recessive inheritance. For instance, in consanguineous families, homozygous (i.e., autozygous) mutations are commonly found, whereas in outbred families segregating an autosomal recessive disease,

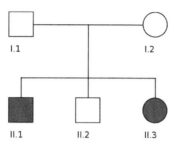

FAM1	I.1	0	0	1	1
FAM1	I.2	0	0	2	1
FAM1	II.1	I.1	I.2	1	2
FAM1	II.2	I.1	I.2	1	1
FAM1	II.3	I.1	I.2	2	2

Figure 20.3. Autosomal recessive inheritance. An example pedigree representing a family segregating an autosomal recessively inherited disease. Here, the son II.1 is affected, the son II.2 is healthy (and thus can be heterozygous for maximally one mutation), and the daughter II.3 is affected. The parents can be assumed to be heterozygous mutation carriers, but this is not shown in the pedigree.

compound heterozygous mutations are more common. In these cases, it is possible to filter specifically for these mutations (see below).

20.2.3 Autosomal dominant pedigrees

Autosomal dominant inheritance is characterized by the following characteristics.

1. A heterozygous mutation (one of the two copies of the gene is affected) causes the disease.

2. Vertical transmission (an affected parent is heterozygous for the disease-causing mutation and can transmit the mutation to a child).

3. Inheritance is independent of the sex of the child.

4. Risk for offspring of an affected person: 50%.

A family segregating an autosomal dominant disease is shown in Figure 20.4.

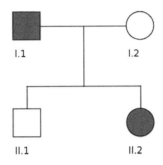

FAM2	I.1	0	0	1	2
FAM2	I.2	0	0	2	1
FAM2	II.1	I.1	I.2	1	1
FAM2	II.2	I.1	I.2	2	2

Figure 20.4. Autosomal dominant inheritance. An example pedigree representing a family segregating an autosomal dominantly inherited disease. The father and his daughter are affected by the disease.

20.2.4 X-chromosomal recessive pedigrees

X-chromosomal recessive inheritance, which is often referred to as X–linked inheritance, shows the following inheritance pattern.

1. Father to son transmission is impossible.

2. All daughters of an affected father inherit the mutation.

3. Healthy males cannot be carriers.

4. Risk for sons of female carriers is 50%.

5. Risk for daughters of female carriers is 50% to be carriers themselves.

A family segregating an X-chromosomal recessive disease is shown in Figure 20.5.

20.2.5 Other modes of inheritance

The most common modes of inheritance encountered in Mendelian disease are autosomal recessive, autosomal dominant, and X–linked recessive inheritance. X–linked dominant inheritance refers to the situation

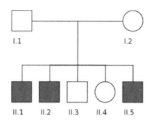

FAM3	I.1	0	0	1	1
FAM3	I.2	0	0	2	1
FAM3	II.1	I.1	I.2	1	2
FAM3	II.2	I.1	I.2	1	2
FAM3	II.3	I.1	I.2	1	1
FAM3	II.4	I.1	I.2	2	1
FAM3	II.5	I.1	I.2	1	2

Figure 20.5. X–linked inheritance. An example pedigree representing a family segregating an X-chromosomal recessive disease. The mother I.2 must be a heterozygous mutation carrier. Each of the affected sons is hemizygous for the mutation (this is generally called as homozygous ALT in VCF files). Unaffected sons must display the wildtype sequence, whereas daughters might be heterozygous for the mutation or may have the wildtype sequence.

where one variant on the X chromosome causes a trait or disorder. The pedigrees may look superficially similar to those observed with autosomal dominant inheritance because the trait is passed "vertically" from generation to generation. However, since males pass their (only) X chromosome to daughters and their Y chromosome to sons, all daughters of an affected male are themselves affected, and none of the sons are. It is for this reason that one instance of father-to-son transmission of a trait is required to confirm autosomal dominant inheritance based on pedigree analysis. The most well-known disease with X chromosomal dominant inheritance is X–linked hypophosphatemia.

Y–linked inherited traits occur in all male descendants of an affected male and never occur in females. There are not many Y–linked diseases and some of the well characterized ones involve conditions such as spermatogenic failure that result in infertility and thus are not transmitted to sons.[3]

The sperm does not contribute mitochondria to the fertilized egg, and therefore all of the mitochondria in the body descend from the mitochondria present in the original oocyte (egg cell) at the time of conception. Therefore, mitochondrial mutations can only be transmitted by the mother, but they are transmitted to all children. Because individual cells contain multiple mitochondria and different proportions of mutated mitochondria may be transmitted to different children, the disease severity may vary widely with mitochondrial disease. We will not discuss other modes of inheritance including anticipation and imprinting here, but refer readers to a textbook of human genetics.

At present, Jannovar screens VCF files for autosomal recessive and dominant as well as X-chromosomal recessive and dominant inheritance.

20.3 INTEGRATING PEDIGREE ANALYSIS WITH ANNOTATION OF VCF FILES USING JANNOVAR

We will now show how to perform simple pedigree filtering of WES/WGS data using Jannovar (see Chapter 14 for instructions on how to install and use Jannovar). If Jannovar is run in the `annotate-vcf` mode with the `--pedigree-file <pedfile>` argument, it will annotate the VCF file as described in Chapter 15. Additionally, it will add annotations to variants that have been inherited in a manner that is compatible with one or more of autosomal dominant (`AD`), autosomal recessive (`AR`), X-chromosomal dominant (`XD`), and X-chromosomal recessive (`XR`). The command used to perform pedigree analysis is the same as that for annotating VCF files as explained in Chapter 14 except that additionally a pedigree file is passed to Jannovar using the `--pedigree-file` flag.

```
$ java -jar Jannovar.jar annotate-vcf \
  -d hg38_refseq.ser \
  --pedigree-file <sample.ped> \
  -i sample.vcf \
  -o out.vcf
```

---

[3] A Y-linked form of the eye disease retinitis pigmentosa has been published but the gene has not been identified to date.

We can run the inheritance filter on our test single-sample VCF file using a PED file with just one entry for NA12878.

```
FAM1    NA12878    0    0    2    2
```

This line states that NA12878 is an affected daughter in family FAM1 and that her parents are not represented in the pedigree. We run Jannovar as follows

```
$ java -jar jannovar.jar annotate-vcf \
    -d data/hg38_refseq.ser \
    -i NA12878_filtered.vcf \
    --pedigree-file NA12878.ped \
    -o NA12878.ped.vcf
```

Jannovar adds the following line to the header of the output VCF file

```
##INFO=<ID=INHERITANCE,\
    Number=.,\
    Type=String,\
    Description="Compatible Mendelian inheritance modes \
    (AD, AR, XD, XR)">
```

See Chapter 13 for information about the VCF format. Jannovar adds INHERITANCE information to the INFO field of all variants/genes showing inheritance that is compatible with one of these four modes of inheritance. For instance, a homozygous variant c.683-44C>T was found in *DUSP1* and annotated as being compatible with autosomal recessive inheritance:

```
INHERITANCE=AR
```

A heterozygous variant in the same gene c.107A>G|p.(Asp36Gly) was annotated as being compatible with autosomal dominant inheritance.

```
INHERITANCE=AD
```

If a gene has multiple heterozygous variants, they can in principle be consistent with autosomal dominant inheritance (if only one variant is pathogenic) or autosomal recessive inheritance (if both are pathogenic), and thus Jannovar annotates them with both possibilities

```
INHERITANCE=AD,AR
```

Similar comments apply to X–linked recessive (XR) and X–linked dominant (XD) inheritance. Based on these annotations, we can later remove variants that are not compatible with the assumed modes of inheritance ("hard filtering"). We will show how this might be done with a script in the Exercises. We note that this kind of filtering is much more useful for VCF files with multiple samples from a family, which we will demonstrate further below.

Jannovar's strategy for performing pedigree analysis is to first group variants according to the gene they are located in. It then performs a filtering operation on the variants and rules out variants or genes if they demonstrate criteria that are not compatible with a given mode of inheritance. In WES/WGS experiments with multiple samples, it is possible that some genotypes are missing because of low coverage or other reasons. Although one might be tempted to exclude such variants from further consideration, this would risk discarding causal mutations. Jannovar therefore follows the strategy of retaining variants unless there is evidence against them.

We note that the inheritance filter is not intended to be used alone to flag variants as pathogenic. Rather, it is a necessary but not sufficient precondition, and should be used in conjunction with other assessments of pathogenicity. For instance, in a typical exome dataset, hundreds of genes might be flagged as having variants compatible with autosomal dominant inheritance. Computational pathogenicity prediction and other prioritization methods would be used subsequently to this step to decide if any of the variants actually represents a strong candidate. On the other hand, if a variant does not cosegregate with the disease, then it can (usually) be filtered out at this step in order to reduce the search space.

20.3.1 Autosomal Recessive Inheritance

Jannovar uses different rules to check whether a gene has variants compatible with autosomal recessive inheritance depending on whether a single sample VCF file or a VCF file with multiple samples from a family is being investigated. For simplicity, we will use the abbreviations ALT for the alternate allele (the variant), REF for the reference allele, HOM-REF for a call of a homozygous reference genotype, HET for a call of a heterozygous genotype, and HOM-ALT for a call of homozygous alternate (variant) genotype.

In the single-sample case, Jannovar checks each gene with called variants. If one of the following two criteria are met, the variants pass the autosomal recessive inheritance filter, and the variants involved are marked with the annotation `INHERITANCE=AR` in the `INFO` field.

(i) The variant has a HOM-ALT call.

(ii) The gene has two or more HET calls.

In the first case, the variant is compatible with autosomal recessive inheritance because it is homozygous. In the second case, each pair of heterozygous variants could be causative, compound heterozygous variants.

For the multisample case, Jannovar again performs separate tests for homozygous and compound heterozygous variants. If at least one category of variant is found to be compatible with `AR` inheritance, the gene passes the filter.

20.3.1.1   Autosomal recessive homozygous inheritance

For the homozygous test, variants are ruled out if any (unaffected) parent is homozygous or if any other unaffected members of the pedigree are homozygous for the variant. The affected members of the pedigree are then investigated. If any affected person is HOM-REF or HET for a variant then the variant is ruled out. Although variants are not ruled out if an affected person has an incomplete genotype (e.g., ./.), at least one affected person in the pedigree must have a HOM-ALT genotype for the variant to be considered a candidate.

20.3.1.2   Autosomal recessive compound heterozygous inheritance

The rules are more involved for compound heterozygous candidates. For each gene with called variants, Jannovar checks whether candidate pairs of heterozygous variants are compatible with autosomal recessive compound heterozygous inheritance. The initial check investigates genotypes in all affected persons and (if available) their parents.[4] For each variant in the candidate pair, Jannovar checks if the variant is HET in one parent and HOM-REF in the other parent. The fact that a genotype is not observed in one or two persons of the trio does not rule out a variant, but a candidate variant pair must be observed in at least one affected person to be considered. The step identifies candidate pairs of variants in each of the affected persons in the pedigree.

---

[4]In this context, the child and the two parents are usually referred to as a "trio".

The next step iterates across all candidate pairs identified in any affected person. For each candidate pair, Jannovar checks whether the pair of variants is compatible with compound heterozygous inheritance with all affected persons in the pedigree. If a variant pair is incompatible for any affected person (see above), it is excluded from further consideration. Finally, Jannovar checks the unaffected persons in the pedigree. If any unaffected person is HOM-ALT for either of the mutations or is compound heterozygous for the mutations, the pair is rejected. If a candidate pair of mutations survives all these tests, both variants are reported as being compatible with **AR** inheritance

To perform pedigree analysis with the PED file in Figure 20.3 (**fam1.ped**), we run Jannovar as follows:[5]

```
$ java -jar Jannovar.jar annotate-vcf \
  -d data/hg38_refseq.ser \
  --pedigree-file fam1.ped \
  -i sample-GRC38.vcf \
  -o fam1.vcf
```

Genes and variants whose segregation is consistent with one of the modes of inheritance are marked with the corresponding annotation in a new field in the **INFO** column. In this case, two compound heterozygous variants in the *CDHR2* gene and one homozygous variant in the *GML* gene receive the annotation **INHERITANCE=AR**. Genes with no variant(s) that are compatible with one of the modes of inheritance do not receive an annotation (see Table 20.2).

20.3.1.3 *Autosomal dominant*

Jannovar's autosomal dominant filter is simple for single-sample VCF files with one affected individual. Any HET variant is assessed as being compatible with autosomal dominant inheritance and will be marked with **INHERITANCE=AD**.

If there is more than one person in the pedigree then there must be at least one compatible call, meaning that all of the following three criteria must be met

1. At least one affected person has a HET call for this variant.

---

[5]The VCF file **sample-GRC38.vcf** and the PED files used for this and the following examples can be downloaded from the book's website.)

Table 20.2. The *CDHR2* is compatible with autosomal recessive inheritance.

Gene	Variant	I.1	I.2	II.1	II.2	II.3	Compatible?
CDHR2	p.V424A	ref	het	het	ref	het	Y
CDHR2	p.M465L	het	ref	het	het	het	Y
WWC1	p.R250C	ref	het	het	ref	het	N
WWC1	p.E865del	ref	het	het	ref	het	N
HK3	p.Q156H	ref	het	het	het	het	N
HK3	p.Q134R	het	ref	het	het	het	N

Note: The VCF file has two variants in a gene that are compatible with autosomal recessive inheritance because they are compound heterozygous in both affected children (II.1 and II.3), and the unaffected child (II.2) is heterozygous for only one of the variants. Each parent is heterozygous for a different variant. The *WWC1* gene is not compatible with `AR` inheritance, because the affected children have inherited both variants from the mother (I.2). Finally, the *HK3* gene has two variants that are found not to be compatible with autosomal recessive inheritance because the unaffected child is compound heterozygous. The column `Compatible?` contains a 'Y' if the gene is compatible with `AR` inheritance and an 'N' otherwise.

2. No affected person has a HOM-REF or HOM-ALT call.

3. No unaffected person has a HET or HOM-ALT call.

We note that very rarely, homozygous mutations are seen with autosomal recessive diseases (e.g., [21]). Often, but not always, the affected individuals show more severe clinical findings. In these cases, the inheritance pattern resembles autosomal recessive inheritance, and this is what Jannovar would report. It is good to be aware of exceptions like this, but in practice they will be rarely encountered.

To perform pedigree analysis with the PED file in Figure 20.6 (fam2.ped), we run Jannovar as follows:

```
$ java -jar Jannovar.jar annotate-vcf \
  -d data/hg38_refseq.ser \
  -i sample-GRC38.vcf \
  --pedigree-file fam2.ped \
  -o fam2.vcf
```

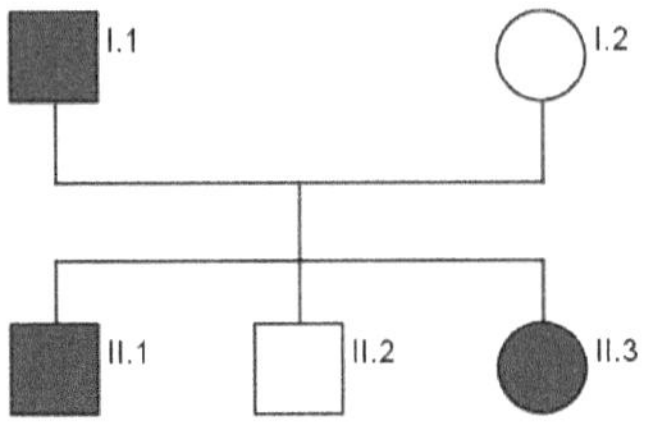

FAM2	I.1	0	0	1	2
FAM2	I.2	0	0	2	1
FAM2	II.1	I.1	I.2	1	2
FAM2	II.2	I.1	I.2	1	1
FAM2	II.3	I.1	I.2	2	2

Figure 20.6. An example pedigree representing a family segregating an autosomal recessive disease. The family corresponds to the `fam2.ped` pedigree file used with Jannovar to perform a pedigree analysis on an autosomal dominant pedigree.

The *FGFR4* and the *WWC1* genes are identified as candidates because they each have a single variant that is heterozygous in all affected persons (`p.P136L` in the case of *FGFR4*, and `p.R49C` in the case of *WWC1*). There is another variant in the *ZC3H3* gene (`p.F149Y`) that is not identified as compatible with autosomal dominant because the unaffected child II.2 is also heterozygous (Table 20.3).

Table 20.3. The *FGFR4* and the *WWC1* genes but not the *ZC3H3* gene are compatible with autosomal dominant inheritance.

Gene	Variant	I.1	I.2	II.1	II.2	II.3	Compatible?
WWC1	p.Arg250Cys	het	ref	het	ref	het	Y
FGFR4	p.Pro136Leu	het	ref	het	ref	het	Y
ZC3H3	p.F149Y	het	ref	het	het	het	N

Note: Even though all affected persons are heterozygous for the variant `p.F149Y`, the unaffected child II.2 is also heterozygous.

20.3.1.4 X-chromosomal recessive

Jannovar again uses a system of rules to determine whether the variant in question is compatible with X-chromosomal recessive (XR) inheritance. Only variants on the X chromosome are considered. With XR, the most common situation is that affected males are born to healthy mothers. In this case, if a single sample from the affected male is sequenced, we expect to see a homozygous variant. (Males have only one X chromosome, and a mutation on a male X-chromosomal gene is called hemizygous. However, current variant callers do not distinguish between sex chromosomes and autosomes and such mutations are usually called as homozygous.) Jannovar therefore will flag any variant called heterozygous or homozygous as compatible with XR if a single affected male is sequenced. For completeness, homozygous or compound heterozygous variants in X-chromosomal genes will be called XR compatible if only one sample from an affected female is sequenced.

Small pedigrees can also be analyzed for potential XR inheritance. In this case, affected males must have HET or HOM-ALT calls.[6] Affected persons cannot be HOM-REF (but can have incomplete data). Unaffected mothers of affected males can be either HET or HOM-REF (in the latter case, the pedigree would be compatible with a *de novo* mutation in the affected male). The father cannot have the variant. Some additional checks are performed to cover rarely observed situations that are described in detail in the online documentation.[7]

To perform pedigree analysis with the PED file in Figure 20.7 (fam3.ped), we run Jannovar as follows:

```
$ java -jar Jannovar.jar annotate-vcf \
  -d data/hg38_refseq.ser \
  -i sample-GRC38.vcf \
  --pedigree-file fam3.ped \
  -o fam3.vcf
```

None of the variants described in the previous sections are identified as compatible with X chromosomal inheritance, simply because the genes are not located on the X chromosome. The following variant is identified as compatible with X-chromosomal inheritance (Table 20.4).

Finally, a variant in the *MAGEB3* gene (p.I112T) is correctly not called as compatible with X-chromosomal inheritance because the son II.2 has the reference sequence, and another variant in the *FAM47A*

---

[6]HET calls are allowed, even though males only have one X chromosome, so as not to rule out candidates called heterozygous as the result of a sequencing error.

[7]http://jannovar.readthedocs.io/

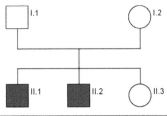

FAM3	I.1	0	0	1	1
FAM3	I.2	0	0	2	1
FAM3	II.1	I.1	I.2	1	2
FAM3	II.2	I.1	I.2	1	2
FAM3	II.3	I.1	I.2	2	1

Figure 20.7. X-chromosomal inheritance. An example pedigree representing a family segregating an X-chromosomal recessive disease. The family corresponds to the `fam3.ped` pedigree file used with Jannovar to perform a pedigree analysis on an X-chromosomal recessive pedigree.

Table 20.4. The variant in the *TCEANC* gene is compatible with X-chromosomal recessive inheritance.

Gene	Variant	I.1	I.2	II.1	II.2	II.3
TCEANC	p.W351L	ref	het	hom	hom	ref

Note: The mother is heterozygous and all affected sons are called as homozygous (actually, they are hemizygous, but this is how variant callers typically report hemizygous variants on the male X chromosome).

gene (`p.E507Q`; `chrX:g.34148877C>G`) is correctly not called as compatible with X-chromosomal inheritance because both the father and the mother are heterozygous for the variant and the healthy daughter is homozygous for it.

20.4 LINKAGE ANALYSIS

Rule-based pedigree analysis suffices to perform simple pedigree filtering of WES/WGS data in nuclear families. Currently, WES studies are commonly run as trios (parents and affected child) or with samples from the parents and several affected and unaffected children. Especially for larger pedigrees, classical linkage analysis is more powerful because it incorporates genetic map and allele frequency information and permits variable penetrance, non-parametric analysis and formal haplotype inference. If large pedigrees are available, linkage analysis can be performed prior to exome sequencing, and the linkage interval can be used to filter the exome results. Alternatively, linkage can be performed using SNP genotypes extracted from exome data [405]. In other cases, homozygosity mapping can be used to filter exome data if consanguinity is suspected [273, 332]. We have also published an approach that uses Hidden Markov Model inference of regions of identity-by-descent(IBD)=2 that can be used to identify regions compatible with autosomal recessive inheritance with unrelated parents [226, 376].

FURTHER READING

Ott J, Wang J, Leal SM (2015) Genetic linkage analysis in the age of whole-genome sequencing. *Nat Rev Genet* **16**:275–84.

20.5 EXERCISES

For the following exercises, download the files `sample-GRCh38.vcf`, `fam1.ped`, `fam2.ped`, and `fam3.ped` from the book's website.

Exercise 1

Perform pedigree analysis for the `sample-GRCh38.vcf` using the PED file `fam1.ped`, which represents the pedigree shown in Figure 20.3 (autosomal recessive inheritance). Explain why the variants found in the *FGFR4* gene do not segregate with the disease if we assume autosomal recessive inheritance in this family. Also analyze the segregation of the variant in the *GML* gene.

Exercise 2

Write a PED file for a family in which the parents are both unaffected. There are four children, a son, two daughters, and a son (in order of birth). The first son and the second daughter are affected by a disease. Also draw the corresponding pedigree.

Exercise 3

For this exercise, use the `trio.vcf` file as described on page 250 (or another trio VCF file of your choice). Now create a PED file in which both the father and the child are affected and annotate the VCF file using Jannovar and this PED file. Create a second PED file in which only the child is affected and perform Jannovar analysis. Examine the list of candidates for `AR` and `AD` in both analyses. Are the candidate lists different? If so, why? Try to understand the reasons for the pedigree filter results based on the genotypes and the PED files.

CHAPTER 21

Intersection Analysis and Rare Variant Association Studies

M ANY early WES studies examined cohorts of patients with the same diagnosis in order to search for the causative gene. As we will see in the following chapters, each individual harbors a relatively large number of rare, predicted pathogenic[1] variants that are not related to a Mendelian disease. Therefore, if we "prioritize" an individual WES dataset solely on the basis of rarity and predicted pathogenicity, typically we will obtain up to many hundreds of "candidates". If we are performing exome sequencing to search for a mutation in a Mendelian disease gene, then at maximum one gene harbors disease-causing mutations, and predicted pathogenic variants in other genes can be regarded as false-positives. If unrelated patients are sequenced, then these false-positives should be independent of one another, and we should not see more overlap than expected by pure chance.

---

[1]The word "pathogenic" means disease-causing, whereas the term "deleterious" is more general and implies that a variant has some negative effect on protein function (which only in some cases can lead to disease). The phrase "pathogenicity prediction" has become standard in the literature, and we will use it here instead of the more correct "deleteriousness prediction", but it is important to note that any WES/WGS sample will have numerous variants predicted to be pathogenic that do not cause disease. Therefore, the term "pathogenic" is not always intended to be equivalent to "disease-causing".

21.1 HOW MANY CANDIDATES ARE OBSERVED IN "TYPICAL" EXOMES?

Exome sequencing and the computational analysis of WES data has been a moving target since WES was first used for the analysis of Mendelian disease in 2010 [307]. The kits used to capture exons and other sequences have been updated and improved multiple times, and the databases used to investigate the variants have grown in volume and sophistication. Also, as we have seen in Chapter 12, different variant calling pipelines yield different sets of called variants. Depending on the depth of sequencing, more or less variants may be called in the flanking intronic regions that are "off-target" but still get captured by exonic probes. Therefore, it is not easy to answer the question "How many variants does one expect to see in a typical exome and how many of these variants are novel?" An excellent review article published in 2012 presented a comprehensive review of the literature and came to the conclusion that a typical exome identifies 20,000 to 50,000 variants [134]. Following removal of low-quality variant calls, roughly 5000 variants remain that are located within the coding portions of exons and are predicted to have some effect on the protein.[2]

A key assumption in WES analysis is that a common variant cannot be the cause of a rare disease.[3] According to this assumption, it makes sense to filter out variants whose population frequencies exceeds some threshold (see Chapter 22 for details). This step typically reduces the number of candidate variants by about 90–95%, leaving 150–500 "private" (not seen in any database) or very rare non-synonymous or splice-site variants [134].

Our in-house experience is that one currently sees up to about 100,000 variants (including off-target variants) in a typical exome, and that up to several hundred candidates remain after the above mentioned filtering steps. It is likely that the number of "private" variants

---

[2]We note that the assumption that "synonymous" variants cannot have a deleterious effect on protein function is not true. For instance, some "synonymous" variants affect exonic splicing enhancers and lead to missplicing and disease, e.g., [314]. It is not currently possibly to predict such mutations reliably, and we will not cover this topic further in this book.

[3]This assumption is generally true, but there are exceptions. For instance, TAR syndrome was found to be caused by a rare multi-gene deletion on one haplotype in conjunction with a relatively common SNP in the regulatory region of one of the involved genes on the other haplotype. The minor allele frequencies of the SNPs were estimated to be ~0.4% and ~3% [11].

will go down with time as databases improve, but it should be noted that current databases do not provide even coverage of all population groups, and the available data for individuals of European descent is more comprehensive than for some other population groups.

As the community transitions from exome to genome sequencing, candidate variants in the non-coding parts of the genome will need to be assessed. There is little published experience, but our initial results with the Genomiser software [403] (see also Chapter 27) suggests that current pathogenicity prediction tools will flag several thousand candidates per genome.

21.2 INTERSECTION FILTERING

A simple strategy can be used to search for novel disease-associated genes in cohorts of unrelated individuals with the same disease. The assumption is that each WES contains numerous variants unrelated to the disease being studied that are not removed by the filtering steps described above. Although each individual may have hundreds of candidate variants, the variants that are unrelated to the disease are likely to be distributed "at random" in the in the cohort.[4] If we examine the intersection between a sufficient number of multiple unrelated individuals, only the disease gene itself will display a mutation in all individuals.[5] For autosomal dominant diseases, each candidate gene must show at least one variant per individual, and for autosomal recessive diseases, candidate genes must have either homozygous or compound heterozygous mutations [374].

Obviously, this strategy is highly susceptible to false-negative and false-positive results if applied naively. Some genes such as the mucins are found to have rare, probably non-deleterious variants in almost all individuals. Thus even if many samples are intersected, some such genes may remain in the intersection. On the other hand, if a mutation is located in a poorly covered exon it may escape detection. If several of the mutations among the sequenced individuals are located in poorly covered exons, the candidate gene would falsely be removed from further consideration.

What do we do therefore if most, but not all of the patients in our

---

[4]For this analysis, we will ignore the fact that some genes show a higher frequency of rare variation than others in the population.

[5]For the moment, let's ignore locus heterogeneity, i.e., the fact that some Mendelian diseases are caused by a mutation in one of several disease genes.

cohort are found to have a mutation in a certain gene? One way of estimating the statistical significance in this scenario compares the number of observed "overlaps" with the number expected by chance. Humans have roughly $M=20{,}000$ protein-coding genes that are interrogated by exome sequencing. We find a mean of m rare, predicted pathogenic mutations in each exome, whereby m is a number such as 200–1000 that depends on the exome kit, the average depth of sequencing, and the thresholds, databases, and parameters used to define whether a given variant is rare and predicted pathogenic.

We can model the event that such a variant occurs in a specific gene as a Bernoulli-distributed random variable X_i, where $X_i = 1$ denotes that a rare predicted pathogenic variant is observed in gene i, and $X_i = 0$ denotes that no variant is observed. The probability of the occurrence of a mutation in a specific gene can be estimated as

$$P(X_i = 1) \approx \frac{m}{M} \tag{21.1}$$

This is simply a Bernoulli distribution with parameter $p = m/M$ [486].

Let us apply this approach to data published in a study on Kabuki syndrome, a rare, autosomal dominant disorder, whose causative disease gene was discovered by exome sequencing in 2010. The authors identified mutations in a gene called *MLL2* in 7 of 10 persons with Kabuki syndrome [306].

It is natural to model the chance of finding a certain number of mutations in the *MLL2* gene in the cohort using a Binomial distribution, which is equivalent to the distribution of the sum of n Bernoulli trials [52]. We will denote the total number of identified mutations in gene i as C_i.

The authors of the Kabuki study identified an average of 753 rare, predicted pathogenic mutations in a person with Kabuki, and therefore, we estimate the probability $P(X_i = 1)$ as $m/M = 753/20000$.

The probability of finding *MLL2* mutations in exactly 7 of 10 patients is then

$$
\begin{aligned}
P(C_i = 7) &= Bin(7, 10, p = 753/20000) \\
&= \binom{10}{7}\left(\frac{753}{20000}\right)^7\left(1 - \frac{753}{20000}\right)^3
\end{aligned}
$$

In R, we calculate the result as

```
> p <- 753/20000
> dbinom(7,10,p)
[1] 1.146926e-08
```

In order to calculate the statistical significance of this result, we need to calculate the probability of observing a result that is at least as extreme.

$$P(C_i \geq 7) = \sum_{k=7}^{10} \binom{10}{k} \left(\frac{753}{20000}\right)^k \left(1 - \frac{753}{20000}\right)^{10-k} \qquad (21.2)$$

Given that we also are implicitly testing all 20,000 protein coding genes, we will also perform a correction for multiple-testing by the Bonferroni procedure. We can perform all steps at once in R

```
> sum(dbinom(7:10,10,p))*20000
[1] 0.0002327798
```

The p-value of $p = 0.0002$ is significant, suggesting that the observation of 7 mutations in a single gene in a cohort of ten patients was not due to chance. The authors of the study went on to prove by various means that *MLL2* is the disease gene in Kabuki syndrome.[6]

21.3   RARE VARIANT ASSOCIATION TESTING

In the initial years of exome sequencing, a number of studies were performed using the intersection strategy described above. More recently, it seems that there are fewer large cohorts of patients available for this approach. More sophisticated strategies have been developed to apply more robust statistical approaches. For instance, the Variant Annotation Analysis and Search Tool (VAAST) exploits allele-frequency information and pathogenicity predictions in a likelihood ratio test framework [469], and was used to identify *NAA10* as the disease gene responsible for Ogden syndrome [378]. A number of methodological questions remain, such as the choice of the best control group [490],

---

[6]This simplified statistical evaluation assumes a uniform probability of observing a rare, predicted pathogenic variant in any given gene. In reality this assumption does not hold because the probability depends on factors such as gene length and genic intolerance to functional variation (see Section 24.1).

what criteria to use for pathogenicity prediction, and how many samples are required for statistical power [223]. Currently, much of the human genetic community is converging around the MatchMaker Exchange platform (MME), which is a network of federated databases of genotypes and phenotypes that allows users to search for additional samples by comparing their genotypic and phenotypic profiles [331]. Sophisticated approaches are being developed that use machine-learning to identify "fuzzy" HPO-encoded[7] phenotypic profiles by phenotypic similarity regression [143], an approach that has already supported the identification of several disease genes [247].

FURTHER READING

Zhu N, Heinrich V, Dickhaus T, Hecht J, Robinson PN, Mundlos S, Kamphans T, Krawitz PM (2015) Strategies to improve the performance of rare variant association studies by optimizing the selection of controls. *Bioinformatics* **31**:3577–83.

Krawitz P, Buske O, Zhu N, Brudno M, Robinson PN (2015) The genomic birthday paradox: how much is enough? *Hum Mutat* **36**:989–97.

Zhi D, Chen R (2012) Statistical guidance for experimental design and data analysis of mutation detection in rare monogenic mendelian diseases by exome sequencing. *PLoS One* **7**:e31358.

21.4 EXERCISES

Exercise 1

An exome sequencing study is performed on 7 individuals diagnosed with a certain rare disease. An average of 400 rare, predicted pathogenic variants is found per sample. Determine the p value of finding mutations in a given gene in k individuals, where k ranges from 0 to 7. Which values of k are statistically significant?

---

[7]HPO is the Human Phenotype Ontology that we will discuss in Chapter 26.

CHAPTER 22

Variant Frequency Analysis

D ELETERIOUS variants are expected to have lower allele frequencies than non-deleterious variants because of negative selection. Therefore, allele rarity is often used as a criterion for predicting pathogenicity in exome and genome sequencing studies. Obviously, a common variant is also not expected to be the cause of a rare disease.

Whole genome sequences (WGS) typically contain well over 4 million variants as compared with the genome reference sequence. Up to roughly 10,000 of these variants are predicted to lead to non-synonymous amino acid substitutions (missense mutations), alterations of conserved splice site residues, or represent small insertions or deletions (NS/SS/I). Depending on the ethnic background of the proband and other factors, up to about 90–95% of these variants can be found in databases of common variants such as dbSNP, the 1000 Genomes Project, and in-house databases. Based on the assumption that variants that are common in the population are not likely to be the cause of rare Mendelian diseases, such variants are typically filtered out before further analysis.

In this chapter, we will show how to use Jannovar to annotate the variants in a VCF file according to their frequency using the ExAC database (Section 22.2). At the time of this writing, ExAC was available only for the previous version of the human genome assembly, GRCh37. For simplicity, we have therefore used a VCF file that was generated using data from the GRCh37 build.

In practice, bioinformatics labs often have their own resources for evaluating variants according to frequency that include data from multiple sources such as dbSNP [393], ESP [309], ExAC [381], Kaviar [136], as well as local, "in-house" databases—often, local populations display variants common in that population that are not yet in global databases. Therefore, the analysis presented in this chapter is not sufficient in itself for the analysis of variant frequencies in WES/WGS data, but it should help to explain the main concepts and goals of the analysis.

We will begin with a brief review of some of the current resources for variant frequency.

22.1 NCBI'S DATABASE FOR SHORT GENETIC VARIATIONS

A key aspect of research in genetics is associating sequence variations with heritable phenotypes. The most common variations are single nucleotide polymorphisms (SNPs), which occur approximately once every 100 to 300 bases. dbSNP[1] is NCBI's variation database, which serves as a central repository for both single base nucleotide substitutions and short deletion and insertion polymorphisms. As of build 149 (available November 2016), dbSNP had amassed over 582 million submissions representing more than 174 million distinct polymorphisms for 18 organisms, including human, mouse, rat, and bee.[2]

A variety of queries can be used for searching: a refSNP number ID ("rs"), a gene name, an experimental method, a population class, a population detail, a publication, a marker, an allele, a chromosome, a base position, a heterozygosity range, a build number, or a strain. Searches return refSNP number IDs that match the query term and a summary of the available information for that refSNP cluster.

22.2 EXAC

The Exome Aggregation Consortium (ExAC) has collected exome sequencing data from a wide variety of large-scale sequencing projects. Currently (2017, version 0.3.1) the dataset made available by ExAC comprises 60,706 unrelated individuals sequenced as part of various disease-specific and population genetic studies [246].[3] This substan-

---

[1]http://www.ncbi.nlm.nih.gov/snp/

[2]However, starting from September 2017, NCBI will phase out support for non-human organism data in dbSNP and dbVar.

[3]Recently, an extended version of ExAC called the Genome Aggregation Database (gnomAD) was released at http://gnomad.broadinstitute.org/.

tially extends previous efforts such as the NHLBI GO Exome Sequencing Project (ESP), which collected exome data on 6,515 individuals [127].[4]

ExAC data has proven to be extremely useful for estimating the population frequency of variants found in exome studies. Filtering on ExAC reduces the number of candidate protein-altering variants and is most powerful when the highest allele frequency in any one population ('popmax') is used rather than the average ('global') allele frequency. At a 0.1% allele frequency filter, filtering with the ExAC data left an average of 154 variants for analysis, compared to 1090 after filtering against ESP [246]. The variants in typical exomes and genomes display a wide range of allele frequencies (Figure 22.1).

22.3 USING JANNOVAR TO ANNOTATE VARIANTS WITH THEIR POPULATION FREQUENCY

Recent extensions of Jannovar enable it to use the ExAC data to annotate VCF files. First download the current version of the ExAC datafile from the ExAC website:[5]

```
ExAC.r0.3.1.sites.vep.vcf.gz
```

The reference genome to which the ExAC file refers (currently, ExAC data is only available for the GRCh37 build) needs to be passed to the program because normalization of variants (see Section 13.8) is performed on the fly. Download the GRCh37 genome file (`hs37d5.fa`) as described in Chapter 8. We will use a GRCh37-mapped VCF file from the Corpasome (see Chapter 4) but any VCF file mapped to GRCh37 can be used.

```
$ java -jar Jannovar.jar annotate-vcf \
  -d data/hg19_refseq.ser \
  -i Sons_VCF_file.vcf \
  -o Corpas_exac.vcf.gz \
  --exac-vcf ExAC.r0.3.1.sites.vep.vcf.gz \
  --ref-fasta hs37d5.fa
```

This command produces a compressed, annotated VCF file. In addition to the annotations described in Chapter 15, the variants are

---

[4]http://evs.gs.washington.edu/EVS

[5]http://exac.broadinstitute.org

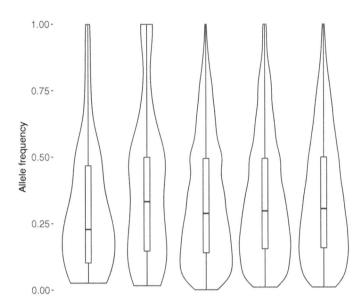

Figure 22.1. ExAC allele frequencies. Violin plots of the allele frequency distribution (`EXAC_AF_ALL`) associated with five variant categories observed in the Corpas exome are shown. The median frequency and a box with the interquartile range are shown, together with the density of data at each frequency.

annotated with the population frequencies and allele counts found in the ExAC dataset. For instance, the variant `chr1:1021346A>G` (`rs10907177`) is annotated with

`EXAC_AC_AFR=2542;EXAC_AC_ALL=20408;EXAC_AC_AMR=1013;(...)`

Jannovar uses abbreviations defined by ExAC but prepends them with `EXAC_`. The full list, together with explanations, is placed in the header section of the annotated VCF file, and further details are available at the ExAC browser. For instance, a variant might be annotated with `EXAC_AC_FIN=936` (936 allele counts in the Finnish samples), and `EXAC_AF_FIN=0.143` refers to the allele frequency with respect to all observed alleles at this position in the Finnish samples.

22.4 HOW TO PERFORM FREQUENCY FILTRATION

As a general rule, a frequency cutoff of 0.1% is chosen, and all variants with a population frequency above that are removed ("filtered out") from further consideration. It seems reasonable to take the maximum frequency observed in any population under the assumption that a variant that is relatively common in some population is unlikely to be causative for a rare disease. While this may be reasonable in some situations, it can be too stringent for autosomal recessive conditions where the carrier frequency of some disease-causing variants may be higher than 0.1% (especially in isolated populations [73, 118]). Therefore, for suspected autosomal recessive diseases, it may be prudent to increase the frequency threshold to 0.5% or even higher. As always, there is a trade off between noise and the risk of false-negative results because of overzealous filtering. The Exomiser (Chapter 27) and other software packages allow variants to be filtered at different levels of stringency, and users may want to repeat the analysis with different cutoffs if no obvious candidates can be found using the standard 0.1% cutoff.

22.5 GENOME DATA

At the time of this writing, the ExAC group has just initiated a Genome Aggregation Database (gnomAD).[6] The Kaviar database [136] is currently the largest database available for WGS frequency data.

FURTHER READING

Lek M, ..., MacArthur DG; Exome Aggregation Consortium (2016) Analysis of protein-coding genetic variation in 60,706 humans. *Nature.* **536**:285–91

22.6 EXERCISES

Exercise 1

In Exercises 1–7 we will get familiar with dbSNP. Go to the dbSNP website[7] and search for the entry `rs25458`. Use the entry to determine where the SNP is located. What chromosome? What chromosome position? Is this SNP located within a gene? If so, what is the name and gene symbol of the gene? The SNP is reported here in the opposite

---

[6]http://gnomad.broadinstitute.org/
[7]http://www.ncbi.nlm.nih.gov/snp

orientation to the genomic strand. Why? Could this have something to do with the orientation of the gene? How many nucleotides is `rs25458` distant from the next nearest SNP?

Exercise 2

Learn how to assess the quality and depth of the evidence for the existence of the SNP represented by an individual dbSNP entry. Things to look out for are (i) whether the SNP was validated by multiple, independent submissions; (ii) whether all alleles (including especially the minor allele) were observed in at least two chromosomes; (iii) whether the SNP was explicitly validated by submitter confirmation; (iv) whether the SNP was validated by a large-scale project such as HapMap or the 1000 Genomes Project; and (v) whether the SNP has been annotated as suspect, for instance because it stems from a region that has a paralogous sequence elsewhere in the genome. Coming back now to the SNP from Exercise 1: What evidence is there for the existence of `rs25458`? Do you believe it truly exists or could it be an artifact?

Exercise 3

For exome sequencing in medical genetics, we are most interested in filtering out high-frequency variants. Many, but not all, of the entries in dbSNP were detected in the course of the 1000 Genomes Project, and it is instructive to go to the 1000 Genomes browser and explore the data. Follow the link on the dbSNP website to the 1000 Genomes entry for `rs25458` (Hover over the symbol of a magnifying glass in the `Integrated Maps` section of the dbSNP webpage.[8] Now have a look at the various populations that were investigated in the 1000 Genomes project. In what population is the frequency of `rs25458` highest? What about lowest? This variation of allele frequency according to population is typical of many SNPs.

Exercise 4

Now go to the dbSNP section entitled **Gene View** and click on the **GO** button. You should see a long list of variants that are located within the coding sequence of the *FBN1* gene, some denoted as likely pathogenic,

---

[8]this function currently seems to work correctly only for the GRCh37 reference genome

some as likely benign, etc. Estimate what proportion of variants in the list are likely pathogenic. It is typical of many genes that numerous presumed neutral variants are common in the population. Also look at the columns heterozygosity and minor allele frequency.[9] Note that this information is available only for some of the variants in dbSNP. That is, it is still difficult to get comprehensive information about the population frequency of many variants in the human population.

Exercise 5

Submitter SNPs (ss), with alleles and flanking sequences on which a reference SNP is based, are summarized in a table below the Sequence Viewer display. The ssIDs link to submitter records providing additional details. How many sequences support `rs25458`?

Exercise 6

Take a look at the bottom of the page to see the frequency of the alleles of `rs25458` in various populations.

1. What was Craig Venter's genotype at `rs25458`?

2. Compare the genotype and allele frequencies for `rs25458` in the `AFD_AFR_PANEL`. Are they more or less what you would expect given Hardy–Weinberg equilibrium?[10] What about `AFD_AFR_PANEL`, is there a closer match?

3. What is the mean overall heterozygosity?[11] Is this what you would expect for a mutation in a genetic (Mendelian) disease?

Exercise 7

Initially, dbSNP was intended to contain neutral population variation, but it now contains ever more disease-related mutations. Search for

---

[9]Minor allele frequency (MAF) refers to the frequency at which the least common second most frequent allele occurs in a given population. For a SNP with only two alleles this is the least common allele. When there are more than two alleles, MAF uses the second most frequent allele because it is a better indicator of how frequent alternative alleles can be found in the population.

[10]At Hardy–Weinberg equilibrium, the allele frequencies $f(A)$ and $f(a)$ of two alleles A and a are directly related to the genotype frequencies: $f(AA) = f(A)^2$, $f(Aa) = 2 \cdot f(A) \cdot f(a)$, and $f(aa) = f(a)^2$.

[11]This can be seen near the bottom of the page below the `population diversity` table.

rs121908912. What disease is this variant associated with? There is a substantial difference in the amount of information available about the population frequency of **rs121908912** and of **rs25458**. Why do you think that is?

See hints and answers on page 501.

CHAPTER 23

Variant Pathogenicity Prediction

Dominik Seelow and Peter Robinson

A N exome typically harbors tens of thousands of DNA variants, among them are thousands of missense variants and hundreds of nonsense, splice-site, and other probably loss of function variants. Typically, hundreds of these variants are rare or even "private".[1] While we can at least hazard a guess that a stop gained ("nonsense") variant has a potential for being pathogenic, it is much harder for missense variants. The consequences of non-coding variants are even more difficult to predict.

The existence of different transcripts poses another obstacle for a precise calculation of a variant's effect. Some DNA variants may have severe consequences in one transcript of a gene (such as a frameshift in the case of an indel variant) but none in other transcripts, in which they might be located outside the coding region. In some cases, a complete loss-of-function mutation in one transcript may be tolerated because other, unaffected transcripts can compensate for the loss. If, however, the other transcripts are not abundant in certain cell types or tissues or lack crucial functional domains, transcript-specific mutations can be disease-causing.

---

[1] A variant is said to be "private" if it is found only in a proband and the immediate relatives of the proband, but not in variant databases. Obviously, as variant data accrues, variants that were once classified as private may be found in other unrelated individuals.

ClinVar is an archive of interpretations of variants in medically relevant genes [235]. As of January 2017, 45,478 of 154,612 variants in the ClinVar database were listed as of uncertain significance (VUS, variant of unknown significance), corresponding to 29.4% of all variants. A further 27,880 variants (18%) were listed as "likely" benign or pathogenic, meaning that their clinical significance was not entirely clear. A total 9746 variants (6.3%) had conflicting interpretations. Thus, we are able to make a confident interpretation of only less than half of the variants in ClinVar, and the proportion of variants for which we can make a reliable, evidence-based interpretation is substantially less in WES or WGS data (Figure 23.1). This means that it is at present impossible to perform a database lookup to evaluate the medical relevance of many of the variants found in WES/WGS data. There is thus a need for computational approaches to predict the clinical relevance (pathogenicity) of variants.

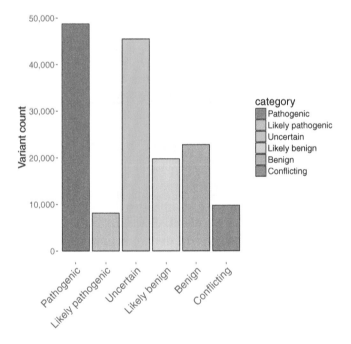

Figure 23.1. Distribution of variant interpretation categories at ClinVar.

In this chapter, we will concentrate on the computational pathogenicity prediction for variants within transcripts of protein-coding genes. We will first present some of the main features that many of the programs use for their prediction [424].

23.1 CRITERIA FOR ASSESSING THE DELETERIOUSNESS OF A VARIANT

Pathogenicity prediction exploits certain attributes of the affected genes, transcripts, transcript positions, and the sequence context of the nucleotide and amino-acid alterations associated with variants in order to assess whether a given variant is disease-causing (pathogenic) or not. For most programs, the predictions are best interpreted as attempting to predict not disease causality but deleteriousness of the variant for protein function: a variant may completely abolish the function of a gene that is simply not associated with any disease.[2] A good example is the *ABO* gene, in which homozygous loss of function variants, although being deleterious at a biochemical level, do not cause any disease but instead are associated with blood type O. We will present some background information about some of the most important attributes of variants that are used for pathogenicity prediction, and will then present one approach, MutationTaster, in detail.

23.1.1 Effects of variants on the protein

Variants that affect protein-coding genes are mainly assessed with reference to their predicted effects on the amino acid sequence, structure, or function of the protein or in some cases with respect to potential changes in the amount of protein produced. A thorough understanding of the amino acids is therefore essential for medical bioinformatics. The most important characteristics of amino acids are **hydrophobicity**, **size**, and **charge** [267], but there are several other important properties, that together define partially overlapping groups of amino acids. Amino acids do not fall neatly into classes — the amino acids have many different combinations of small/large, charged/uncharged, and polar/nonpolar properties (Figure 23.2).

The following list indicates the most important properties of amino acid sequences that are taken into account in pathogenicity assessment.

---

[2]Despite this, the phrase "pathogenicity prediction" is standard in the literature, and we will use it throughout this book.

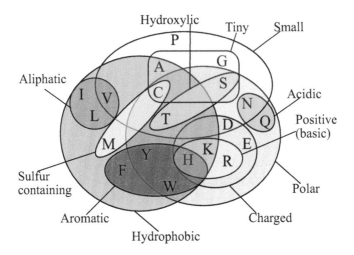

Figure 23.2. Physicochemical properties of the amino acids.

Chemical similarity of substituted amino acids.

One should consider whether an observed amino acid substitution is likely to change the function of the protein. For instance, glycine is the smallest and most flexible of all amino acids, while proline is not only much larger but also the least flexible amino acid because of its reduced rotational freedom. Intuitively, one would expect that a substitution of glycine by proline would substantially alter the structure of the affected protein chain. On the other hand, a substitution of leucine by isoleucine might be expected to have less of an effect, because they are both aliphatic, branched hydrophobic amino acids. Therefore, early attempts to understand the effects of missense variants on proteins often focused on pairwise comparisons of the physicochemical characteristics between the wild-type and variant amino acids. From an initial perusal of the properties of the amino acids, it may not be immediately clear how to calculate a numerical similarity between two different amino acids. The Grantham score attempts to capture the overall distance between any pair of amino acids based on differences in side chain atomic composition, polarity, and volume between two amino acids [141], but changes of the Grantham score do not correspond well with the effect on protein function or the disease potential of such a substitution [202].

Conservation of substituted amino acids.

This involves the comparison between wildtype and the variant amino acid and the evolutionarily tolerated amino acid range of variation at its position in the protein by applying conservation matrices such as BLOSUM. The BLOSUM matrices were derived from the frequencies with which the 20 amino acids are observed to substitute for each other in multiple sequence alignments of related proteins in different species. As with the Grantham score, the average BLOSUM62 score is often lower for pathogenic changes than for neutral replacements. However, the comparison of the substitutions found in disease mutations with those found in polymorphisms reveals that there is no clear correlation between BLOSUM62 scores and functional tolerance. A likely explanation is that some substitutions can have a drastic effect on the protein when they occur in certain domains (e.g., in structural domains such as α helices) but are tolerated elsewhere.

Evolutionary conservation of the sequence context.

A high level of conservation of amino acids surrounding the variant site may indicate a functional relevance of the protein site.

Protein structural considerations.

Disease-causing mutations often affect a critical amino acid residue of the protein in a way that changes its function. If, for instance, a glycine located in an α helix is replaced by another amino acid, the helix might be disrupted. Substitutions of amino acids by proline residues may also prove deleterious to protein function because of its conformational rigidity. A clinical example is the *KRAS* oncogene which is mutated in numerous kinds of cancer. KRAS (as well as the related proteins HRAS and NRAS) are GTPases that function as molecular switches—they are active if bound to GTP and inactive if bound to GDP, and possess an intrinsic GTPase that converts bound GTP to GDP and thereby switches off signaling. Mutations in KRAS can favor GTP binding and lead to constitutive activation with consequent increased cell proliferation. Knowledge of this fact allows us to better interpret mutations located at particular structures of the protein (Figure 23.3).

Protein domains.

The substitution of amino acids within functional or structural protein domains may have more drastic consequences than else-

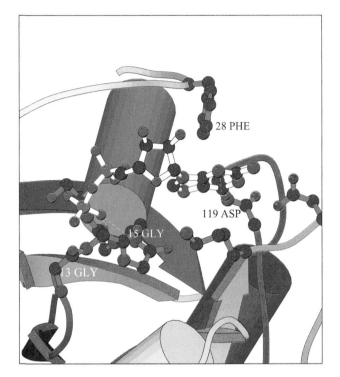

Figure 23.3. Positions of selected oncogenic mutations in KRAS. p.Gly13Ser lies within the GTP nucleotide binding domain of the KRAS protein and inhibits GTPase activity of KRAS [295]. p.Gly15Ser also lies within the GTP-binding domain, and has found to be mutated in sequencing studies. p.Asp119Asn, another mutation that lies within the GTP-binding region, decreases GTPase activity of KRAS resulting in increased activation of the downstream MAPK signaling pathway [201]. p.Phe28Leu lies within the switch II region of the KRAS protein and confers a loss of function. The residues surround the GDP moiety (the two phosphate groups with the purple P atoms as well as the ribose and guanine moieties can be recognized). In general, location of a residue affected by a variant close to an active/functional site of a protein is a hint (but by no means proof) that the variant may be deleterious. Information about the mutations was retrieved from the JAX Clinical Knowledge Base [188]. The figure was generated with MolScript [221] by modifying a script available at the MolScript Website.

where in the protein. If, for example, a cysteine which is part of a disulfide bond, is replaced by another amino acid, the disulfide bond will be lost. For instance, many missense mutations in Marfan syndrome affect cysteine residues in one of the 43 cbEGF domains of fibrillin-1 and thereby disrupt the structure of the protein. Each cbEGF domain has six cysteine residues which form three disulfide bridges which are important for the structure of cbEGF domains. All missense variants affecting cysteine residues in cbEGF domains in fibrillin-1 that have been described to date are pathogenic, and therefore it is plausible that this category of variant is pathogenic in general.

Premature termination codon variants.
Nonsense variants and other variants such as frameshifts associated with premature termination codons (PTC) either lead to a truncated protein or to nonsense-mediated decay (NMD). NMD occurs when a PTC is located at least 50 bases upstream of the last exon/intron boundary in the coding sequence, and lowers the concentration of the variant mRNA.

Shifted start ATGs.
Mutations of the wildtype start codon or in the Kozak consensus sequence that surrounds the start codon can cause the translational machinery to start from another codon; if this codon is out of frame, the mutation will lead an early frameshift. In the latter case, no functional protein will be produced at all.

Substitution frequencies.
Similar to calculating evolutionary tolerated amino acid substitutions, it is also possible to compare the substitution frequencies observed in common polymorphisms and disease mutations. These frequencies often vary drastically from those obtained by evolutionary comparisons.

23.1.2 Effect of variants on DNA and RNA

Splice site analysis.
The loss or substitution of bases located within or nearby a splice site may abolish the splice site, thereby leading to effects such as exon skipping. This class of mutation often not only removes crucial parts of the protein but can also cause a frameshift, usually resulting in a premature stop codon and nonsense-mediated de-

cay of the mRNA. DNA variants, either intronic or exonic, can also introduce new, "cryptic" splice sites.

DNA conservation.

A cross-species comparison of the DNA motifs at a given site often reveals a higher degree of conservation than might be expected, even though single bases may be subject to changes. This indicates a functional relevance of the site. Such data is provided by scores such as phastCons [397], phyloP [340], or GERPP++ [97].

Functional elements.

Nucleotide substitutions in functional elements such as polyadenylation signals, transcription start sites, or Kozak consensus sequences may alter or abolish their function and lead to disease.

We will take MutationTaster as an example to further explore different tests which provide a means to assess the deleteriousness of a variant, because it includes more tests than many other variant prediction tools and has often been shown to outperform other tools in terms of accuracy [202, 251, 280].

23.2 MUTATIONTASTER

MutationTaster was the first pathogenicity prediction tool able to handle more than only missense variants. In contrast to tools such as PolyPhen [6], SIFT [305], or PROVEAN [71], MutationTaster is not restricted to analysis at the protein level but also includes tests on the DNA level such as splice site analysis or multi-species conservation of the genomic sequence. It can therefore analyze non-synonymous as well as silent and non-coding intragenic variants. In addition to SNVs, it predicts the disease potential of short indels. Another difference is that MutationTaster displays all test results in a detailed output, thereby allowing researchers to draw their own conclusions about the variant's disease potential.

The software is explicitly transcript-based and considers the effect of a variant on each affected protein-coding transcript. Variants can be entered by their position within the coding sequence, the mature mRNA sequence or within the gene. Alternatively, variants can be entered within a sequence snippet as obtained by Sanger sequencing,[3] or

---

[3]For instance, the snippet `ACTGTC[A/T]GTGTF` indicates an A→T substitution

by chromosomal position. In the latter case, all protein-coding transcripts affected by the variant will be analyzed. Another option is the upload of complete VCF files (see Chapter 13) containing all genotypes from a deep sequencing project. This will lead to the automatic analysis of all variants included in the file in a highly parallel fashion — analysis of a complete WES or sample including all polymorphisms and all non-coding variants takes less than 20 minutes.

MutationTaster performs many different tests on each variant, whereby the actual analyses chosen depend on the nature of the variant. Tables 23.1, 23.2, and 23.3 illustrate the tests.

All test results are used as input for a Naive Bayes classifier which has been trained on all intragenic disease mutations from HGMD Pro 2013 [409], excluding variants also found in polymorphism databases in homozygous state and several million variants from the 1000 Genomes Project [1] which were found in at least 20 healthy individuals in homozygous state. MutationTaster discriminates between missense variants, more complex changes in the amino acid sequence, and synonymous or non-coding variants. As each type of variant requires different tests (e.g. amino acid conservation cannot be tested in synonymous variants), three different classifiers adapted to the specific tests are employed.

MutationTaster does not include all tests suggested in the beginning of this chapter. Instead of considering Grantham or BLOSUM scores to evaluate the deleteriousness of missense variants, MutationTaster evaluates the actual amino acid substitution(s) with a classifier trained with all available exchanges found in harmless polymorphisms and known disease mutations. These intolerance scores do not correspond well with either Grantham or BLOSUM scores.

The software makes use of the available knowledge on common polymorphisms and disease mutations: MutationTaster integrates data from the 1000 Genomes Project, ExAC, and ClinVar to automatically annotate polymorphisms or confirmed disease mutations as such. Polymorphisms are considered benign if they occur in homozygous state in at least 5 healthy individuals; for the study of dominant traits, the software also provides an option to filter for allele carriers while uploading the VCF file.

---

within the given sequence context. Together with a transcript ID, sequence snippets can uniquely identify the position of a variant.

Table 23.1. MutationTaster: Tests for all variants (on DNA level).

Test	Source	Details and Result
Known polymorphism	ExAC [246]	**Details**: homozygous in 40+ individuals **Result**: automatic classification as 'polymorphism'
Known polymorphism	1000G[1]	**Details**: homozygous in 5+ individuals **Result**: automatic classification as 'polymorphism'
Known disease mutation	ClinVar[235]	**Details**: annotated as 'pathogenic' **Result**: automatic classification as 'disease mutation'
DNA conservation	phastCons[397]	**Details**: determines conservation of the affected and the flanking bases **Result**: score
DNA conservation	phyloP[340]	**Details**: determines conservation of the affected and the flanking bases **Result**: score
Splice site loss	MutationTaster	**Details**: variants located directly within a splice site are assessed as abolishing the splice site **Result**: score
Potential splice site loss	nnsplice[359]	**Details**: variants located close to a splice site annotated for this transcript are considered to decrease the splice site likelihood of the variant compared to wildtype **Result**: score
Potential splice site gain	nnsplice[359]	**Details**: splice site likelihood of the variant and wildtype are compared to determine splice site gains **Result**: score
Regulatory DNA elements	ENCODE[115], JASPAR[284]	**Details**: tests whether a variant is located within a regulatory element **Result**: score

Table 23.2. MutationTaster: Special tests for variants within the coding sequence.

Test	Source	Details
Conservation	bl2seq [423]	Conservation of variant/lost amino acid(s) in 10 species (by blasting non-human and mutated sequences against reference protein isoform sequence
Protein domains	UniProt/ Swiss-Prot	Potential disruption of protein domain(s) by modified/lost amino acids within the domain or upstream changes of splice sites
Shifted start ATG		Test whether a variant destroys the start ATG
Frameshift		Test whether an indel or a shifted start ATG causes a frameshift
Loss of Kozak consensus sequence		See in Table 23.3
Nonsense mutations		Finds premature stop codons introduced by DNA variants (including indels) and considers the differences between nuclear and mitochondrial genetic code
Nonsense-mediated decay		Checks if a premature stop codon is located at least 50 bases 5' to the last coding intron/exon boundary and might therefore lead to nonsense-mediated decay
Stop codon lost		Test whether a stop codon is lost, resulting in an elongated protein
Tolerance of substitution		Compares the frequency of tolerated substitutions (from polymorphisms) vs. deleterious substitutions (from disease mutations)

Table 23.3. MutationTaster: Special tests for variants within the UTRs.

Test	Source	Details
Loss of poly-adenylation signal	polyadq [416]	Compares strength of polyadenylation signal in wildtype and variant sequence
Loss of Kozak consensus sequence		MutationTaster checks whether the consensus sequence is weakened by the variant

23.3 OTHER PATHOGENICITY PREDICTION PROGRAMS

PolyPhen is probably the most prominent representative of the class of variant pathogenicity prediction programs. Its name stands for "Polymorphism Phenotyping". The program was first published around the turn of the millennium and has seen some major improvements since then, leading to the current version PolyPhen-2 published in 2010 [6]. PolyPhen focuses on amino acid conservation and potential changes in the three-dimensional structure of a protein to predict the deleteriousness of a variant. The prediction is generated by a Naive Bayes classifier.

Several tools do not run their own tests but instead combine the scores obtained by different prediction programs, e.g., CONDEL [137], KGGSeq [258], CADD [197], and IMHOTEP [202]. Most of them use their own training data set to assign weights to the scores made by the different prediction programs, reflecting both the scoring scheme (scale or Boolean) and the accuracy of the program in predicting the effect of variants on which the original programs disagree. While such an approach has the potential to compensate wrong predictions, it also bears the risk of mixing predictions for different data: The purely amino-acid based-tools such as PolyPhen-2 or SIFT have one prediction for each amino acid substitution while a DNA-based tool such as Mutation-Taster gives many different predictions because this substitution may be the result of different base exchanges. The problem increases when different transcripts are involved, which is the usual situation in human genes.

Another weakness is that the different programs may utilize different gene annotation systems. NCBI, UCSC, and Ensembl often disagree in the number and the exact sequence of protein-coding transcripts of a gene. The combination of scores obtained from different gene annotations may lead to questionable results if a genomic coordinate is assigned to a coding base by one tool but to an intronic base in another.

23.4 ACCESSING THE PREDICTION SCORES

Most predictors and combination score providers offer web interfaces that allow the query of single variants on the fly. Some also provide batch analyses, either in special formats (e.g., PolyPhen-2) or for standard VCF files (e.g., MutationTaster).

Another convenient way to integrate predictions is the download of dbNSFP [265], which includes precomputed analyses of many different prediction programs as well as a variety of combination scores for all possible non-synonymous variants.

23.5 SO HOW GOOD ARE CURRENT PATHOGENICITY PREDICTION PROGRAMS?

Many assessments of pathogenicity prediction programs have been published. The prediction programs often "disagree" on individual variants, and no one program can correctly predict all variants. The predictive values are generally estimated at about 75 to 90% [248, 430, 447]. The classical variant prediction tools can only assess the effect of the variant on the protein or gene product. They do not take the phenotypic relevance of the gene affected into account, and a prediction of deleterious does not equate with a prediction of disease causality. Given the huge number of DNA variants, including synonymous variants, detected by WES, the prediction of the variant's effect on the protein or gene product alone is often not sufficient to identify the causal mutation — users need to take other evidence into account.

Table 23.4 shows the application of different variant prediction programs to the exome of a "healthy" person who doesn't have a severe monogenic disease. Any homozygous variant which is predicted to be disease-causing must hence be a false positive. Even though MutationTaster is the only one of the tested programs which is able to predict the effect of non-coding variants and therefore covers many more variants, it has the lowest number of false positives — this is partly due to the fact that MutationTaster exploits data about common polymorphisms.

Table 23.4. False-positive rate of different variant predictors.

Case	Homozygous non-synonymous, synonymous, and non-coding				Only homozygous missense			
	MT2	PPH	SIFT	PROVEAN	MT2	PPH	SIFT	PROVEAN
	all predictions				*all predictions*			
FP	103	464	353	462	6	376	295	331
TN	7,714	943	6,623	6,436	2,771	776	2,482	2,446
FPR	1.3%	33.0%	5.1%	6.7%	0.2%	32.6%	10.6%	11.9%
	variants analyzed by all tools				*variants analyzed by all tools*			
FP	7	401	286	308	6	376	274	290
TN	1,224	830	945	923	1,146	776	878	862
FPR	0.6%	32.6%	23.2%	25.0%	0.5%	32.6%	23.8%	25.2%

Note: FP, false positives; TN, true negatives; FPR, false positive rate = FP / (FP + TN); PPH: Polyphen-2 (HumDiv Model); MT2: MutationTaster2. All predictions that assumed pathogenicity (including "possibly damaging" and "probably damaging") were counted as false positives.

FURTHER READING

Adzhubei IA, Schmidt S, Peshkin L, Ramensky VE, Gerasimova A, Bork P, Kondrashov AS, Sunyaev SR (2010) A method and server for predicting damaging missense mutations. *Nat Methods* **7**:248–9.

Knecht C, Mort M, Junge O, Cooper DN, Krawczak M, Caliebe A. (2017) IMHOTEP-a composite score integrating popular tools for predicting the functional consequences of non-synonymous sequence variants. *Nucleic Acids Res* **45**:e13.

Liu X, Wu C, Li C, Boerwinkle E (2016) dbNSFP v3.0: A One-Stop Database of Functional Predictions and Annotations for Human Nonsynonymous and Splice-Site SNVs. *Hum Mutat* **37**:235–41.

Schwarz JM, Cooper DN, Schuelke M, Seelow D (2014) MutationTaster2: mutation prediction for the deep-sequencing age. *Nat Methods* **11**:361–2. (http://www.mutationtaster.org/info/statistics.html)

23.6   EXERCISES

Exercise 1

Study the effect of a missense mutation on the protein.

Choose a known disease mutation leading to the substitution of a single amino acid from ClinVar.[4] A good example is the H47R mutation in the *SOD1* gene which leads to amyotrophic lateral sclerosis type 1 (ALS1, OMIM #105400). To find the mutation, type the gene symbol into the search field of the ClinVar homepage. The result lists all variants found in this gene with a ClinVar entry, regardless of their pathogenicity. To filter for disease mutations, set the *Clinical significance* to *Pathogenic* (on the left). To narrow the search to missense variants, select *Missense* (under *Molecular consequence*).

ClinVar uses the three-letter code, so search for *His47Arg* and click on the hyperlink. On the next page, you will find some basic information about the variant, such as its position in the current genome build(s), the substitution in different GenBank entries (gene, mRNA, and protein), a link to OMIM, the UniProt ID of the protein, and, last but not least, the dbSNP ID. All variants stored in ClinVar have dbSNP IDs — a dbSNP ID does not mean that a variant is benign!

The *Assertion and evidence details* section below indicates why this variant is considered to be a disease mutation. To learn more about

---

[4]https://www.ncbi.nlm.nih.gov/clinvar/.

it, you can follow the links to PubMed — you will notice that this mutation was assigned to position 46 in the older literature, and hence was previously termed "H46R".

On the ClinVar webpage, follow the dbSNP link to get more data about the variant (clicking on it will not directly open dbSNP's page for this SNP but another hyperlink which will then lead to the correct page).

The *HGVS Names* panel lists different ways to annotate the mutation; here the variant is shown on each the chromosomal (NC), gene (NG), mRNA (NM), and protein (NP) level. Please note that NM and NP clearly name one transcript/isoform of the gene — the position may be very different in other isoforms. There are currently two chromosomal annotations (`NC_000021.8:g.33036170A>G` and `NC_000021.9:g.31663857A>G`). The format includes the chromosome ('21') and the genome version ('.8' for GRCh37/hg19 and '.9' for GRCh38/hg38).

In the following exercises, you will be asked to investigate the mutation from Exercise 1 with the prediction tools SIFT, MutationTaster, and PolyPhen-2.

Exercise 2

Investigate the pathogenicity of the mutation from Exercise 1 with SIFT/PROVEAN.[5] Enter the mutation in the correct format (examples are presented on the webpage of SIFT) and click on submit.[6] Check the prediction and try to find out how this mutation impairs protein function.

Exercise 3

Investigate the pathogenicity of the mutation from Exercise 1 with MutationTaster.[7] Enter the mutation on the mRNA level. For this, type the gene symbol (SOD1) into the *Gene* field and click on the *show available transcripts* link. Now choose the correct transcript. MutationTaster is based on Ensembl but includes some major NCBI Genbank transcripts as well — you will find the Genbank ID from

---

[5] http://sift.jcvi.org/

[6] Various tools are available on the SIFT/PROVEAN Website. For the SIFT Human SNPs tool it is important to note that the gene is located on the forward (1) strand.

[7] http://mutationtaster.org/

the dbSNP page (NM_000454) right of one of the Ensembl transcripts. Our position is coding-sequence-based as indicated by the c. in NM_000454.4:c.140A>G. Therefore *Position / snippet refers to* must be set to *coding sequence*. Now enter the position and the new base under "single base exchange by position" and click on submit. Check the prediction and try to find out how this mutation impairs protein function.

Exercise 4

Investigate the pathogenicity of the mutation from Exercise 1 with PolyPhen-2.[8] PolyPhen-2 is protein-based and restricted to missense variants. Enter the UniProt identifier of the protein (P00441, available from ClinVar or MutationTaster's results), chose the original and the replaced amino acid, enter the position (47) and click on submit. Check the prediction and try to find out how this mutation impairs protein function.

---

[8]http://genetics.bwh.harvard.edu/pph2/

VI

Prioritization

CHAPTER 24

Variant Prioritization

S TUDIES suggest that about 25% of individuals with a rare disease wait 5–30 years and see at least three doctors to get a diagnosis, with the initial diagnosis being wrong in at least 40% of cases [116, 296]. The use of Next-Generation Sequencing (NGS) in diagnostics and research has grown enormously since the introduction of NGS a little over a decade ago. Research programs such as the 100,000 Genomes Project in England, the Undiagnosed Diseases Network of the National Institutes of Health (NIH), the Kids First Pediatric Research Program of the NIH, as well as Chinese, French, American and British programs for genomic and precision medicine are likely to generate over a million genomes of individuals with suspected rare, genetic disease by 2020. In addition, NGS is becoming increasingly important for personalized therapy decisions in cancer treatment, and a number of large-scale programs have been founded to advance the field of precision oncology, including the NIH's Precision Medicine Initiative, INdividualized Therapy FOr Relapsed Malignancies in Childhood (INFORM) in Germany, and the Personalized OncoGenomics (POG) Program in Canada.

Despite this, NGS-based diagnostics has not been a panacea for diagnostics in rare genetic disease or cancer. In fact, the overall diagnostic yield of WES in large cohorts of individuals with suspected Mendelian disease can be surprisingly low at about 11% to 25% [98, 356, 473, 474, 491].

Even if WES is combined with methods such as Array Comparative Genomic Hybridization (aCGH), the overall yield of both methods in a large cohort of autistic children was only 18% [418]. WGS may be able to improve overall diagnostic yield due to more uniform coverage of the exome and better capabilities to identify structural variation, and a pilot study using WGS on children with severe intellectual disability achieved a diagnostic rate of 42% [133]. However, it is impossible to know if this finding will generalize or if the higher yield was related to the patient cohort, especially given that an analysis of a large WGS cohort with varying disease categories showed a yield of only 21% [426].

The essential problem is that, while it has become relatively easy to determine the genome sequence of individual patients, it can still be extremely difficult to interpret the findings. A typical exome contains up to 140,000 variants [88]. The analysis of the called variants generally begins with the removal of common variants (Chapter 22), the annotation of variants to transcripts and to categories such as missense, nonsense, etc. (Chapter 15) and the assessment of their potential pathogenicity (Chapter 23). If a patient cohort is being analyzed, a rare variant association test can be performed (Chapter 21), and if a family is available, pedigree-based filtering or linkage analysis can be performed (Chapter 20). Nevertheless, in many situations, these steps do not identify a single candidate but instead yield long lists of candidate variants.

There are many reasons why a simple analysis of the variants identified by WES/WGS is not sufficient. Firstly, about 200 or more never-before-seen missense variants can be identified in a single exome. Second, with the exception of a few comprehensively investigated variants that are associated with well understood diseases, it can be difficult to decide whether any given variant is truly disease-causative or not. Often conflicting information is available about genes or mutations, and current databases may contain a substantial amount of erroneous information [35, 139]. It is not always possible to retrieve all evidence for or against the pathogenicity of a variant, which would be necessary for truly evidence-based diagnostics [51].

Since the medical or biological assessment of each candidate can take a substantial amount of time, it is advantageous to place the best candidates towards the top of the list. If the algorithms work, they not only save time but also increase the likelihood that a novel disease gene will actually be found, since researchers and clinicians might not have the time to look at a list of tens or hundreds of candidate genes and variants with the necessary attention to detail.

Computational methods for rearranging candidate lists in order to place the "best" candidates near the top of the list are referred to as "gene prioritization" [297, 334].

In this book, we have distinguished between *filtering steps*, in which individual variants are removed or down-weighted owing to character-istics of the variant, and *gene prioritization*, in which the genes and the variants located in them are ranked according to their estimated likelihood of being associated with the disease. Not all literature on this topic makes this distinction, and some authors consider what we call filtering to be prioritization.

In previous chapters, we have shown how variants are filtered ac-cording to a number of criteria.

- Quality thresholds (Chapter 16)

- Population frequency of variants (Chapter 22)

- Functional category of variation (Chapter 15)

- Status as *de novo* variant (Chapter 18)

- Cosegregation (pedigree) analysis (Chapter 20)

- Predicted pathogenicity (Chapter 23)

In the following chapters, we will provide detailed presentations of several strategies for prioritization:

- Searching for genes close to known disease genes in the protein-protein interaction network (Chapter 25).

- Phenotype-driven genomic analysis (Chapter 26).

- Integration of model-organism phenotype data (Chapter 26).

- Prioritization on the basis of medical knowledge (Chapter 28).

There are many other methods that can be used to prioritize can-didate genes that we will not cover in detail. This chapter will provide short reviews of some of the most important such methods.

24.1 GENIC INTOLERANCE TO FUNCTIONAL VARIATION

Humans have roughly 20,000 protein-coding genes. The overwhelming majority of Mendelian diseases result from a mutation in a protein-coding gene, but it is *not* the case that each gene in our genome is associated with its own Mendelian disease. Currently, roughly 3,500 genes have been identified as being associated with Mendelian diseases.

It seems likely that many thousands of additional Mendelian disease genes remain to be discovered, but it is certain that there are many genes that can be deleted or otherwise altered without causing disease. Intuitively, if we know that a gene exhibits a high frequency of rare variants in the (more or less healthy) general population, then it become less likely that any given rare variant in the gene is the cause of a specific rare disease. We can use this information to prioritize variants in an exome or genome.

It is commonly observed in exome sequencing studies that certain genes seem to have an excess of variants. For instance, mucin genes are often found to have variants because they represent recent gene duplication events with many paralogous alignments, which in turn give rise to systematic false positive variant calls (see Figure 17.12 in Chapter 17).

One approach to this problem is to make a list of genes such as the mucins and the HLA genes that frequently are found to have variants. The list could be used to simply remove variants in the genes prior to further analysis. However, if done naively, this could filter out true disease genes that also show a relatively high degree of variability such as the titin gene, mutations in which can cause forms of cardiomyopathy [377].

The genic intolerance approach provides a method of flagging genes that simply show a higher degree of variability than one would expect. These genes can be flagged as such in evaluating pipelines (or removed if desired), and thereby the physicians or biologists who are evaluating the exome results will be warned not to believe too easily that these genes are causative of disease.

Petrovski and colleagues introduced a method to calculate a rare-variant intolerance score (RVIS) for each gene in the human genome [329]. The intuition behind the score was to distinguish between genes with very few rare, functional mutations in the general population and genes that often carry non-conservative amino acid substitutions and stop mutations at high frequencies yet trigger no clinical diagnosis. To suggest causation in a gene that is commonly mutated in the healthy general population, one demands a substantially higher amount of evidence. We will refer to the original publication for details on the method [329]. The authors have recently extended the framework to non-coding variants [328] and protein domains [151]. Others have developed similar schemes for assessing copy number variation [381].

24.2 GENE EXPRESSION

It is intuitive that, with some exceptions, if a Mendelian disease causes a functional or morphological defect in some tissue, then the associated disease gene should be expressed in that tissue. For instance, one study reported using expression in brain tissue to prioritize candidate genes for the neurobehavioral disorder autism [82]. One of the authors used spatial gene expression patterns from the human and mouse brain to prioritize candidate genes for X–linked mental retardation (XLMR) and genetic (generalized) epilepsy with febrile seizures plus (GEFS+) [335, 336, 337]. Recently, many resources for investigating tissue-specific expression have become available, such as the dataset generated by the Genotype-Tissue Expression (GTEx), which comprises RNA-seq expression data for more than thirty distinct human tissues [147].

24.3 PATHWAYS

The interrelationships between functional pathways and disease are poorly understood, but since gene mutations alter the functions of genes which in turn participate in cellular pathways, a number of groups have used pathway information to assess candidate genes. For instance, the pathogenesis of Marfan syndrome and Loeys-Dietz syndrome is known to be related to dysfunction of the TGFβ pathway; therefore, it was predicted that the gene associated with Shprintzen-Goldberg syndrome, which shares many clinical manifestations with the former diseases, would also be involved in the TGFβ pathway, which ultimately helped to identify mutations in TGFβ repressor *SKI* as the disease-associated gene [109].

A number of resources have therefore included pathways and other genomic data sources as a component of prioritization algorithms. For instance, SPRING (Snv PRioritization via the INtegration of Genomic data) integrates six functional effect scores to assess variants as well as data derived from Gene Ontology, protein-protein interactions, protein sequences, protein domain annotations, and gene pathway annotations [465].

24.4 INTEGRATIVE METHODS

Many of the best performing methods for prioritization combine multiple predictors using statistical or machine-learning techniques. For instance, ENDEAVOUR ranks genes based on their similarity with a set of known training genes, resulting in one prioritized list for each data source. These rankings are then combined into a single ranking based on order statistics [7]. A recent study provided an extensive analysis of disease-gene associations using network integration and fast kernel-based gene prioritization methods [435].

24.5 SUMMARY

Exome analysis combines a series of variant filtration and assessment steps with prioritization methods that estimate the likelihood that variants in a gene are disease-causative in a particular case or study. This is still a very active area of bioinformatics research, and new methods are certain to be developed in coming years. The methods presented in this book are currently among the most powerful, but clearly there is still a lot of room for improvement.

FURTHER READING

Moreau Y, Tranchevent L-C (2012) Computational tools for prioritizing candidate genes: boosting disease gene discovery. *Nat. Rev. Genet.* **13**:523–36.

Petrovski S, Wang Q, Heinzen EL, Allen AS, Goldstein DB (2013) Genic intolerance to functional variation and the interpretation of personal genomes. *PLoS Genet.* **9**:e1003709.

Piro RM, Di Cunto F (2012) Computational approaches to disease-gene prediction: rationale, classification and successes. *FEBS Journal* **279**:678–696.

CHAPTER 25

Prioritization by Random Walk Analysis

P ROTEINS interact with other proteins in order to fulfill their functions. Defining the binding partners of a protein therefore helps
to understand its activity. Protein-protein interactions (PPIs) can be
measured by a wide range of high-throughput assays, such as yeast two-
hybrid systems, bimolecular complementation methods, proximity ligation assays, co-immunoprecipitation approaches, and others [53]. Many
databases have been developed for PPI data including HIPPIE [386],
which currently lists 273,153 interactions with 16,737 human proteins,
and STRING, which offers experimental PPIs and other functional interactions between proteins [415].

Genes associated with phenotypically similar diseases tend to be located close to each other in the interaction network [319]. The random-
walk with restart (RWR) method provides a method of assessing the
vicinity of any two genes in the PPI network to one another based on
an analysis of the global network structure. In this chapter, we will
introduce the basic concepts behind the method and then present ExomeWalker, a tool we developed to prioritize genes in WES data based
on RWR analysis of the vicinity of WES candidates to members of
disease-gene families.

25.1 DISEASE GENE FAMILIES

The basic idea of network-based disease gene prioritization is that a
sought-after (novel) gene is likely to be located in the same neighbor-
hood of the PPI network as other (currently known) genes involved in

the same or similar diseases. For instance, the ciliopathies include diseases such as Bardet Biedl syndrome that are clinically highly similar and caused by mutations in genes encoding proteins that interact with one another within the cilia.

A disease-gene family is therefore defined as a group of genes characterized by the fact that a mutation in any one of the genes leads to a clinically similar disorder. The Online Mendelian Inheritance in Man (OMIM) database[1] [17] provides "phenotypic series", which can be used to derive disease gene families. For instance, phenotypic series PS127550 currently comprises 14 forms of dyskeratosis congenita, each of which is caused by a mutation in one of 9 different genes. Many other methods can be used to compile disease gene families including simply review of the medical or scientific literature or computational methods based on measures of disease similarity [333].

25.2 DIRECT PROTEIN-PROTEIN INTERACTIONS AND SHORTEST PATHS

The earliest method for using PPIs to prioritize genes looked for direct protein interactions between known disease genes and sets of candidate genes [7, 232]. This approach was used for example in studies where a novel disease locus had been identified in a certain genomic interval, and the candidate genes were chosen for Sanger sequencing based on whether they had direct interactions to known disease genes (Figure 25.1).

A more sophisticated approach ranked the candidate genes according to the shortest path between known disease genes and genes in the linkage interval in order to prioritize candidates [132]. However, in this case a relatively high probability that randomly chosen pairs of genes will be connected by a relatively short path, which can lead to false-positive results.[2] Also both of these methods are not able to choose between multiple genes with a given shortest path length (Figure 25.2).

---

[1]https://www.omim.org

[2]Protein interaction networks have been found to follow a scale free distribution, which entails a ""small-world" behavior, which means that in general, the maximum number of links separating any two proteins is small [29].

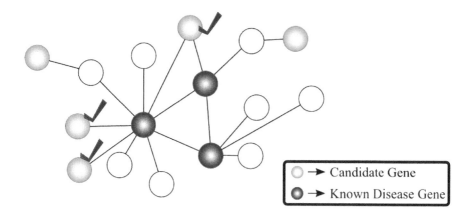

Figure 25.1. Direct protein-protein interactions. An analysis of direct PPIs would search for all candidate genes (e.g., those genes with a predicted pathogenic variant in a WES study), and flag those that have a direct PPI with a gene that is already known to be associated with the disease in question (represented by red check marks). Figure adapted from ref. [207].

25.3 REPRESENTING PPIS AS AN ADJACENCY MATRIX

The PPI network can be displayed as an undirected graph with nodes representing the genes and edges representing the mapped interactions of the proteins encoded by the genes. The graph in turn can be represented as a matrix $\mathbf{A}_{n \times n}$ in which proteins are numbered $1, \ldots, n$, and protein i is Represented by row i and column i. If there is an interaction between protein i and j, then the i,jth entry of the matrix, a_{ij} is equal to 1, otherwise $a_{ij} = 0$. This matrix is known as an adjacency matrix. Formally, the nodes of graph G are represented as $V(G) = \{v_1, v_2, \ldots, v_n\}$. The matrix $\mathbf{A}$ is called the adjacency matrix of G if

$$a_{ij} = \begin{cases} 1 & \text{if } (v_i, v_j) \in E(G) \\ 0 & \text{otherwise} \end{cases} \qquad (25.1)$$

Here, $E(G)$ refers to the set of edges of the graph. Figure 25.3 shows a network and the corresponding adjacency matrix for a toy example with 6 proteins.

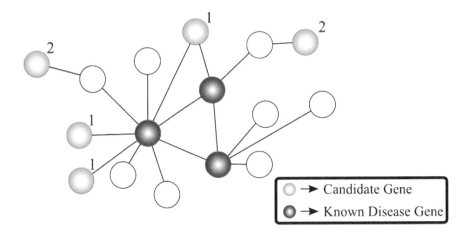

Figure 25.2. Shortest path PPI analysis. A shortest path analysis of protein-protein interactions would search for all candidate genes that are connected by any path to one of the known disease genes, and then rank the candidates according to distance (number of edges along the path). Figure adapted from ref. [207].

25.4   RANDOM WALK ON THE PPI NETWORK

The random walk on graphs is defined as an iterative walker's transition from its current node to a randomly selected neighbor starting at a given source node, s. The distribution of this walk will allow us to

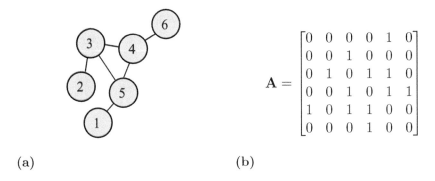

(a)   (b)

Figure 25.3. A small protein-protein interaction network. A small example protein-protein interaction network with six proteins is shown. The corresponding adjacency matrix is shown on the right.

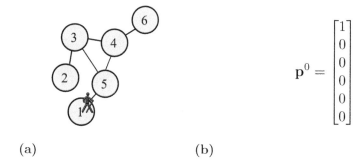

$$\mathbf{p}^0 = \begin{bmatrix} 1 \\ 0 \\ 0 \\ 0 \\ 0 \\ 0 \end{bmatrix}$$

(a)   (b)

Figure 25.4. At the initial time point ($t = 0$), the walker is located at node 1.

calculate a similarity value between the known disease genes (which we call the "seed genes") and any other gene in the network.

To perform the random walk analysis, we first column-normalize the adjacency matrix $\mathbf{A}$ so that the sum of all the entries in each column (which represent the transition probabilities) is equal to 1. We call the column-normalized adjacency matrix $\mathbf{W}$.

$$\mathbf{W} = \begin{bmatrix} 0 & 0 & 0 & 0 & 1/3 & 0 \\ 0 & 0 & 1/3 & 0 & 0 & 0 \\ 0 & 1 & 0 & 1/3 & 1/3 & 0 \\ 0 & 0 & 1/3 & 0 & 1/3 & 1 \\ 1 & 0 & 1/3 & 1/3 & 0 & 0 \\ 0 & 0 & 0 & 1/3 & 0 & 0 \end{bmatrix}$$

We represent the position of the walker at time t as the column vector $\mathbf{p}^t$. If the walker is located (completely) at node i at time t, then the ith entry of the vector is equal to one, and the other entries are equal to zero. We will start the walker off at node 1 at time zero (Figure 25.4).

Taking a step in the network is implemented mathematically by matrix multiplication. It is easy to see that after multiplying the position vector $\mathbf{p}^0$ by the normalized adjacency matrix $\mathbf{W}$, we cause the walker to advance from node 1 to node 5 (Figure 25.5).

Continuing the walk, we see that at time $t = 2$, the walker has been "divided" across three nodes each with weight $1/3$.

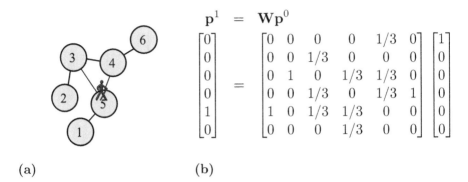

(a)   (b)

Figure 25.5. At the next time point $(t = 1)$, the walker is located at node 5.

$$\mathbf{p}^2 = \mathbf{W}\mathbf{p}^1$$

$$\begin{bmatrix} 1/3 \\ 0 \\ 1/3 \\ 1/3 \\ 0 \\ 0 \end{bmatrix} = \begin{bmatrix} 0 & 0 & 0 & 0 & 1/3 & 0 \\ 0 & 0 & 1/3 & 0 & 0 & 0 \\ 0 & 1 & 0 & 1/3 & 1/3 & 0 \\ 0 & 0 & 1/3 & 0 & 1/3 & 1 \\ 1 & 0 & 1/3 & 1/3 & 0 & 0 \\ 0 & 0 & 0 & 1/3 & 0 & 0 \end{bmatrix} \begin{bmatrix} 0 \\ 0 \\ 0 \\ 0 \\ 1 \\ 0 \end{bmatrix}$$

If the walker is allowed to walk around the network in an unconstrained fashion, then the probability weight will diffuse away from the original seed genes. For this reason, the actual random walk is performed with a restart of the walk at the original seed nodes in order to achieve a balance between the walker exploring pathways in the network and not diffusing too far away from the seed genes. At each time step, the walker "restarts" at the seed node s with probability r. For exome analysis, we choose $r = 0.7$ [400]. Formally, the random walk with restart is defined as:

$$\mathbf{p}^{t+1} = (1 - r)\mathbf{W}\mathbf{p}^t + r\mathbf{p}^0 \tag{25.2}$$

The random walk algorithm then iterates this walk until convergence as judged by $\frac{1}{n}\|\mathbf{p}^t - \mathbf{p}^{t+1}\|_1 < 10^{-6}$, where the L1 norm $\|\mathbf{x}\|_1$ is defined as $\|\mathbf{x}\|_1 = \sum_{i=1}^{n} |x_i|$.

For gene prioritization the initial probability vector $\mathbf{p}^0$ assigns equal probabilities to the nodes representing members of the disease-gene family, with the sum of the probabilities equal to 1. This is equivalent to letting the random walker begin from each of the known disease genes with equal probability. Candidate genes are ranked according to the values in the steady-state probability vector $\mathbf{p}^\infty$. While it is possible to obtain $\mathbf{p}^\infty$ by explicitly calculating Equation (25.2) until convergence, we instead solve the equation

$$\mathbf{p}^\infty = (1-r)\mathbf{W}\mathbf{p}^\infty + r\mathbf{p}^0 \qquad (25.3)$$

to obtain

$$\mathbf{p}^\infty = r\left(\mathbf{I} - (1-r)\mathbf{W}\right)^{-1}\mathbf{p}^0 \qquad (25.4)$$

By precalculating the matrix $r\left(\mathbf{I} - (1-r)\mathbf{W}\right)^{-1}$, we can perform random walk analysis as a simple matrix multiplication of the vector $\mathbf{p}^0$ in $\mathcal{O}(n^2)$ time, where n is the number of genes in the network. Therefore, denoting $r\left(\mathbf{I} - (1-r)\mathbf{W}\right)^{-1}$ by $\mathbf{R}$, we can calculate the result of the random walk analysis as:

$$\mathbf{p}^\infty = \mathbf{R}\mathbf{p}^0 \qquad (25.5)$$

We can further simplify the calculations by noting that most of the elements of the vector $\mathbf{p}^0$ are zero, with only the elements representing the m seed genes having the non-zero value $1/m$. Denoting the set of the indices of these elements as $\{j'\}$, it is easy to see that only the corresponding columns of $\mathbf{R}$ contribute to the final values of $\mathbf{p}^\infty$, whose i^{th} element can be given as

$$\mathbf{p}^\infty[i] = \frac{1}{m}\sum_{j\in j'}\mathbf{R}[j,i] \qquad (25.6)$$

That is, to get element i in $\mathbf{p}^\infty$, we need only to take the sum of the products of the non-zero elements of $\mathbf{p}^0$ with the corresponding elements of column i of $\mathbf{R}$. The computational complexity of the random walk analysis in Equation (25.2) is dominated by the matrix-vector multiplications in each step, which is $\mathcal{O}(n^2)$ for an $n \times n$ matrix. In contrast, our method requires precomputation of one matrix inversion, but the actual calculation of $\mathbf{p}^\infty$ is $\mathcal{O}(mn)$ with $m \ll n$, as there are $\mathcal{O}(m)$ operations to calculate Equation (25.6), which has to be done for each of the n elements of $\mathbf{p}^\infty$. $\mathbf{p}^\infty$ is a probability vector, with each entry representing the probability of the walker being located at node i in the network after an infinite number of steps. The value of $\mathbf{p}^\infty$ at i (p_i) can be interpreted as the similarity between the start genes (the known disease genes) and gene i.

$$\mathbf{p}^\infty = \begin{bmatrix} p_1 \\ p_2 \\ \vdots \\ p_N \end{bmatrix} \qquad \text{where} \qquad \sum_{i=1}^{N} p_i = 1$$

The random walk with restart (RWR) analysis as defined above represents a global distance measure in networks that integrates all of the paths in the graph. Intuitively, the RWR algorithm calculates the similarity between two genes, i and j, on the basis of the likelihood that a random walk through the interaction network starting at gene i will finish at gene j, whereby all possible paths between the two genes are taken into account. In our initial implementation of the RWR algorithm, we addressed the challenge of linkage analysis (in pre-NGS days), which flagged genomic intervals of 0.5–10 cM containing up to 300 genes. In this scenario, a previously unknown disease gene would be sought by linkage analysis. For instance, if a genetically heterogeneous disease such as retinitis pigmentosa[3] (RP) was diagnosed in a family and mutations in the known RP genes had been ruled out, then linkage analysis could be performed using samples from affected and unaffected members of that family to identify the chromosomal region where the disease gene must be located. The region identified by linkage analysis often contained 100–300 genes depending on the size of the family or families being investigated. In the heyday of linkage analysis, selected genes in the interval would be subjected to Sanger sequencing, which was a time-consuming and expensive process. The goal of gene prioritization for linkage analysis was to order the genes in the linkage interval according to the estimated likelihood of the gene being the causal disease-associated gene [297, 334]. However, this idea does not require candidate genes to be located within the same chromosomal region and can therefore be also be of use for prioritizing candidate genes from WES/WGS studies [334].

---

[3]A hereditary eye disease characterized by gradual deterioration of vision.

Our method for linkage analysis, GeneWanderer [207], lets the random walk start with equal probability from each of the known disease-gene family members in order to search for an additional family member in the linkage interval (Figure 25.6).

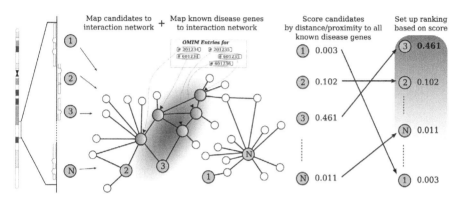

Figure 25.6. Gene Wanderer. All candidate genes contained in the linkage interval are mapped to the interaction network, as are all previously known disease genes associated with the disease gene family. Our method then assigns a score to each of the candidate genes based on the relative location of the candidate to all of the known "disease genes" as determined by a global network-distance measure, namely RWR. The genes in the linkage interval are ranked according to the score in order to define a priority list of candidates for further biological investigation. Figure adapted from ref. [207].

The advantage of the RWR method over direct-interaction or shortest path analysis can be best explained with a toy example. Figure 25.7 shows three subnetworks with a different configuration but consisting of the same number of nodes. The global distance between a hypothetical disease gene (x) and a candidate gene (y) is different in each case. In the left panel, proteins x and y are connected via a hub node with many other connections, so that the global similarity between node x and node y is less than in the middle panel, where x and y are connected by a protein with fewer connections than those of the hub. In the right panel, x and y are connected by multiple paths and therefore are assigned a higher similarity score than in the other two panels. Note that the shortest path between x and y is identical in each case, so that a shortest path approach cannot differentiate between these three types of connection. Even worse, the approach taking only direct interactions with gene x into account would identify gene y as a candidate in none of the three cases. In the original publication, we demonstrated the superiority of the RWR approach using simulations [207].

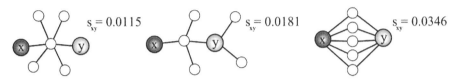

Figure 25.7. Example subnetworks. A hypothetical disease gene (x) is connected by a distance of two edges (PPIs) to a candidate gene (y). The structure of the rest of the network influences the similarity score between x and y (see text). Figure adapted from ref. [207].

25.5 EXOMEWALKER

In the exome case, we not only have information about the PPIs of genes, we additionally have the information about variants found in the WES data [400].

ExomeWalker performs an analysis of the variants found in an exome to identify predicted pathogenic variants, and then performs a random walk analysis that searches for genes with predicted pathogenic variants that are also near to the seed genes in RWR analysis. Each gene with one or more variants is assigned a combined ExomeWalker score, which is a combination of the random walk score and the score for the best scoring variant in that gene. In the case of AR inheritance under a compound heterozygous model, the variant score is taken to be the average of the two highest scoring variants (Figure 25.8).

The ExomeWalker application has been incorporated into the Exomiser suite of WES/WGS software, and we will explain how to run the ExomeWalker algorithm with the Exomiser in Chapter 27.

FURTHER READING

Köhler S, Bauer S, Horn D, Robinson PN (2008) Walking the interactome for prioritization of candidate disease genes. *Am J Hum Genet* **82**:949–58.

Smedley D, Köhler S, Czeschik JC, Amberger J, Bocchini C, Hamosh A, Veldboer J, Zemojtel T, Robinson PN (2014) Walking the interactome for candidate prioritization in exome sequencing studies of Mendelian diseases. *Bioinformatics* **30**:3215–22.

25.6 EXERCISES

Exercise 1

Use STRING[4] [415] to find genes that interact with the disease gene for Bardet Biedl syndrome type 1 (BBS1). Restrict your search to Homo sapiens. Are any of the other genes associated with other forms of Bardet Biedl syndrome? Consult OMIM[5] for information about the genes if necessary.

Exercise 2

Examine PPI subnetworks for a disease-gene family. Go to OMIM and get the genes belonging to a phenotypic series. For instance, the genes *TERC*, *TERT*, *NOLA2*, *TINF2*, *NOLA3*, *ACD*, *PARN*, *WRAP53*, and *RTEL1* are associated with dyskeratosis congenita. The phenotypic series can be found as links on OMIM pages that describe individual diseases. Other families include Primary coenzyme Q10 deficiency (see OMIM:607426), Lissencephaly (OMIM:607432), and Noonan syndrome (OMIM:163950), but there are many more. Go back to the STRING database, and use the "multiple proteins" option to show protein interactions between the members of the disease gene family you chose. What do you observe? Try the same exercise with a "random" list of genes.

Exercise 3

We will explain how to run ExomeWalker in Chapter 27. After reading that chapter, run the example dataset for ExomeWalker analysis included in the Exomiser distribution.

---

[4]http://string-db.org/
[5]https://www.omim.org/

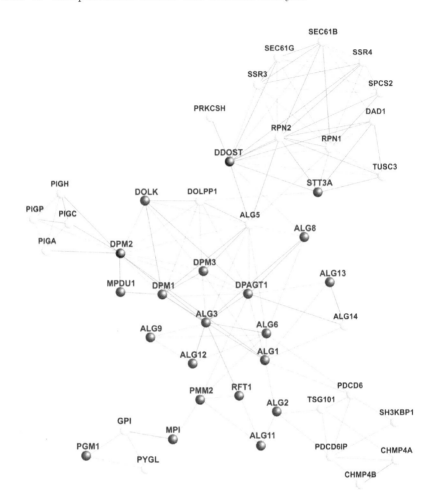

Figure 25.8. ExomeWalker. The figure shows a protein-protein association network generated for a simulated exome that was analyzed by ExomeWalker. The group of congenital disorders of glycosylation, type I (CDG-I) were used to define the seed genes. ExomeWalker has flagged two candidate genes *DDOST* and *DMP2* (shown in blue). These genes interact with multiple other CDG-I genes (shown as red nodes in the network) via paths of length one and two. *DDOST* has only two direct interactions with CDG seed genes and is at some distance from the others, leading to it only being ranked 23rd. However, *DPM2* has multiple direct and second-degree interactions with CDG genes leading to it being ranked as the top-ranked candidate (Figure taken with permission from [400]).

CHAPTER 26

Phenotype Analysis

Peter Robinson and Sebastian Köhler

P HENOTYPE-based analysis of WES/WGS data is one of the most powerful approaches towards gene prioritization [402]. In this chapter, we will present the Human Phenotype Ontology (HPO) and explain how HPO-based computational phenotype analysis works. The word "phenotype" can be used with multiple different meanings. In biology, phenotype is often used to mean the set of observable characteristics of an organism, including morphological, physiological, and behavioral traits.[1] In medicine, however, the word phenotype is usually used to describe a deviation from normal morphology, physiology, or behavior. This is the definition we will follow in this book. Although the clinical process is not usually described in these terms, one of the most important tasks of a clinician is to determine the phenotype of patients and to use the phenotypic information to make a correct diagnosis. This is done by taking a medical history, performing a physical examination of the patient, and in some cases by performing laboratory tests, ordering procedures such as a chest X-ray or magnetic resonance imaging (MRI) of the brain, and so on. Making a diagnosis can sometimes be easy, but is often an enormously challenging task, especially with rare diseases. One of the essential approaches towards differential diagnosis has been described as deep phenotyping — the precise and comprehensive analysis of the individual phenotypic abnormalities observed in a patient. The concept of deep phenotyping stands in contrast

---

[1]e.g., the length of a bird's wing, mouse myocardial wall function, or honey bee colony pollen collection rates.

to intuitive approaches to the differential diagnosis in which the overall "gestalt" enables the physician to recognize the diagnosis — a process which can be beset with errors [87].

26.1 THE HUMAN PHENOTYPE ONTOLOGY (HPO)

The HPO (http://www.human-phenotype-ontology.org) is an ontology of phenotypic abnormalities that can be observed in human disease. The HPO provides comprehensive bioinformatic resources for the analysis of genes, phenotypes, and diseases, and represents a computational bridge between genome biology and clinical medicine. The HPO was originally published in 2008 with the goal of integrating phenotypic information in different scientific and medical disciplines and databases [372]. Since this time, the scope and uptake of the HPO have grown substantially [146, 209, 214].

The HPO is structured as four independent subontologies for Phenotypic abnormality, Mode of inheritance, Mortality/Aging, and Clinical Modifier. The HPO currently (February 3, 2017) contains 12,243 terms that are connected by 15,911 subclass relations. The HPO is structured as a directed acyclic graph in which individual terms are connected by subclass (`is_a`) relations (Figure 26.1). Each term contains a set of attributes including human-readable and computational definitions, synonyms, cross-references, and other items (Table 26.1).

In addition to the attributes shown in the table, HPO terms have computer-readable "logical definitions" such as the following for *Atrial septal defect*.

```
'has part' some
  ('closure incomplete'
    and ('inheres in' some 'interatrial septum')
  and ('has modifier' some abnormal))
```

The definitions are formulated as Web Ontology Language (OWL)[2] classes that refer to multiple other ontologies for data on anatomy, biochemistry, histology, proteins, gene function, and pathology. The definitions are key components of algorithms for ontology quality control and cross-species phenotype mapping. We refer readers to the original publications for more details [208, 210, 401].

---

[2]https://www.w3.org/OWL/

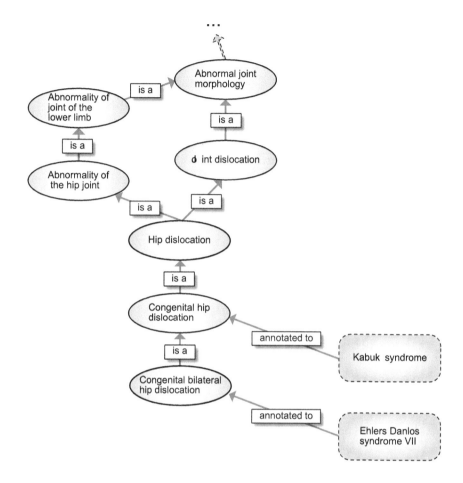

Figure 26.1. Human Phenotype Ontology. Excerpt of the HPO. The ovals represent HPO terms (e.g., *Hip dislocation*). HPO terms are used to describe phenotypic abnormalities that characterize diseases or that are observed in individual patients. For example, individuals with Kabuki syndrome are often observed to have congenital hip dislocation. HPO terms can be linked to one another by is_a (subclass) relations. For instance, every individual with *Congenital hip dislocation* can be said to have a *Hip dislocation*, the more general term.

26.2 ANNOTATIONS

Individual HPO terms do not describe diseases. Instead, in order to create a computational model of a disease, we associate a set of one

Table 26.1. Anatomy of an HPO term.

Item	Example	Comment
id	HP:0001631	Accession number for this term
name	Atrial septal defect	Preferred label
synonym	ASD; Atrial septum defect	Synonyms
definition	Atrial septal defect (ASD) is a congenital abnormality of the interatrial septum that enables blood flow between the left and right atria via the interatrial septum.	Human-readable definition
xref	ICD-10:Q21.1	Cross reference to synonymous concepts in other databases
is_a	Abnormality of cardiac atrium (HP:0005120); Abnormality of the atrial septum (HP:0011994)	Link(s) to one or more "parent terms", i.e., more general terms that are located directly above the current term in the hierarchy of the HPO

Note: The table shows the most important attributes of the HPO term *Atrial septal defect*.

or multiple HPO terms to the disease. We refer to the associations as "annotations". Currently, the HPO provides 123,724 annotations of HPO terms to genetic diseases in Orphanet, Decipher and OMIM [214]. The initial focus of the HPO was genetic disease, but the HPO is being extended to other areas of medicine. Recently, 132,620 annotations were derived by a text-mining approach for 3,145 common, complex diseases [146].

26.3 INTEGRATION OF THE HPO WITH OTHER DISEASE DATABASES

The HPO differs from other clinical terminologies and ontologies such as SNOMED, whose major purpose is the support of hospital IT systems, billing, and administration, by having been developed as a bioinformatic tool for differential diagnosis and translational research. The HPO enables the computational integration with numerous other bioinformatic resources for research such as the Gene Ontology, as well as phenotype data of tens of thousands of genetically modified mouse and zebrafish models of human disease. The HPO offers an extremely broad and deep coverage of phenotypic abnormalities. The entire Unified Medical Language System (UMLS), which imports numerous other terminologies such as MeSH and MedDRA, covered only 54% of the concepts in the HPO; SNOMED CT covered only 20% [349]. The UMLS, a terminology integration tool of the US National Library of Medicine, has therefore completely integrated the HPO starting with the 2015AB edition. The HPO is being used by numerous databases in the field of human genetics to annotate patient data and to enable phenotype-based searching and analysis. This approach has already contributed to the discovery of several novel disease genes (e.g., [9, 413, 434]). Table 26.2 shows a selection of publicly accessible databases that are using the HPO.

Table 26.2. A selection of databases and projects that use HPO for annotating and analyzing patient phenotype data.

Name	Reference
PhenomeCentral	[55]
DDD (Deciphering Developmental Disorders)	[464]
DECIPHER (DatabasE of genomiC varIation and Phenotype in Humans using Ensembl Resources)	[49]
ECARUCA (European Cytogeneticists Association Register of Unbalanced Chromosome Aberrations)	[445]
The 100,000 Genomes Project	[283]
Geno2MP	[72]
NIH UDP (Undiagnosed Diseases Program)	[41]
NIH UDN (Undiagnosed Diseases Network)	[350]
Phenopolis	[341]
GenomeConnect	[198]
FORGE Canada & Care4Rare Consortium	[34]
RD-Connect	[428]

The HPO aims to be not only a standard for the exchange of phenotype data, but also a computational foundation for algorithms and applications. Probably the most important class of application exploits the calculation of a numerical similarity value between any two HPO-encoded phenotype profiles, enabling patient to patient, patient to disease, and disease to disease similarity calculations that can be used in diagnostic, translational, and a wide range of medical systems biology algorithms. The concept of information content is essential to understand how these algorithms work.

26.4 INFORMATION CONTENT IN ONTOLOGIES

Ontologies consist of well-defined concepts that are connected to one another by semantic relations. This has inspired a number of algorithms that exploit these relationships in order to define similarity measures for terms or groups of terms in ontologies. Semantic similarity measures have been used in computational biology as a way of validating results of gene expression clustering, prediction of molecular interactions, disease gene prioritization, and for clinical diagnostics, among other things. A large number of different methods for calculating semantic similarity have been developed since the seminal paper of Resnik [360].

In 1995, Resnik introduced a method for evaluating the semantic similarity between two concepts (terms) in an ontology with is_a relations [360, 361]. Resnik's idea was to associate probabilities with the terms of the ontology. Let $\mathcal{T} = \{t_1, t_2, \ldots, t_n\}$ be the set of terms in the ontology, permitting multiple inheritance. Define a function $p : \mathcal{T} \longrightarrow [0, 1]$, such that $p(t_i)$ is the probability of encountering an instance of class $t_i \in \mathcal{T}$ (usually this is defined as the frequency of t_i in the entire database).

Resnik defined the *information content* of a concept (ontology term) as the negative logarithm of its probability:

$$\mathrm{IC}(t) = -\log p(t) \tag{26.1}$$

This definition makes intuitive sense. As the probability of a concept increases, its information content decreases (we gain less new information from an observation that something common has happened than from an observation that something rare has happened). Moreover, the information content associated with the root term, which subsumes all concepts in the ontology and thus has a probability of 1, is zero because $-\log(1) = 0$. Indeed, because *all* terms are also instances of the root term, the root term conveys no information at all about a particular instance.

A key concept for the understanding of how ontologies work is subsumption. Annotations within the ontologies such as the Gene Ontology and the HPO follow the so-called Annotation Propagation Rule.[3] If a gene (in Gene Ontology) or a disease (with the HPO) is annotated to a given ontology term, then it is also annotated to all of the is_a ancestors of the term [371]. For instance, the Gene Ontology currently shows that 20 human genes are annotated to the Gene Ontology *DNA topoisomerase activity* (GO:0003916). This term is a child term of *isomerase activity* (GO:0016853) in the Gene Ontology. The true path rule implies that the 20 genes annotated to *DNA topoisomerase activity* are also annotated to *isomerase activity*. The term *isomerase activity* has a total of 18 child terms, each of whose annotations are transferred to *isomerase activity*. Additionally, there are 7 annotations made directly to the term *isomerase activity* in the current set of GO annotations, so that *isomerase activity* in total has 269 annotations. Currently, 19381 proteins are annotated in the GO annotation file, so that the frequency of annotations to *DNA topoisomerase activity* is 20/19381, and the frequency of annotations to *isomerase activity* is 269/19381. Thus, p is a monotonically increasing function as we move up the ontology to more general concepts: if t_i is_a t_j, then $p(t_i) \leq p(t_j)$.

We will illustrate the basic ideas using a toy wine ontology (Figure 26.2). Imagine the ontology terms have been used to annotate 1,000 items from a wine catalog. If we pick a wine at random and learn it is annotated to the term *Rioja*, this gives us some information about the wine, as reflected in the positive information content of $-\log(250/100) = 1.386$. But there are many specific types of Rioja wine. We get much more information if we learn the wine is a *Reserva*, because there are only 5 such wines in the catalog; correspondingly, the information content of *Reserva* is $-\log(5/1000) = 5.298$. Note that the term *Rioja* has 250 annotations. Assuming the annotations to the three child terms are disjoint, the $40 + 5 + 15 = 60$ annotations of *Crianza*, *Reserva*, and *Gran Reserva* are propagated to the term *Rioja*, which thus must have 250-60=190 direct annotations. On the other hand, the number of annotations of the term *Riesling* is equal to the sum of annotations of its children, meaning that *Riesling* has only propagated, but no direct annotations.

Resnik used his definition of information content to define the semantic similarity between two terms in an ontology. The more infor-

---

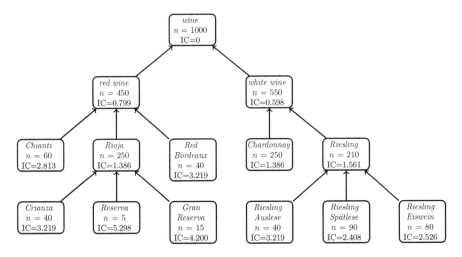

Figure 26.2. Information content of wine ontology terms. The figure shows a toy wine ontology. Each term is shown with its frequency among annotations for 1000 items of a wine catalog, and the corresponding information content (IC).

mation two terms share in common, the more similar they are. The information shared between two terms is indicated by the information content of their *most informative common ancestor* (MICA). This is defined using the function $\texttt{Anc}(t)$, which returns all of the ancestors of the term t in the ontology, including the term t itself.

$$t_{MICA(t_1,t_2)} = \underset{t \in \texttt{Anc}(t_1) \cap \texttt{Anc}(t_2)}{\arg\max} -\log p(t) \qquad (26.2)$$

Equation (26.2) identifies the term t with the maximum information content that is an ancestor of both t_1 and t_2. We can now define the similarity of two terms t_1 and t_2 as being equal to the information content of their MICA.

$$\text{sim}(t_1, t_2) = -\log p\left(t_{MICA(t_1,t_2)}\right) = \underset{t \in \texttt{Anc}(t_1) \cap \texttt{Anc}(t_2)}{\max} IC(t) \qquad (26.3)$$

For example, the MICA of terms *Chianti* and *Gran Reserva* in the ontology of Figure 26.2 is the term *red wine*. Therefore, noting that the term *red wine* has the highest information content of any of the common ancestors of the terms *Chianti* and *Gran Reserva* (Figure 26.3), we conclude that:

$$\text{sim}(\text{Chianti}, \text{Gran Reserva}) = \text{IC}(\text{red wine}) = 0.799$$

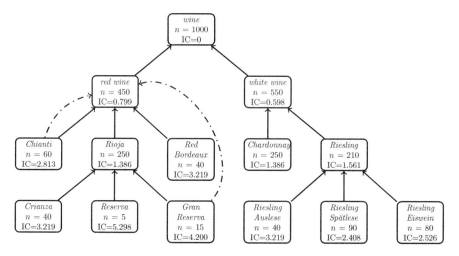

Figure 26.3. Node-based semantic similarity of ontology terms. The semantic similarity of the terms *Chianti* and *Gran Reserva* of the ontology of Figure 26.2 is calculated by finding their MICA and calculating the information content of the MICA.

On the other hand, the MICA (and the only common ancestor) of the terms *Chianti* and *Riesling* is the root of the ontology (Figure 26.4), meaning that they have a semantic similarity of zero.

26.5 SEMANTIC SIMILARITY OF ITEMS ANNOTATED BY ONTOLOGY TERMS

In practice, one is generally more interested in semantic similarity between items that are annotated by one or more ontology terms than in the semantic similarity of the terms themselves. We will show below how this can be used for clinical differential diagnostics, but for didactic purposes we will return again to the topic of wine. The toy ontology shown in Figure 26.2 can be regarded as a "domain ontology" that describes various types of wine. Each of the terms corresponds to a category of wines. In contrast, an "attribute ontology" describes the characteristics of the wines (Figure 26.5). Imagine that we are searching for a dry white wine with a citrony taste, but we do not know

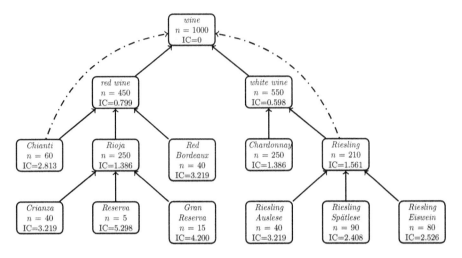

Figure 26.4. Node-based semantic similarity of ontology terms. The semantic similarity of the terms *Chianti* and *Riesling* of the ontology of Figure 26.2 is calculated by finding their MICA, which in this case is the root of the ontology. The information content of the root of an `is_a` ontology (which annotates all items) is $-\log(1) = 0$.

how the wine is called — therefore, we cannot order a wine with these characteristics in an online catalog until we find the name of a wine annotated to these three ontology terms (or related terms). Search procedures could exploit the structure of the wine attribute ontology to perform a kind of "fuzzy search" to retrieve wines with related characteristics. For instance, suppose the wine Petrocelli Chardonnay is annotated with the terms *yellow, lemony*, and *dry*. Only one of the three attributes matches exactly, but because *yellow* wine is a particular kind of *white* wine, and wine with a *lemony* taste has a particular kind of *citrony* taste, this Chardonnay is actually a good match. We will see below that phenotype-driven differential diagnostics is similar — diseases are the domain items we are trying to find, and the attributes are the phenotypic features. But first we will explain how one of the computational algorithms for searching in attribute ontologies works.

In the following equations, we define I_i to be the set of terms t used to annotate wine w_i; thus $t \in I_i$ returns the set of all annotations for w_i.

One commonly used similarity measure iterates over each term t

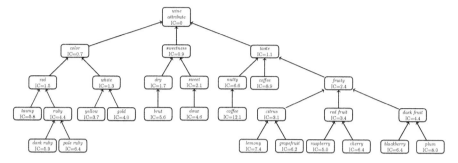

Figure 26.5. Wine attribute ontology. In contrast to the wine ontology, the wine attribute ontology describes the characteristics of wine. Specific types of wine can be annotated with one or several terms of this ontology. For instance, Contador, Rioja, 2014, might be annotated to the terms *dark ruby*, *blackberry*, and *dry*.

that annotates wine w_1 and finds the best matching term that annotates wine w_2, taking the average of the similarity scores of these matches:

$$\text{sim}(w_1, w_2) = \frac{1}{|I_1|} \sum_{t_1 \in I_1} \max_{t_2 \in I_2} \text{sim}(t_1, t_2) \qquad (26.4)$$

This similarity metric is not symmetric, but it is easy to create a symmetric version by taking the average of both directions:

$$\text{sim}^{\text{symmetric}}(w_1, w_2) = \frac{1}{2}\text{sim}(w_1, w_2) + \frac{1}{2}\text{sim}(w_2, w_1) \qquad (26.5)$$

For example, to calculate the similarity between Rioja and Cabernet Sauvignon (see Table 26.3), we find the best matches for their annotations. Figure 26.6 shows that their similarity can be calculated as 3.33. In contrast, if we calculate the similarity of Montepulciano with Pinot grigio, the similarity would be calculated based on the average of the information content of the terms *color* (best match between *red* and *white*), *taste* (best match between *coffee* and *lemony*), and *dry*, or $(0.7 + 1.1 + 1.7)/3 = 1.167$. Thus, Montepulciano and Pinot grigio are much less similar to one another than Rioja and Cabernet Sauvignon, as we would expect given their characteristics as shown in Table 26.3.

Table 26.3. Attributes of wines.

Wine	Characteristics
Rioja	*dark ruby, blackberry, dry*
Cabernet Sauvignon	*dark ruby, cherry, dry*
Malbec	*red, plum, brut*
Merlot	*red, red fruit, dry*
Lambrusco	*pale ruby, cherry, sweet*
Pinot grigio	*white, lemony, dry*
Montepulciano	*red, coffee, dry*
Nebbiolo	*ruby, raspberry, dry*
Eiswein	*golden, lemony, sweet*
Riesling	*white, fruity, dry*
Chenin blanc	*golden, fruity, dry*

Note: Wine types have been classified with three attributes each for a hypothetical online wine catalog.

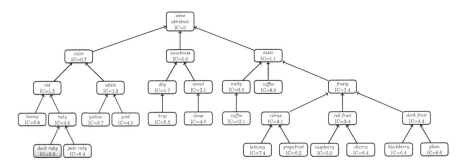

Figure 26.6. Similarity between Rioja and Cabernet Sauvignon. Both wines share the attributes *dark ruby*, and *dry*. The term *blackberry* for (Rioja) has as its nearest match *cherry* in Cabernet Sauvignon, and the similarity between these terms is the information content of *fruity*. The similarity is calculated as $(2.4+5.9+1.7)/3 = 3.33$ according to Equation (26.5).

26.5.1 Applications of Semantic Similarity for Clinical Diagnostics

The above-mentioned applications of semantic similarity were designed to identify similar items in databases. It is also possible to use semantic similarity as a way of processing arbitrary user queries to a database, in which the user enters one or more desired attributes (encoded as ontology terms) and searches for the best matches amongst items in the database that are annotated to the terms of the ontology. We applied this approach to the realm of diagnostics in clinical genetics [213].

Making the correct diagnosis is arguably the most important role of the physician, and is required to plan the correct treatment, to discuss prognosis and natural history of a disease, and to schedule appropriate surveillance examinations to avoid disease complications. Medical diagnostic procedures include a history and physical examination, blood tests and other laboratory investigations, and imaging techniques such as X-rays or computer-assisted tomography scans. The differential diagnostic process attempts to identify candidate diseases that best explain a set of clinical features. This process can be complicated by the fact that the features can have varying degrees of specificity, and by the presence of features unrelated to the disease itself. Depending on the experience of the physician and the availability of laboratory tests, clinical abnormalities may be described in greater or lesser detail.

One obvious advantage of capturing phenotypic information in the form of an ontology is that search routines can be designed to exploit the semantic relationships between terms. For instance, the search procedure can be designed such that a search on *Abnormality of the cardiac septa* will not just return all diseases annotated to this term, but also all diseases annotated to subclasses of this term such as *Ventricular septal defect* and *Atrial septal defect.*[4]

Semantic similarity metrics can be adapted to measure phenotypic similarity between queries and hereditary diseases annotated using the HPO. The importance of a clinical finding for the differential diagnosis depends on its specificity. As described above, in ontologies the specificity of a term is reflected by its information content. For medical diagnostics, the physician will enter the various abnormalities observed upon physical and laboratory examination of the patient using terms of the HPO. The information content of each HPO term in this application is defined as $-\log p_t$, where p_t is the frequency of term t among all of the diseases in the database. For instance, if *Atrioventricular block* is used to annotate three diseases among a total of 4,813 diseases, its IC would be calculated as $-\log(3/4,813) = 7.38$. The more general term *Abnormality of the musculoskeletal system* pertains to 2,352 diseases, so its IC is $-\log(2,352/4,813) = 0.72$.

---

[4] A search browser for HPO terms with this functionality is available at the HPO website, http://www.human-phenotype-ontology.org/.

We implemented a score to calculate the similarity between the query terms entered by the physician and the terms used to annotate the diseases in a database. That is, for each of the query terms the "best match" among the terms annotated to the disease is found, and the average over all query terms is calculated. This is defined based on Equation (26.4) as the similarity:

$$\text{sim}(\mathcal{Q} \to \mathcal{D}) = \text{avg}\left[\sum_{t_1 \in \mathcal{Q}} \max_{t_2 \in \mathcal{D}} IC(MICA(t_1, t_2))\right] \quad (26.6)$$

Figure 26.7 provides an overview of the approach, whereby the query is made up of the HPO terms *Downward slanting palpebral fissures* (a downward slanting of the line defined by the meeting of the eyelids) and *Hypertelorism* (widely spaced eyes), both of which are features that can be observed in many different hereditary diseases.

Equation (26.6) will return a high score if a good match is found for each term in the query, but it does not take into account the possibility that there could be a number of terms annotated to the syndrome in addition to those used for the maximum match. For instance, this would be the case if a specific query is compared to two syndromes, both of which are annotated by terms that exactly match the query, but one of the syndromes is annotated by a number of additional terms. Using the one-sided formula (Equation 26.6), both syndromes would receive the same score. It is also possible to define a symmetric version of Equation (26.6) in which the similarity of the query to the disease is averaged with the similarity of the disease to the query:

$$\text{sim}^{\text{symmetric}}(\mathcal{D}, \mathcal{Q}) = \frac{1}{2}\text{sim}(\mathcal{Q} \to \mathcal{D}) + \frac{1}{2}\text{sim}(\mathcal{D} \to \mathcal{Q}) \quad (26.7)$$

26.6 STATISTICAL SIGNIFICANCE OF SEMANTIC SIMILARITY SCORES

One drawback of the methods presented above is that it is difficult to assign a meaning to any particular semantic similarity score. For instance, even if the phenotypic abnormalities entered by the physician do not correspond to any of the diseases in the database, a result will be returned in which the diseases with the best scores are shown. How does one determine a cutoff below which a semantic similarity score is not meaningful? How trustworthy is any given semantic similarity score?

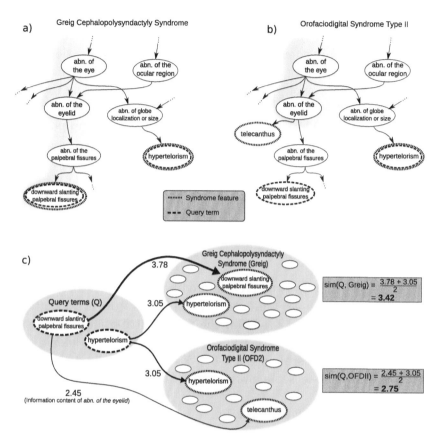

Figure 26.7. Searching the HPO with semantic similarity. Two of about 5,000 diseases annotated with HPO terms are shown. In **a)**, two of the 33 annotations for Greig Cephalopolysyndactyly syndrome [OMIM:175700] are shown. In **b)**, 2 of the 32 annotations for orofacial digital syndrome type II [OMIM:252100] are shown. In part **c)**, we see the two query terms entered by the physician: *downward slanting palpebral fissures* and *hypertelorism*. There is an exact match to the terms of Greig Cephalopolysyndactyly syndrome, which is reflected in the similarity score of 3.42; The term *hypertelorism* is also used to annotate the disease orofacial digital syndrome type II, but the term *downward slanting palpebral fissures* is not. Therefore, the information content of the most informative common ancestor of both terms, *abnormality of the eyelid*, is used for the calculation of the semantic similarity.

In the setting of medical diagnostics, the null hypothesis would be that the terms entered by the physician are not related to a disease in the database, meaning that the semantic similarity score obtained for the disease is due only to chance. We can simulate the semantic similarity score distribution for queries of n terms for each of the diseases in the database by drawing n terms at random and calculating the resulting similarity score many times. We can estimate the probability of obtaining a given score by chance by simply calculating the proportion of random scores that are equal to or higher than the score, and use this probability as an empirical p-value. For instance, if we perform 100,000 random queries against the disease *Marfan syndrome* we can store the values obtained for each of these queries. If a real query then has a score that is less than that of only 10 of the random queries, then we can estimate the p-value of this query as $10/100,000 = 10^{-4}$. Figure 26.8 shows the distribution of scores for random queries for one disease.

The score distribution is then calculated for each of the diseases in the database. For each query, the semantic similarity score is calculated and a p-value is generated based on the score distribution. The best matches are returned ranked according to the p-value (which should be corrected for multiple testing against the N diseases in the database), and a threshold can be set against some significance level (usually $\alpha = 0.05$).

This approach now allows a p-value to be assigned to the results of a query. Intuitively, if the highest scoring candidate diagnosis has a significant p-value, this would indicate to the clinician that this syndrome is a likely differential diagnosis and should be considered further. If on the other hand the highest scoring candidate does not have a significant p-value, this would indicate that the combination of phenotypic abnormalities entered by the physician is not specific enough to allow a diagnosis, or perhaps that the combination of features pertains to a clinical entity that is not present in the database being queried.

Although the idea of estimating the distribution of semantic similarity scores by means of simulations is simple enough, it requires several hundred thousand simulations for each of the diseases in the database and cannot be adjusted dynamically for queries against only a subset of diseases in the database.

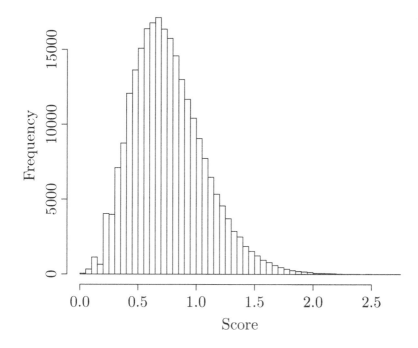

Figure 26.8. Sampled distribution of semantic similarity scores for the single target set for the Currarino syndrome. Random queries of size 10 were generated and their score was determined according to Equation (26.6).

These methods were implemented in a freely available web application called the Phenomizer (Figure 26.9).[5]

26.7 PHENOGRAMVIZ

Similar methods can be used to perform phenotype-driven evaluation of genomic variation. We will explore the use of this strategy for exome and genome sequencing in Chapter 27. Here, we will present a tool called PhenogramViz,[6] which was designed to support the phenotype-driven evaluation of copy-number variants (CNV). Methods such as array comparative genomic hybridization (CGH), SNP genotyping array

---

[5]http://compbio.charite.de/phenomizer/
[6]https://phenomics.github.io/software-phenoviz.html

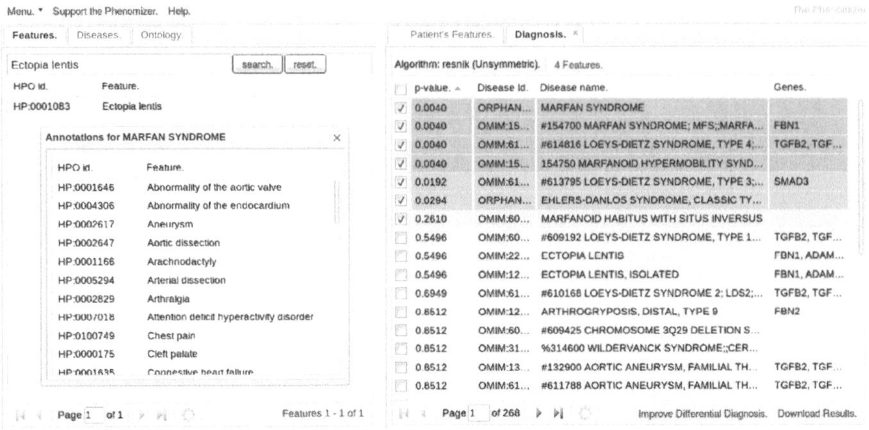

Figure 26.9. Phenomizer. The user has entered a set of four HPO terms that characterize some of the manifestations seen in individuals with Marfan syndrome. The Phenomizer returns a differential diagnosis sorted according to probability. A number of additional features are offered to help users explore the possibilities. Right-click on "show annotations" opens a window with all of the HPO annotations for the disease in question. The website has a detailed tutorial.

and genome sequencing enable the genome-wide detection of structural variants. Array CGH typically reveals from 20 to over 100 duplications and deletions. Most CNVs are relatively small in size, but some CNVs observed by array CGH may span up to several megabases. Each CNV can either represent neutral polymorphic variation or convey clinical phenotypes by inducing gene dosage effects or dysregulation of genes, and the interpretation and classification of rare CNVs remains difficult [482].

PhenogramViz supports phenotype-driven interpretation of aCGH findings based on multiple data sources, including the integrated cross-species phenotype ontology Uberpheno, in order to visualize gene-to-phenotype relations [212]. See Figure 26.10 for an example.

FURTHER READING

Köhler S, Schulz MH, Krawitz P, Bauer S, Dölken S, Ott CE, Mundlos C, Horn D, Mundlos S, Robinson PN (2009) Clinical diagnostics in human genetics with semantic similarity searches in ontologies. *Am J Hum Genet* **85**:457-64.

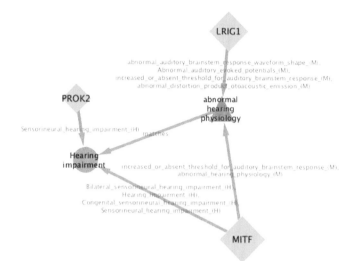

Figure 26.10. PhenogramViz. In this example, a hypothetical patient with hearing impairment and a deletion of the genes *MITF*, *LRIG1*, and *PROK2* is shown. The phenotypic feature of the patient (*hearing impairment*) is shown as a blue circle, and the deleted genes are shown as green diamonds. PhenogramViz draws edges (arrows) between these nodes if the gene involved in the CNV can provide a plausible explanation for the phenotype(s) observed in a patient because mutation in the human (H), mouse (M), or zebrafish (Z) ortholog of the gene is associated with a related phenotype. This visualization is intended to be an aid for the interpretation of CNVs in diagnostic or research settings. For instance, knockout of the gene *Lrig1* in the mouse leads to the phenotype *abnormal auditory evoked potentials*. This can be related to the patient phenotype *Hearing impairment* because of a common ancestor in the multispecies phenotype ontology (*abnormal hearing physiology*). The genes *MITF* and *PROK2* are related in an analogous fashion to *Hearing impairment*.

Köhler S, Schoeneberg U, Czeschik JC, Doelken SC, Hehir-Kwa JY, Ibn-Salem J, Mungall CJ, Smedley D, Haendel MA, Robinson PN (2014) Clinical interpretation of CNVs with cross-species phenotype data. *J Med Genet* **51**:766–72.

26.8 EXERCISES

Exercise 1

Using the wine ontology shown in Figure 26.2, calculate the semantic similarity of the terms *Chardonnay* and *Riesling Eiswein* using Equation (26.5).

CHAPTER 27

Exomiser and Genomiser

Julius O.B. Jacobsen, Damian Smedley, and Peter Robinson

E XOMISER and Genomiser are software suites developed by our group to enable start to finish phenotype-driven analysis of WES and WGS data. Genomiser expands the Exomiser by adding functions to analyze and prioritize non-coding variants [373, 403]. Both algorithms make use of many of the filtering and prioritization methods we have discussed in the previous chapters, and additionally use phenotype analysis as discussed in Chapter 26 to reorder the candidate list. In this chapter, we will present the algorithms used by Exomiser and Genomiser together with a series of practical tutorials. The software for both applications can be freely downloaded.

27.1 INTRODUCTION

The Exomiser is designed to analyze VCF files containing the called variants from an individual with suspected Mendelian disease and, optionally, other affected and unaffected family members.

The Exomiser analyzes these VCF files to first filter the variants and then to prioritize the remaining candidates to help identify the causative variant or variants. The filtering step is critical in order to reduce the 100,000+ variants seen in a typical exome to a more manageable size, and comprises frequency filters similar to those described in

Chapter 22. By default, Exomiser removes synonymous and off-target[1] variants (intergenic, upstream, downstream or intronic). Exomiser will remove variants whose minor allele frequency (MAF) is above a user-defined threshold (e.g., 0.1%), under the assumption that a common variant cannot be the cause of a rare disease. Other optional filters allow the user to remove variants that are below a particular quality in the QUAL column of the VCF file or to ignore those that are not in a predefined set of genes or within a genomic interval. Another optional filter is to remove variants that do not fit the expected inheritance pattern. The Exomiser uses Jannovar for pedigree filtering according to the rules described in Chapter 20. Exomiser requires a PED file for pedigree analysis, and currently only works with PED files that represent single families.

The Exomiser suite contains a number of different methods for variant prioritization based on protein-protein interactions and/or phenotype comparisons between a patient and existing human disease databases and model organisms.

27.1.1 PHIVE

The original implementation of Exomiser used the PhenoDigm algorithm using OWLTools [401] to calculate the phenotypic similarity between a patient's clinical signs and symptoms and observed phenotypes in mouse mutants associated with each candidate gene in the exome. The rationale behind this approach is that if a mouse model exists for the gene containing the disease-associated mutation, then it is likely to exhibit phenotypic similarity to the clinical phenotypes. For instance, the human phenotype *Otosclerosis* (HP:0000362), an abnormal growth of bone in the middle ear that can cause hearing loss, can be inferred to be similar to *fusion of middle ear ossicles* (MP:0003740) by ontological analysis of the Human Phenotype Ontology (HPO) [214] and the Mammalian Phenotype Ontology (MPO) [404]. The algorithm behind the original version of the Exomiser was termed PHIVE (*PHenotypic Interpretation of Variants in Exomes*), and works by first filtering variants according to rarity, location in or adjacent to an exon, and compatibility with the expected mode of inheritance, and then ranking all remaining genes with identified variants according to the combination of variant score (frequency and pathogenicity of the variant[s]) and

---

[1]i.e., not located in the targeted exons.

the phenotypic relevance score. In essence, our method searches for a phenotypically relevant gene that also has deleterious exome sequence variants, taking advantage of the voluminous data available for model organisms.

The variant score is calculated from those variants that are both rare and predicted to be pathogenic. The estimated frequency of variants was derived from the 1000 Genomes Project Consortium [1], from the Exome Server Project (ESP) [309] and (since version 7), the ExAC consortium [246]. Variants can be removed from further consideration if their population frequency exceeds a defined threshold (1% by default). Any variants remaining after filtering are assigned a frequency factor that rewards rarity of the variant. The predicted pathogenicity of the variants is calculated by a score that combines the predictions of SIFT [305], PolyPhen2 [6], and MutationTaster [390] (see Chapter 23) for missense variants and assigns a heuristic score to other categories of variant such as nonsense and splicing variants.

The phenotypic relevance score is calculated based on the semantic similarity of a human disease (the HPO annotations entered to represent the clinical manifestations of the individual being sequenced) and the phenotypic manifestations observed in a mouse model (the MPO annotation). OWLTools was used to calculate the phenotypic similarity between each of the HPO-annotated OMIM disease records and MPO-annotated mutant mice, resulting in a phenotypic relevance score for the corresponding mouse genes and their human orthologs. Our semantic matching approach allows similar but nonexact phenotypes to be detected and a score to be generated for how similar the two phenotypes being considered are and how specific the match is.

Finally, PHIVE ranks candidate genes according to the average of variant scores and phenotypic relevance scores.

27.1.2 hiPHIVE

The original PHIVE algorithm was extended to hiPHIVE, which stands for "human/interaction PHIVE". The hiPHIVE method calculates phenotypic similarity not only for mouse, but also for zebrafish and human data. The zebrafish data originate from the Zebrafish Model Organism Database (ZFIN) [170] and human phenotype data derived from the HPO database (see Chapter 26). This allows known (human) disease-associated genes to be flagged as potential diagnoses, and novel candidate genes can still be prioritized as in the original PHIVE algorithm. A prioritizer is used that assesses the match of the genotypes of candidate genes to the inheritance mode of the known disease associated with the genes. Additionally, for genes with no phenotype data from any of these sources, a random walk with restart algorithm (Chapter 25) is used to score how close the candidate is in the a protein-protein association network to genes with strong phenotypic similarity to the patient.

27.2 RUNNING THE EXOMISER: A TUTORIAL

The Exomiser is a self-contained software package that was designed to be able to run efficiently on standard consumer laptops with modest hardware requirements. It requires a 64 bit Windows, MacIntosh, or Linux operating system with 4–12Gb RAM, at least 50 GB free disk space, and Java 8 or newer.

The new version of the Exomiser is available at the Monarch Initiative [288, 300] website:

```
http://monarch-exomiser-web-dev.monarchinitiative.org/exomiser/
```

Source code can be downloaded here:

```
https://data.monarchinitiative.org/exomiser/
```

For this book, we have used version 7.2.1. Download the zipped files for the Exomiser distribution and data, and optionally the sha256 checksum file.

```
exomiser-cli-7.2.1-data.zip
exomiser-cli-7.2.1-distribution.zip
exomiser-cli-7.2.1.sha256
```

The checksum can be used to check the integrity of the downloaded files. It is calculated by means of a hash function, and if the downloaded file is identical to the file one the server, than the results should be identical. We can check this as follows:

```
$ sha256sum -c exomiser-cli-7.2.1.sha256
exomiser-cli-7.2.1-distribution.zip: OK
exomiser-cli-7.2.1-data.zip: OK
```

If there was some problem, then FAILED would appear instead of OK for one or both files. Unzip the files.[2]

---

[2]On Windows you can use 7-zip: http://www.7-zip.org/.

```
$ unzip exomiser-cli-7.2.1-distribution.zip
$ unzip exomiser-cli-7.2.1-data.zip
```

If installation was successful, then the Exomiser is ready to go. To test it, we perform whole-exome sequencing (WES) analysis on a test dataset that is provided in the Exomiser installation. The WES VCF file was produced by adding a published mutation associated with Pfeiffer syndrome to a publicly available WES VCF file from an individual without known Mendelian disease (see Chapter 4). Pfeiffer syndrome can be caused by heterozygous mutations in the *FGFR2* gene and is characterized by craniosynostosis (premature fusion of skull bones in infants), wide thumbs and big toes, and other features. The mutation that we added to the VCF file, NM_022970.3:c.1697A>C, leads to a missense mutation in the tyrosine kinase 1 (TK1) domain of the fibroblast growth factor receptor 2, p.(E566A), and was found in a severely affected individual with Pfeiffer syndrome [480]. To run the Exomiser, we change to the Exomiser directory created after unpacking the archive, and run the following command.

```
$ cd exomiser-cli-7.2.1
$ java -Xms2g -Xmx4g -jar exomiser-cli-7.2.1.jar \
    --analysis test-analysis-exome.yml
```

This command causes the Exomiser to perform an analysis with the parameters indicated in the yml settings file. We will explain this in a moment, but first let us examine the HTML file produced by the Exomiser with a summary of the results. The file is written to the **results** subdirectory and is called **Pfeiffer-hiphive-exome-SPARSE.html**. It can be opened with any Web browser. This file has five main sections.

Analysis settings

This section shows the parameters set in the file passed to Exomiser via the --**analysis** argument (in this case, **test-analysis-exome.yml**), whereby the comments are not shown. This is useful to document the way in which a result was reached.

Filtering summary

This section provides an overview of the filtering procedure. In this case, the Target Filter removes variants that were off the exome target (e.g., intergenic variants) and variants judged to have a low pathogenic potential (e.g., synonymous variants). For our test exome, a total of 7925 variants passed this filter, and 29,375 failed and were excluded from further analysis. Similarly, the frequency filter shows that 593 variants in the exome had an annotated population frequency (minor allele frequency) of 1% or less. The data sources used to make this assessment were shown under `frequencySources` in the Analysis Settings section. The Exomiser optionally uses a Pathogenicity Filter to discard variants predicted to be non-pathogenic. In this case, the settings file specified that this filter was not to be applied, and thus no variants were discarded in this step. A total of 593 variants were used for gene prioritization.

Variant type distribution

This section provides an overview of the categories of variants found in the VCF file being analyzed. In the example exome, for instance, there were 5328 missense variants, 17,882 coding-transcript intron variants, and so on. It is a good idea to become familiar with the typical distribution of variant categories in exome and genome data so as to be able to recognize deviations from the norm, which may indicate technical problems (for instance if the capture procedure did not work well, there may be fewer on-target variants than otherwise expected).

Prioritised genes

This section displays a list of genes ranked according to the estimated likelihood of the gene being responsible for the disease. Let us examine the first gene in the list, *FGFR2* (Figure 27.1).

The full entry for *FGFR2* also shows that there was a phenotypic similarity of 0.804 to a mouse mutant involving *Fgfr2* as well as proximity to SALL4 in the protein interactome; SALL4 is assessed to be of potential relevance because mutations in *SALL4* are associated with Duane-radial ray syndrome, which displays a number of phenotypic similarities; For instance, the HPO term *Broad thumb* (HP:0011304) was matched to *Triphalangeal thumb* (HP:0001199) in Duane radial ray syndrome. The Exomiser prioritizes FGFR2 because it has a first degree interaction with SALL4, which is phenotypically relevant. The assumption of this method is that mutations in genes that encode proteins that are close to one another in the interaction network (first-degree and other protein protein interactions) are likely to be associated with phenotypically similar diseases (see Chapter 25).

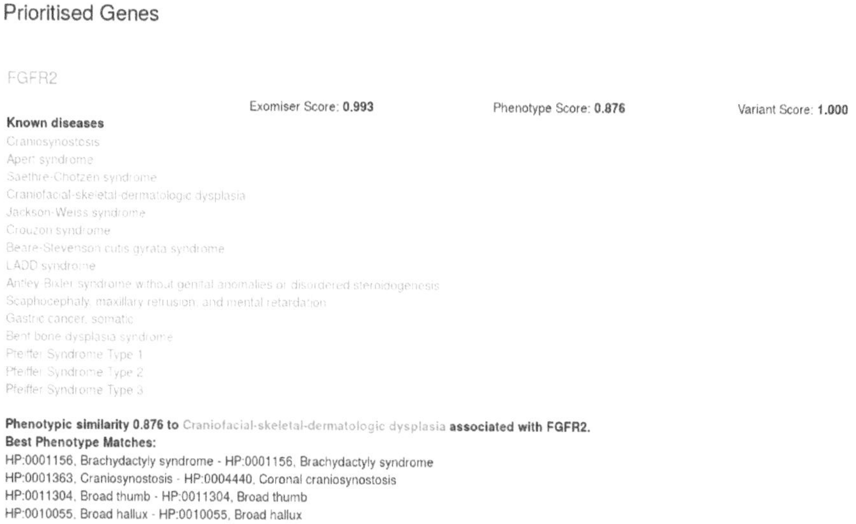

Prioritised Genes

FGFR2

Exomiser Score: **0.993**	Phenotype Score: **0.876**	Variant Score: **1.000**

Known diseases
Craniosynostosis
Apert syndrome
Saethre-Chotzen syndrome
Craniofacial-skeletal-dermatologic dysplasia
Jackson-Weiss syndrome
Crouzon syndrome
Beare-Stevenson cutis gyrata syndrome
LADD syndrome
Antley-Bixler syndrome without genital anomalies or disordered steroidogenesis
Scaphocephaly, maxillary retrusion, and mental retardation
Gastric cancer, somatic
Bent bone dysplasia syndrome
Pfeiffer Syndrome Type 1
Pfeiffer Syndrome Type 2
Pfeiffer Syndrome Type 3

Phenotypic similarity 0.876 to Craniofacial-skeletal-dermatologic dysplasia **associated with FGFR2.**
Best Phenotype Matches:
HP:0001156, Brachydactyly syndrome - HP:0001156, Brachydactyly syndrome
HP:0001363, Craniosynostosis - HP:0004440, Coronal craniosynostosis
HP:0011304, Broad thumb - HP:0011304, Broad thumb
HP:0010055, Broad hallux - HP:0010055, Broad hallux

Figure 27.1. Exomiser. Screenshot of the Exomiser showing the top ranked gene, *FGFR2*. The Exomiser score is a composite score derived by logistic regression from the Phenotype score and the Variant score. Those genes that both contain a variant predicted to be pathogenic (variant score) and are associated with human diseases or mouse or zebrafish models that are phenotypically similar to the phenotypic profile entered by the user get high Exomiser scores. In this case, the top-ranked gene, *FGFR2*, is associated with multiple human diseases including Pfeiffer syndrome as well as craniofacial skeletal dermatologic dysplasia, which was assigned a high phenotype matching score on the basis of the HPO terms shown in the figure.

The Exomiser output includes a link to the STRING [415] website, in which the interactions leading to the prioritization result can be visualized (Figure 27.2). The STRING interaction network is defined by high-confidence (>0.7) interactions from STRING (Search Tool for the Retrieval of Interacting Genes/Proteins), version 9.05. The high-confidence interactions include direct (physical) and indirect (functional) protein–protein interactions, as well as associations transferred by orthology from other species or obtained through text mining.

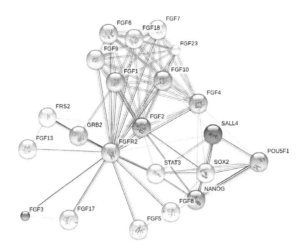

Figure 27.2. FGFR2 is near to SALL4 in the protein interaction network (STRING screenshot [415]).

27.3 OUTPUT FILES

The Exomiser can be run as a stand-alone application that will output an HTML page that summarizes the results of the analysis. Exomiser also can be made to output tab-separated value (TSV) and VCF files containing details on the analysis results in a form that can be easily used as input for larger analysis pipelines. Table 27.1 shows data from the `genes.tsv` file taken from the line for FGFR2 in our example exome. The Exomiser gene phenotype score represents the phenotypic relevance score for FGFR2 (0.8763), being the maximum score from human, mouse and fish phenotype score. The three fields for human, mouse and fish phenotype evidence show the source for the corresponding score. For instance, the human score is derived from a match for `Craniofacial-skeletal-dermatologic dysplasia` (`OMIM:101600`):

```
Brachydactyly syndrome(HP:0001156)-Brachydactyly syndrome(HP:0001156),
Craniosynostosis (HP:0001363)-Coronal craniosynostosis (HP:0004440),
Broad thumb (HP:0011304)-Broad thumb (HP:0011304),
Broad hallux (HP:0010055)-Broad hallux (HP:0010055).
```

The walker score is derived based on proximity of other genes to the current gene using random walk analysis as described in Chapter 25 [41]. The sources for the walker score are shown in the PPI evidence fields. For instance, the walker score of 0.512 was derived from a match to SALL4, which is a direct neighbor in the interactome and is associated with similar phenotypic abnormalities (as described in the previous section).

Table 27.1. Exomiser TSV output.

Item	Example
Gene symbol	FGFR2
Entrez Gene ID	2263
Exomiser gene-pheno score	0.8763
Exomiser gene-variant score	1.000
Exomiser gene combined score	0.9933
Human pheno score	0.8763
Mouse pheno score	0.8039
Fish pheno score	0.000
Walker score	0.5102
PHIVE all species score	0.8763
OMIM score	1.000
Matches candidate gene	0
Human pheno evidence	
Mouse pheno evidence	(see text)
Fish pheno evidence	
Human PPI evidence	
Mouse PPI evidence	(see text)
Fish PPI evidence	

The `variants.tsv` file contains information about each variant, including the frequency information, and the scores from Mutation-Taster, Polyphen2, and SIFT that are used to derive the Exomiser variant score. The file should be largely self-explanatory, and we will offer several exercises for readers who would like to work more closely with these files.

The main difference between the `variants.tsv` and the VCF file (other then the obvious format difference) is that the TSV file has one line per allele in the case of 1/2 genotypes[3]

---

[3]i.e., heterozygosity for two non-reference alleles.

27.4 RUNNING THE EXOMISER: YAML FILES

The analysis procedures performed by the Exomiser are controlled by the YAML file (it is also possible to use analogous command line arguments to set analysis parameters, but we will not cover this here).

YAML[4] stands for *YAML Ain't Markup Language*, being a simple data serialization language that is intended to be used for configuration files and other applications. YAML is really very simple. We will not attempt to provide a full explanation — for this book it will suffice to say that YAML uses spaces (not tabs) to denote structure, shows lists with a leading hyphen (-), one per line, and associated arrays are represented by showing key value pairs as the key followed by a colon followed by a space followed by the value.

We will use the file `test-multisample.yml` (see below) to illustrate how to set the parameters. This file controls the analysis of a four-sample VCF with simulated data, with two parents, an unaffected son, and an affected daughter. In this simulation, the daughter is the only affected person in the family, and we run the Exomiser using a filter for autosomal dominant diseases. In this simulation, Pfeiffer syndrome has occurred in the daughter as the result of a *de novo* mutation, which is not rare in this disease [444].[5]

```
analysis:
    vcf: data/Pfeiffer-quartet.vcf
    ped: data/Pfeiffer-quartet.ped
    modeOfInheritance: AUTOSOMAL_DOMINANT
    analysisMode: SPARSE
    geneScoreMode: RAW_SCORE
    hpoIds: ['HP:0001156', 'HP:0001363',
      'HP:0011304', 'HP:0010055']
    frequencySources: [
        THOUSAND_GENOMES,
        ESP_AFRICAN_AMERICAN, ESP_EUROPEAN_AMERICAN,
        ESP_ALL, EXAC_AFRICAN_INC_AFRICAN_AMERICAN,
```

---

[4]http://www.yaml.org/

[5]For simplicity, we just show the analysis for autosomal dominant, whereby the search for a de novo mutation is performed because the PED file indicates that only the daughter is affected and we have limited to the mode of inheritance to autosomal dominant in the YAML file. In general, one would consider other modes of inheritance such as autosomal recessive if the parents are healthy and a child is affected.

```
        EXAC_AMERICAN, EXAC_SOUTH_ASIAN,
        EXAC_EAST_ASIAN, EXAC_FINNISH,
        EXAC_NON_FINNISH_EUROPEAN, EXAC_OTHER
        ]
    pathogenicitySources: [POLYPHEN, MUTATION_TASTER, SIFT]
    steps: [
        variantEffectFilter: {remove: [UPSTREAM_GENE_VARIANT,
            INTERGENIC_VARIANT,
            REGULATORY_REGION_VARIANT,
            CODING_TRANSCRIPT_INTRON_VARIANT,
            NON_CODING_TRANSCRIPT_INTRON_VARIANT,
            SYNONYMOUS_VARIANT,
            DOWNSTREAM_GENE_VARIANT,
            SPLICE_REGION_VARIANT]},
        frequencyFilter: {maxFrequency: 0.01},
        pathogenicityFilter: {keepNonPathogenic: false},
        inheritanceFilter: {},
        omimPrioritiser: {},
        hiPhivePrioritiser: {},
    ]
outputOptions:
    outputPassVariantsOnly: false
    numGenes: 10
    outputPrefix: results/PfeifferQuartet
    outputFormats: [TSV-GENE, TSV-VARIANT, VCF, HTML]
```

Assuming this file is stored as `test-multisample.yml`, the Exomiser can be run with the following command.

```
$ java -jar exomiser-cli-7.2.1.jar \
    --analysis test-multisample.yml
```

The YAML files provided with the Exomiser provide a starting point for creating your own files, and are divided into sections. The required options need to be filled in either in the YAML settings file or on the command-line or the Exomiser cannot run. There is currently only one required item, the path to the VCF file that is to be analyzed.

```
vcf: data/Pfeiffer-quartet.vcf
```

The sample data items contain the path to the PED file, which is required only for multi-sample VCF files. For single-sample VCF files, this entry can be left blank (i.e., "`ped: `").

`ped: data/Pfeiffer-quartet.ped`

The other items indicate the HPO terms that describe the clinical features of the patient (`hpoIds`), the sources of population frequency data to be used for frequency filtering (see Chapter 22), and the pipeline to be used for the filters and prioritizers. In this case, Exomiser is instructed to apply the variant effect filter (assessing the predicted pathogenicity of variants), the frequency filter ("`frequencyFilter: {maxFrequency: 0.01}`"), which will remove variants whose population frequency in any of the indicated data sources is 1% or more, the pathogenicity filter (variants assessed as non-pathogenic will be discarded), the inheritance filter (which filters genes according to compatibility with autosomal dominant inheritance according to the PED file — if we had indicated "`modeOfInheritance: UNDEFINED`", then no filtering would be applied), and so-called OMIM filter (which checks whether the mode of inheritance matches that of the diseases), and finally the hiPHIVE prioritizer, which is an improved version of PHIVE (for details of the algorithm, see [41]).

The output options cause Exomiser to generate TSV, VCF and HTML files for the top 10 genes (if this is left blank, all genes will be shown). The output prefix indicates the base name that will be given to the generated files. It can be seen from these files that the *FGFR2* gene was again ranked in top place (the HTML file is probably the easiest to understand at a glance).

The YAML files available with the Exomiser distribution contain copious comments (lines or ends of lines that start with #). The comments help to explain how to customize the Exomiser analysis for specific goals. Several exercises at the end of this chapter will provide practice in how to write own YAML configuration files for the Exomiser. The following sections will briefly present the configurations needed to run different algorithms within the Exomiser framework.

27.5 EXOMEWALKER

The ExomeWalker application was described in Chapter 25. The basic assumption of this application is that we are searching for a novel or known disease gene that belongs to a disease-gene family, i.e., a group of related genes that tend to cluster together in the protein-protein association network. We will again use the Pfeiffer exome for our example, but we will now search for the causative gene (*FGFR2*) by starting from a "family" of related genes: *FGFR1* (Gene id: 2260), *FGF1* (Gene id: 2246), and *FGF8* (Gene id: 2253). The YAML configuration file is as follows:

```
analysis:
    vcf: data/Pfeiffer.vcf
    ped:
    modeOfInheritance: UNDEFINED
    analysisMode: SPARSE
    geneScoreMode: RAW_SCORE
    hpoIds:
    frequencySources: [
        THOUSAND_GENOMES,
        ESP_ALL]
    pathogenicitySources: [POLYPHEN, MUTATION_TASTER,
      SIFT]
    steps: [
        variantEffectFilter: {remove:
            [UPSTREAM_GENE_VARIANT,
            INTERGENIC_VARIANT,
            REGULATORY_REGION_VARIANT,
            CODING_TRANSCRIPT_INTRON_VARIANT,
            NON_CODING_TRANSCRIPT_INTRON_VARIANT,
            SYNONYMOUS_VARIANT,
            DOWNSTREAM_GENE_VARIANT,
            SPLICE_REGION_VARIANT]},
        frequencyFilter: {maxFrequency: 0.01},
        pathogenicityFilter: {keepNonPathogenic: false},
        inheritanceFilter: {},
        exomeWalkerPrioritiser:
          {seedGeneIds: [2260,2246,2253]}
    ]
outputOptions:
    outputPassVariantsOnly: false
    numGenes: 20
    outputFormats: [TSV-GENE, TSV-VARIANT, VCF, HTML]
```

This file is similar to the previous one except that we have indicated the ExomeWalker prioritizer and passed the Entrez Gene identifiers of the three genes mentioned above. The result of the analysis is that the *FGFR2* gene is the only gene to receive a non-zero ExomeWalker score, and the variant again receives the top score, so that the *FGFR2* gene is prioritized in first place.

27.6 PHENIX

The PhenIX application (Phenotypic Interpretation of eXomes) was designed to evaluate the results of a clinical exome (often referred to as a "Mendeliome") that essentially enriches exons as with exome sequencing but restricts the capture probes to genes known to have a disease association. In our initial study we designed a panel with 2741 well established Mendelian disease genes. PhenIX ranks variants based on pathogenicity and assesses the genes according to semantic similarity with Mendelian diseases with known etiology (3991 at the time of publication). PhenIX is thus not intended to be used in research settings to search for novel disease genes, but instead provides an application that can be used in diagnostic settings in which research is not possible or not desired. In a prospective study on 40 individuals who had not received a diagnosis despite intensive medical genetics workup, PhenIX was able to establish a definitive diagnosis in 11 cases [481]. To run PhenIX, this line needs to be uncommented in the YAML line, and the other prioritizers need to be commented out.

```
phenixPrioritiser: {}
```

27.7 GENOMISER

The Genomiser application was designed to provide a method to interpret small non-coding variants (single-nucleotide substitutions and indels less than 25 nt), which still constitutes a major challenge in the application of whole-genome sequencing in Mendelian disease. Although a number of methods have been developed to predict the pathogenicity of non-coding variants [197, 241, 366, 487], none of these methods alone is capable of reliably identifying disease-causing non-coding mutations in WGS data. We adapted our phenotype driven approach to the prioritization of coding and non-coding variants by extending Exomiser to include a novel machine-learning score for non-coding variants and a framework that includes not only the coding portions of genes but also relevant regulatory regions such as promoter, untranslated regions (UTRs) and enhancers, in addition to non-coding RNA genes [403]. We will refer to the original publication for details about the algorithms.

Genomiser is run as an Exomiser application with appropriate settings of the YAML file.

```
analysis:
    vcf: data/NA19722_601952_AUTOSOMAL_RECESSIVE_\
      POMP_13_29233225_5UTR_38.vcf.gz
    ped:
    modeOfInheritance: AUTOSOMAL_RECESSIVE
    analysisMode: PASS_ONLY
    geneScoreMode: RAW_SCORE
    hpoIds: ['HP:0000982',
             'HP:0001036',
             'HP:0001367',
             'HP:0001795',
             'HP:0007465',
             'HP:0007479',
             'HP:0007490',
             'HP:0008064',
             'HP:0008404',
             'HP:0009775']
    frequencySources: [
        THOUSAND_GENOMES,
        ESP_AFRICAN_AMERICAN, ESP_EUROPEAN_AMERICAN,
        ESP_ALL, EXAC_AFRICAN_INC_AFRICAN_AMERICAN,
        EXAC_AMERICAN, EXAC_SOUTH_ASIAN,
        EXAC_EAST_ASIAN, EXAC_FINNISH,
        EXAC_NON_FINNISH_EUROPEAN, EXAC_OTHER
        ]
    pathogenicitySources: [POLYPHEN,
      MUTATION_TASTER, SIFT, REMM]
    steps: [
        hiPhivePrioritiser: {},
        priorityScoreFilter:{priorityType:HIPHIVE_PRIORITY,
          minPriorityScore: 0.501},
        variantEffectFilter: {remove: [SYNONYMOUS_VARIANT]},
        regulatoryFeatureFilter: {},
        frequencyFilter: {maxFrequency: 1.0},
        pathogenicityFilter: {keepNonPathogenic: true},
        inheritanceFilter: {},
        omimPrioritiser: {}
```

```
]
outputOptions:
    outputPassVariantsOnly: false
    numGenes: 50
    outputPrefix: results/POMP
    outputFormats: [HTML, TSV-GENE, TSV-VARIANT, VCF]
```

The overall structure of the YAML file is similar for the Genomiser and the Exomiser. To run the Genomiser, add REMM to the pathogenicitySources. The ReMM score is the machine learning score for non-coding variants developed for the Genomiser [403]. We recommend running the prioritizer followed by a priorityScore-Filter in order to remove genes which are least likely to contribute to the phenotype defined in hpoIds. This will dramatically reduce the time and memory required to analyze a genome. A threshold of 0.501 is a good compromise to select good phenotype matches and the best protein–protein interactions hits from hiPHIVE. The regulatoryFeatureFilter removes all non-regulatory, non-coding variants over 20 kb from a known gene. Also worth nothing that the analysisMode is set to PASS_ONLY which will completely discard any variants that fail an analysis filter from memory. This is also a major contributor to being able to run this on a genome.

This analysis identifies the mutation NM_015932.5:c.-95delC in the *POMP* gene. The simulated patient has the disease keratosis linearis with ichthyosis congenita and sclerosing keratoderma (KLICK; MIM:601952; ORPHA:281201). The phenotypic features of the patient that were passed to Genomiser are shown in Table 27.2. The mutation is located in the 5' UTR of the *POMP* gene, and results in an abnormality of gene regulation that leads to altered distribution of POMP in epidermis and a perturbed formation of the outermost layers of the skin [93]. The *POMP* gene is the best hit in this simulation, meaning that the Exomiser provided a effective prioritization of the 4,333,911 variants in the corresponding VCF file.

FURTHER READING

Robinson PN, Köhler S, Oellrich A, Sanger Mouse Genetics Project, Wang K, Mungall CJ, Lewis SE, et al. (2014) Improved exome prioritization of disease genes through cross-species phenotype comparison. *Genome Res* **24**:340–8.

Table 27.2. HPO annotations for KLICK

ID	Name
HP:0000982	Palmoplantar keratoderma
HP:0001036	Parakeratosis
HP:0001367	Abnormal joint morphology
HP:0001795	Hyperconvex nail
HP:0007465	Honeycomb palmoplantar keratoderma
HP:0007479	Congenital nonbullous ichthyosiform erythroderma
HP:0007490	Linear arrays of macular hyperkeratoses in flexural areas
HP:0008064	Ichthyosis
HP:0008404	Nail dystrophy
HP:0009775	Amniotic constriction ring

Note: The HPO terms used to annotate the simulated patient with keratosis linearis with ichthyosis congenita and sclerosing keratoderma (KLICK; MIM:601952).

Smedley D, Jacobsen JO, Jäger M, Köhler S, Holtgrewe M, Schubach M, Siragusa E, et al. (2015) Next-generation diagnostics and disease-gene discovery with the Exomiser. *Nat Protoc* **10**:2004–15.

Zemojtel T, Köhler S, Mackenroth L, Jäger M, Hecht J, Krawitz P, Graul-Neumann L, et al. (2014) Effective diagnosis of genetic disease by computational phenotype analysis of the disease-associated genome. *Sci Transl Med* **6**:252ra123.

Smedley D, Schubach M, Jacobsen JO, Köhler S, Zemojtel T, Spielmann M, Jäger M, et al. (2016) A whole-genome analysis framework for effective identification of pathogenic regulatory variants in Mendelian disease. *Am J Hum Genet* **99**:595–606.

27.8 EXERCISES

The best way to learn how to use the Exomiser is obviously to use it on exome data. We will assume that at least some readers of this book do not have access to "real-life" cases with exome or genome data and other medical data. We will show here how to simulate datasets based on the medical literature in a way that will allow all readers to try the features of the Exomiser. We will obtain data on mutations and clinical features and disease diagnoses from the medical literature.

Readers can then use the data to run the Exomiser. We recommend trying different parameters and assessing the influence on the rankings (obviously, if the Exomiser works well, then the mutations we used will be prioritized in first place or at least within the top ranks).

In the following exercises, we will present several examples. Readers may want to choose simulated "cases" that are in their own sphere of interest. To simulate a "case", one can obtain an exome file from the 1000 Genomes Project website,[6] and then find a disease-causing mutation from the medical literature. Then, "spike" the mutation into the 1000-Genomes VCF file. Note that you will need to determine the chromosomal location of the mutation, which often is not given in the original publication. We have found that often one can use the Mutalyzer [459] and the UCSC Genome Browser [379] (including the BLAT tool) to identify the chromosomal position of published mutations. Make sure that you use the same genome build for the mutations as is used in the VCF file.

Then, identify the HPO terms corresponding to the description of the patient in the published article. There are many ways of doing this, but you may find the HPO Browser at the HPO website to be useful for this purpose.[7]

Exercise 1

In this exercise, we will simulate a patient with Noonan syndrome. The patient was found to have a *de novo* mutation in the *NRAS* gene, NM_002524.4:c.149C>T, corresponding to p.(Thr50Ile) [78]. The variant is classified as pathogenic in ClinVar [235] (variant ID 13902), and the information in the ClinVar entry makes it easy to determine the chromosomal location of the variant. We will use the GRCh37 genome reference for this exercise, and the location of the mutation is NC_000001.10:g.115256562G>A on chromosome 1 (the *NRAS* gene is on the negative strand of chromosome 1.[8] It is helpful to examine variants in the UCSC Genome Browser [379] to confirm the position, bases, and strand.

We can easily add ("spike in") the variant to a "normal" VCF file. We will show how to do it with the SonsVCFfile.vcf from the

---

[6]http://www.internationalgenome.org/

[7]http://human-phenotype-ontology.github.io/

[8]Hence the chromosome sequence has the substitution G>A at that position, not C>T as reported for the coding sequence

Corpasome, but any VCF file will do (be sure that the genome assembly used for the mutation matches that of the VCF file).

For simplicity, copy a line corresponding to a heterozygous variant with good quality values, and replace the chromosome, position, as well as the reference and alternate bases.

```
1   115256562  .  G  A  421.49  PASS  AC=1;(...)  \
   GT:AD:DP:GQ:PL  0/1:8,17:25:99:451,0,257
```

Add this line as the last line of SonsVCFfile.vcf (Note that the order of the variants in the VCF file does not matter for the Exomiser). Consulting the supplementary table 1 of the original publication, we see that one of the patients found to have a *de novo* T50I mutation was a 14 year-old boy with the phenotypic features shown in Table 27.3. For the exercise, create a YAML settings file (or modify an existing one), and add some of the HPO IDs from the table. Explore the influence of various settings of the prioritizers and the thresholds for the filters. If all goes well, the *NRAS* gene should be prioritized in first place.

Table 27.3. The HPO terms used to annotate the patient with Noonan syndrome and a T50I mutation in the *NRAS* gene.

ID	Name
HP:0010880	Increased nuchal translucency
HP:0001561	Polyhydramnios
HP:0001639	Hypertrophic cardiomyopathy
HP:0011675	Arrhythmia
HP:0004322	Short stature
HP:0000256	Macrocephaly
HP:0000465	Webbed neck
HP:0000028	Cryptorchidism
HP:0001270	Motor delay
HP:0040180	Hyperkeratosis pilaris
HP:0000545	Myopia
HP:0002212	Curly hair
HP:0001762	Talipes equinovarus

Exercise 2

Cystinosis is an autosomal recessive disease characterized by elevated cystine levels, leading to cystine accumulation in tissues and failure to thrive, impaired renal function, ocular damage and hypothyroidism. Most individuals with cystinosis develop end-stage renal disease by the age of 10 years unless therapy is instituted at an early age. The homozygous mutation NM_004937.2:c.414G>A, corresponding to p.(Trp138Ter), in the *CTNS* gene, can cause cystinosis. For this exercise, go to the Human Phenotype Ontology website[9] and find the HPO terms associated with nephropathic cystinosis by using the HPO Browser. The chromosomal coordinates of the mutation are NC_000017.10:g.3558599G>A (on chromosome 17 of the genome assembly GRCh37). Spike a line into a VCF file (note that you will need to use the homozygous ALT genotype, 1/1). Create a YAML settings file, run the Exomiser, and explore the influence of various settings of the prioritizers and the thresholds for the filters as in the previous exercise. If all goes well, the *CTNS* gene should be prioritized in first place.

---

[9]http://human-phenotype-ontology.github.io/

CHAPTER 28

Medical Interpretation

Peter Robinson and Johannes Zschocke

B IOINFORMATICS is rapidly maturing to a powerful discipline that is poised to transform the way genomic medicine is performed. However, current algorithms, programs, and frameworks for bioinformatic analysis of WES/WGS data are far from being able to reliably interpret WES/WGS data without expert human interpretation. Thus, at best, bioinformatics pipelines can be regarded as systems for experts, rather than computational expert systems that replace the judgment of a physician. Therefore, the bioinformatic analysis of medical WES/WGS data should be viewed and discussed in the medical context, taking the clinical features of the investigated individual and the specific relevant questions into consideration. This is best done in close collaboration with the physicians involved in the care of the individual.

Some centers involved in medical genomics have begun to hold "genome conferences" similar in intent to the "tumor conferences" that are a standard part of medical care in oncology. Tumor conferences bring together multidisciplinary panels of physicians (surgeons, medical oncologists, radiation oncologists, pathologists and others) and other healthcare professionals to review each patient's details and to debate the best course of treatment given the diagnostic findings and the personal situation of the patient. We suggest that for optimal care of individuals with suspected Mendelian disease who have undergone genomic diagnostics, analogous conferences are required that bring together bioinformaticians, geneticists, and other relevant clinicians to discuss potential or likely relevance of the called variants in the light of the clinical picture. In our experience, this is an extremely effective way of assessing the list of candidate genes. Even the best computational analysis is not reliable enough to make the diagnosis without very substantial human intervention. Sending DNA samples to remote, centralized laboratories that provide a report without interaction with the clinicians is unsatisfactory and may result in suboptimal medical advice. Considering that regional centers cannot normally afford the expensive technical infrastructure, a possible solution may be training of clinical (molecular) geneticists in regional centers in the evaluation and interpretation of exome data, using the bioinformatic infrastructure of large national or international sequencing service providers.

In this chapter, we will briefly review several situations in which medical knowledge can be essential for the correct interpretation of WES/WGS data. The chapter makes no attempt to provide an exhaustive list of such situations, but may give the readers of this book an idea of the type of situation where medical input can be most helpful.

28.1 ONLY A SINGLE HETEROZYGOUS MUTATION FOR AN AUTOSOMAL RECESSIVE DISEASE (I)

> WES has been performed as part of the diagnostic examination of a child with progressive neurological deterioration. A single heterozygous nonsense mutation in the *SGSH* gene is found. Mutations in *SGSH* cause mucopolysaccharidosis type III (Sanfilippo syndrome). This disease is inherited in an autosomal recessive fashion, and therefore, two heterozygous mutations (i.e., compound heterozygous mutations) or one homozygous mutation are required for the development of disease. The finding of a single heterozygous mutations is not diagnostic for Sanfilippo syndrome.
> **How should we proceed?**

Sanfilippo syndrome is an autosomal recessive lysosomal storage disease with progressive neurological symptoms and impaired cognitive function caused by impaired degradation of heparan sulfate [135]. Since the clinical picture of the child matches well with the disease, one should search for a second mutation that might have been missed by

WES. In fact, one case of a gross deletion of *SGSH* has been reported, which may been missed by WES.[1]

The important lesson to be learned here is that WES/WGS are not able to detect all causative variants, i.e., WES/WGS analysis has a limited sensitivity that is substantially lower than 100%. If there is a pressing clinical suspicion, then additional diagnostic tests may be indicated to determine whether the person being examined is (by chance) a heterozygous carrier for the identified mutation or whether the second mutation has been missed by the analysis.

28.2 ONLY A SINGLE HETEROZYGOUS MUTATION FOR AN AUTOSOMAL RECESSIVE DISEASE (II)

> In our second example, WES has been performed as part of the diagnostic examination of a child with progressive neurological deterioration. A single heterozygous nonsense mutation in the *G6PC* gene is found. Mutations in the *G6PC* gene are known to cause the autosomal recessively inherited disorder glycogen storage disease Ia. **How should we proceed?**

Although the general situation seems similar to our first example, in this case one should look elsewhere to find the cause of the progressive neurological deterioration. Homozygous or compound heterozygous mutations in *G6PC*, the gene for glucose-6-phosphatase, cause the autosomal recessive glycogen storage disease type I (GSDI). This condition is characterized by the inability to release glucose from the cellular glycogen stores, causing severe hypoglycemia with seizures and coma as well as other metabolic abnormalities at short fasting periods. Accumulation of cellular glycogen leads to the massive enlargement of various organs, e.g., liver and kidneys. However, GSDI is not characterized by progressive developmental delay (unless perhaps caused by chronic or recurrent hypoglycemia) [199]. Since the clinical picture does not match, we should search elsewhere for the diagnosis. It is important to realize that it is the norm that any individual in the population is a carrier for multiple mutations in genes associated with autosomal recessive disease. The carrier status does not affect the health of the

---

[1]If a heterozygous deletion encompasses the exon in which a heterozygous variant is located on the other chromosome, the variant would be called as homozygous ALT. However, if the deletion affects other exons, then the deletion can be missed by WES.

individual, and the finding of a heterozygous mutation in genes associated with autosomal recessive disease is not in itself an indication for further medical workup.

The only difference between our first two examples was the clinical analysis — do the clinical findings of the individual being sequenced match with the findings expected for the diseases associated with the genes found to have heterozygous mutations by exome sequencing?

28.3 NOT ALL CATEGORIES OF MUTATION CAN BE DETECTED BY WES/WGS

> We fail to find a mutation despite a relatively clear clinical suspicion of Mendelian disease. **How should we proceed?**

It is important to realize that short-read WES/WGS technologies are limited with respect to their ability to detect certain categories of genomic variation. It is important to consider alternate diagnostic investigations if there is a strong index of suspicion and the findings of WES/WGS were negative. We will briefly discuss four examples, but this list is not exhaustive.

28.3.1 Single-exon deletions

Large deletions that encompass whole exons are possible causative mutations in most Mendelian diseases due to genetic loss-of-function. Identification of such mutations requires quantitative analyses which are not always reliable, especially when the coverage of the respective exon is generally poor. This is a particular problem for autosomal dominant diseases where there is no second mutation that may point towards a particular diagnosis. However, it remains difficult to identify small SNVs in exome data, although recent work shows that about half of single-exon deletions predicted from WES data could be independently confirmed [362], but it is more difficult to determine how many single-exon deletions are missed by routine WES. WGS may be more reliable for the detection of chromosomal rearrangements associated with single-exon deletion.

28.3.2 Mutations in enhancers and regulatory sequences

There is a limb-specific enhancer of the sonic hedgehog gene that is located about one million base pairs away from the SHH transcription start site. Mutations in this enhancer cause skeletal defects such as polydactyly [165]. Regulatory sequences are not represented on current exome designs, and thus variants in enhancers are not assessed by current exome technology.

28.3.3 Pseudogenes and paralogs

Congenital adrenal hyperplasia (CAH) is a group of autosomal recessive diseases that are characterized by defects in the ability of the adrenal glands to produce one or more of the corticosteroids, mineralocorticoids, and androgens (which are related groups of hormones). The most common type of CAH is caused by mutations in the gene for 21-hydroxylase, *CYP21A2*. A related pseudogene *CYP21A2P* is located near this gene; gene conversion events involving the functional gene and the pseudogene account for many cases of 21-hydroxylase deficiency. The pseudogene has a number of differences compared to the actual gene including premature stop codons. The following alignment shows a region where the pseudogene sequence is nearly identical with that of the actual gene (only one of 202 nucleotides is a mismatch).

```
CYP21A2   ATGCTGCTCCTGGGCCTGCTGCTGCTGCTGCCCCTGCTGGCTGGCGCCCGCCTGCTGTGG
          ||||||||||||||||||||||||||||||||||||||||||||||||||||||||||||
CYP21A2P  ATGCTGCTCCTGGGCCTGCTGCTGCTGCTGCCCCTGCTGGCTGGCGCCCGCCTGCTGTGG

CYP21A2   AACTGGTGGAAGCTCCGGAGCCTCCACCTCCCGCCTCTTGCCCCGGGCTTCTTGCACCTG
          |||||||||||||||||||||||||||||||||| |||||||||||||||||||||||||
CYP21A2P  AACTGGTGGAAGCTCCGGAGCCTCCACCTCCTGCCTCTTGCCCCGGGCTTCTTGCACCTG

CYP21A2   CTGCAGCCCGACCTCCCCATCTATCTGCTTGGCCTGACTCAGAAATTCGGGCCCATCTAC
          ||||||||||||||||||||||||||||||||||||||||||||||||||||||||||||
CYP21A2P  CTGCAGCCCGACCTCCCCATCTATCTGCTTGGCCTGACTCAGAAATTCGGGCCCATCTAC

CYP21A2   AGGCTCCACCTTGGGCTGCAAG...
          ||||||||||||||||||||||
CYP21A2P  AGGCTCCACCTTGGGCTGCAAG...
```

It can be appreciated that short-read analysis (WES or WGS) may not be able to distinguish a mutation in *CYP21A2* caused by gene conversion from the analogous sequence in the pseudogene itself, or to distinguish sequences of *CYP21A2* from sequences of two highly similar paralogs *CYP11B1* and *CYP11B2*. Instead, diagnostic laboratories employ strategies such as long PCR with primers that span distinguishing nucleotides, or long-range PCR with allele-specific primers [64].

28.3.4   Repeat expansions

Repeat expansion diseases are caused by an increase in the number of repeats in specific microsatellite sequences. For example, Huntington disease is a neurodegenerative disease that is caused by an expansion of a CAG repeat in the gene encoding huntingtin. The normal gene sequence contains less than 36 CAG repeats, which encode a tract of glutamine residues in the N-terminal part of the huntingtin protein. An increase in the number of repeats leads to disease. Currently, ten neurodegenerative diseases are known to be related to CAG expansions, four to CTG expansions, and a total of eight other nucleotide repeat units have been describe with a variety of neurological diseases [485]. For the most part, expansions such as this are difficult or impossible to detect with current NGS technologies.

28.3.5   Mitochondrial genes

The mitochondrion has its own 16,569-bp closed-circle genome that encodes two rRNAs, 22 tRNAs, and 13 mRNAs coding for proteins of the respiratory chain. Mitochondrial diseases tend to affect tissues that require high amounts of energy such as brain, muscle and eye. Although most mitochondrial proteins are encoded in the cell nucleus, mutations in the mitochondrial DNA (mtDNA) are an important cause of a number of inherited metabolic and neurological diseases. There are specific difficulties that arise with the analysis of genes encoded by mtDNA. The mtDNA itself is not specifically targeted by most exome kits, but because white blood cells can have many hundreds or thousands of copies of the mitochondrial genome, a relatively good overall coverage of about 100X is often observed. Mutations in mtDNA can display heteroplasmy, whereby only a proportion of the mitochondrial genomes carry the mutation. The mixture of mutant and reference sequence is called heteroplasmy and can range from 0 to 100%. With a coverage of 100X, it is estimated that it is possible to detect heteroplasmic mtDNA variants only at a proportion of 10% or more [383]. Bioinformaticians should realize that the analysis of mitochondrial variation requires special attention. Tools such as MitoSeek have arisen to address this need [148].

28.3.6 Structural variation

Structural variants (SVs) such as copy-number variants are an important cause of inherited disease and contribute to both inherited conditions such as intellectual disability and cancer. It remains difficult to detect SVs with WES/WGS, and according to the clinical situation, other (additional) diagnostic modalities such as quantitative DNA arrays (SNP arrays or comparative genomic hybridization) may be indicated (see Chapter 19).

28.3.7 Difficult exons

It is a common observation that not all exons are covered well in typical exome studies. The first exons of many genes are GC rich, which often leads to poor coverage. A mutation which is located in a poorly covered exon may be missed by typical analysis pipelines. For instance, the Shprintzen-Goldberg syndrome is caused by mutations in the *SKI* gene [109]. Most published mutations are located in exon 1 of the SKI gene, which is very GC rich (737 or 70.8% of 1041 nucleotides are G or C). In one in-house case we performed exome sequencing on a sample from an individual with Shprintzen-Goldberg syndrome. Despite a high overall coverage, only one read covered the mutation in exon 1 of the SKI gene — this is not called as a mutation by software such as GATK, and was only discovered after manual inspection of the reads in exon 1 of the *SKI* gene using IGV. Again the lesson is that if there is clinical suspicion for a certain disease it is necessary to check whether all exons of the relevant gene have been adequately covered. The true false-negative rate of exome sequencing is unknown but in our opinion it is likely to be substantial.

28.3.8 Uniparental disomy

Uniparental means "from one parent" (mother or father) and "disomy" means two chromosomes. Normally, children receive one chromosome from each of their parents (with special rules applying to the sex chromosomes). Uniparental disomy (UPD) refers to the situation where an individual receives two copies of the same chromosome from one parent. If that parent was heterozygous (carrier) for a mutation in a gene associated with an autosomal recessive disease and the mutation lies on the chromosome inherited by the child, then the child will be homozygous for that mutation and will develop the disease. Alternatively, if one parent has a large deletion encompassing a certain gene and the other parent is heterozygous (carrier) for a disease-causing mutation in that gene, the mutation is likely to be reported as homozygous by variant callers. These two situations have substantially different consequences for recurrence risk for future siblings (very low in the case of UPD and 25% in the other case). Also, a number of imprinting diseases are caused by UDP or other epigenetic changes that are not easily recognized by WES. Stretches of homozygosity may be observed with UPD. Segregation analysis and assessment by a medical geneticist are advisable especially with nonconsanguineous index cases [407].[2]

28.4 HOW DO WE PROCEED WITH A NEGATIVE RESULT IN WES ANALYSIS?

The answer to the question posed above, **How do we proceed with a negative result in WES analysis?**, is therefore not simple. Since currently the diagnostic yield in large cohorts of individuals analyzed by WES or WGS is only 25–40% [98, 426, 473, 474], we expect that most results will be negative or inconclusive. It may be difficult to know when to stop searching, but it is important to realize that currently WES/WGS analysis is not able to detect all classes of mutations and there are clinical situations where it is important to use other diagnostic modalities to prevent missing diagnostic (and in some cases therapeutic) opportunities.

28.5 KNOWLEDGE IS EVOLVING RAPIDLY

Currently, many hundred novel disease-gene associations are discovered each year. Centers should consider a strategy for periodic reassessment of unsolved cases in the light of new knowledge and improved analysis approaches.

---

[2]Uniparental disomy is a special case of copy-neutral loss of heterozygosity (LOH) which is often also observed in cancer genomes where it can constitutes about 20 to 80% of the LOH events [38, 268].

28.6   THE PROBLEM WITH THE PARAMETERS

We mentioned in Chapter 12 that different variant calling pipelines yield different variant sets. This is true for any of the analyses presented in this book. Perhaps the prioritization algorithms are most sensitive to parameter choice. For instance, ExomeWalker (Chapter 25) searches for genes in exome data that encode proteins with many associations to other known proteins of a disease-gene family. Obviously, ExomeWalker is unlikely to work well if the wrong family is chosen. Many of the parameters for the Exomiser (Chapter 27) depend on medical analysis. For instance, if the wrong presumed mode of inheritance is entered as a filter parameter, the true disease-causing mutation could be mistakenly filtered out of the hit list. Therefore, optimal use of the tools and algorithms presented in this book is highly dependent on input from the physicians caring for a patient or from researchers who are performing a study.

28.7   CLINICAL APPLICATION

Persons with rare genetic diseases wait on average 5 to 30 years to get a diagnosis, and see three or more physicians during this time. The initial diagnosis is wrong in at least 40% of cases [211]. Therefore, one important application of NGS diagnostics for Mendelian disease is simply to make the diagnosis. This can relieve feelings of guilt and uncertainty in affected families, allow more exact calculations of recurrence risk as well as molecular diagnostics in other family members, and can help to find exact information about prognosis and potential complications. Although treatments are not available for most Mendelian diseases, effective treatments are known for a growing number of them, and cases have been published where WES was not only used to make a diagnosis, but also to identify an effective treatment regimen [463]. WES is now being used as a standard molecular diagnostic test in some clinical settings, and additional challenges to setting up analysis pipelines in clinical settings need to be addressed [181]. This chapter has concentrated on the medical application of WES/WGS for persons with rare diseases, but NGS technologies are also quickly gaining relevance for cancer diagnostics as a tool for identifying druggable mutations or predicting treatment outcome.

28.8 SUMMARY

Exome sequencing remains a difficult endeavor. There are many things better left to bioinformaticians, but there are also many aspects where expert clinical or scientific knowledge can make a key difference in the evaluation of exome results. Although bioinformatics has become a truly indispensable part of WES/WGS analysis, the final assessment of the results of WES/WGS is best performed by an interdisciplinary team. Also, simple "eyeballing" of candidate mutations in IGV (Chapter 17) is highly recommended to assess the quality of the variant and search for potential artifacts. If there is one message we would like to leave with our readers, it is the importance of teamwork and of simple common sense in the interpretation of results.

VII

Cancer

CHAPTER 29

A (Very) Short Introduction to Cancer

M OST of what we have discussed and learned in the previous chapters can be directly applied to DNA sequencing data from cancer patients. However, there are important differences and particular characteristics of tumor sequencing data that need to be taken into account to ensure a meaningful interpretation.

A full discussion of DNA sequencing in cancer research and clinical care would require a book of its own. Here, we will introduce several of the quintessential procedures and methods for the analysis of tumor data without any pretense of completeness. Our aim is first and foremost to get readers unfamiliar with the topic started on the right track. This chapter will begin with a high-level overview of cancer biology that will help understand the rationale behind the analysis approaches described in the following chapters.

29.1 TUMORS ARE HETEROGENEOUS TISSUES

Most tumors are highly heterogeneous tissues, a fact that has many important implications for the analysis of tumor data. Tumors are heterogeneous tissues in two senses.

First, tumors are not exclusively made of neoplastic cells. Instead, (solid) tumors additionally contain non-cancerous endothelial and other cells that form blood vessels to support the high energy and nutrient needs [155, 156], fibroblasts and other stromal cells, and often infiltrating immune cells. The boundaries between tumors and adjacent

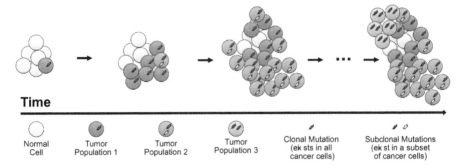

Figure 29.1. Clonal evolution of tumors.

tissues may not be well defined, and surgical samples often contain not only tumor but also surrounding normal tissue.

Although it is in principle possible to purify tumor samples before sequencing their DNA to at least partly remove non-cancerous cells using techniques such as laser capture microdissection (LCM) [114], this is rarely done in practice. In general, the tumor content of a sample is considered to be sufficiently high if 80% of the nuclei are from tumor cells,[1] but differences in sample purity can partly depend on intrinsic properties of the cancers, and in some cases the tumor content may be lower [20].

The second reason for the heterogeneity of tumor samples is an intrinsic inhomogeneity of tumors. Cancer originates from a single cell, and thus at the beginning at least is monoclonal. Over the trajectory that begins with this cell and leads to a malignant cancer, however, additional mutations accrue because of the background mutation rate and in many cases genomic instability or other tumor-promoting mutagenic processes. Tumorigenesis can be regarded to be a stepwise evolutionary process characterized by an accumulation of multiple mutations or non-genetic cancer-promoting factors (e.g., aberrant epigenetic gene regulation) [411]. Indeed, many cancers develop subclones with distinct sets of mutations — this is the substrate for the "Darwinian selection" that operates on cancer cells and allows them to compete for nutrients and adapt to their microenvironment (Figure 29.1) [32, 142, 313]. Distinct subclonal populations of tumor cells may even harbor (at least partly) different sets of driver mutations [107, 142, 281, 304] and dis-

---

[1]See, for example, the tissue sample requirements of The Cancer Genome Atlas (TCGA) project at https://cancergenome.nih.gov/cancersselected/biospeccriteria.

play considerable variability regarding clinically relevant phenotypes such as the capability to metastasize and therapy resistance due to mutations which alter drug metabolism, lead to impaired apoptosis or activate alternative survival pathways [282, 479].

Both the subclonal nature of tumors and the impurity of the samples invalidate one of the basic assumptions made for variants from germline data: heterozygous variants should be found on average on 50% of the covering reads and homozygous variants in most if not all of the covering reads. Instead, true somatic mutations are sometimes present at low variant allele fractions (i.e., a small fraction of the reads)[2] which, in the case of germline data, would easily be confused with noise or false positives.

29.2 THE BASICS OF TUMOR BIOLOGY

Somatic mutations in individual cells occur throughout normal life from infancy and childhood to adulthood. Only rarely does a combination of factors co-occur such that they lead to an initial clonal expansion of the cell with the creation of a benign tumor. Malignancy occurs if tumor cells additionally gain invasive potential and the ability to detach from the primary tumor mass to form metastases elsewhere in the body.

29.2.1 The hallmarks of cancer

The proliferative potential and malignancy of cancer cells is determined mostly by characteristics which are frequently termed the "hallmarks of cancer" [155, 156]:

- **Self-sufficiency in growth signals**: Cancer cells become independent from external growth signals by, for example, constitutively activating internal signaling pathways or overexpressing growth receptors to amplify low external signals.

- **Insensitivity to antigrowth signals**: Cancer cells evade growth suppressors and antigrowth signals.

---

[2]Sometimes the term "variant (or mutant) allele frequency" is used instead. In other contexts, "allele frequency" refers to the frequency of the variant within a population, but in the current context the fraction of reads showing the variant is meant.

- **Evasion of apoptosis**: Cancer cells resist programmed cell death which would normally be caused by extensive DNA damage or external signals.

- **Sustained angiogenesis**: Fast cell proliferation requires nutrients. Therefore, cancer cells stimulate the growth of new blood vessels to support the tumor development.

- **Limitless replicative potential**: Replication is (in most cases) not an unlimited process. Cancer cells must be able to replicate more often then non-cancerous cells.

- **Tissue invasion and metastasis**: Malignancy is associated with the potential to invade surrounding tissues and frequently also with the capability to build new tumor colonies in other organs or parts of the body. For this purpose, individual cancer cells need to detach from the initial tumor mass in order to metastasize.

Importantly, there are different paths which can lead to cancer. That is, these hallmarks do not need to be acquired in one specific, predefined order [155].

29.2.2 Cancer genes

Not all genes affected by somatic mutations are related to tumorigenesis. Indeed, in most cases only a few of the somatic mutations found in a tumor are true "driver" mutations that are causally implicated in oncogenesis. Most mutations are merely random somatic mutations or "passengers", caused by genomic instability or impaired repair control mechanisms of the DNA, without any positive or negative effect on tumor development.

Cancer-associated genes can be divided into two types:

- **Oncogenes**: Genes which positively influence processes such as cell cycle progression and growth signaling may have tumor-promoting effects if over-expressed or mutated. Some signaling molecules, for instance, can become constitutively active by specific gain-of-function mutations.

- **Tumor suppressors**: Genes involved in tumor-suppressing processes like DNA repair or programmed cell death can be affected by loss-of-function mutations, deletions or other mechanisms that negatively affect their expression levels or functions. Tumor suppressor genes can in turn be subdivided into three classes: "gatekeeper genes" which inhibit cell growth or control apoptosis, "caretaker genes" (also termed "stability genes") which are responsible for genomic stability and DNA repair and thus suppress mutations, and "landscaper genes" whose mutations can stimulate the surrounding stromal environment to support unregulated cell proliferation of the tumor cells [101, 195, 291]. If a tumor has a defective stability gene, then the overall rate of mutations can be higher, increasing the probability of *de novo* mutations in oncogenes and other tumor suppressor genes, which tends to accelerate the process of tumorigenesis.

Although many if not most tumors depend on underlying genetic or genomic alterations, non-genetic factors such as deregulated gene expression also play an important role. For example, some tumors largely depend on epigenetic or epigenomic phenomena such as a CpG island methylator phenotype [271].

29.3   HEREDITARY CANCER SYNDROMES

Most cancers are sporadic, meaning that their initiation and development depends on somatic alterations. However, an estimated 5–10% of cases are associated with hereditary mutations that increase cancer predisposition [308].

In 1971, Knudson proposed a two hit model for the development of retinoblastoma [203]. Hereditary retinoblastoma is an autosomal dominant disease in which all cells in the body have one germline mutation in the tumor suppressor retinoblastoma gene (*RB1*). This means that all cells of the retina are susceptible to the development of a retinoblastoma owing to only a single additional somatic mutation which inactivates the second allele of *RB1*. On the other hand, in persons without hereditary retinoblastoma, a sporadic tumor develops if first there is a somatic mutation in the retinoblastoma gene in some cell, and then a second, independent somatic mutation occurs in the second *RB1* allele of the same cell [203, 427] (Figure 29.2). We now know that the development of cancer is a complex, multistep process involving mutations in multiple genes as well as epigenetic alterations.[3] With hereditary forms

---

[3]While Knudson's two-hit hypothesis illustrates an instructive special case of

of cancer, all cells have a heterozygous mutation in a tumor suppressor gene. One might say that the first of many steps in the development of cancer has already occurred in all cells. Other types of hereditary cancer involve germline loss-of-function mutations of the genes *BRCA1* and *BRCA2*, which increase the risk of breast and ovarian cancer in women [270], and the rare Li-Fraumeni syndrome, which is caused by germline mutations in the tumor suppressor *TP53* [189]. The latter is also one of the most frequently mutated genes in sporadic tumors.

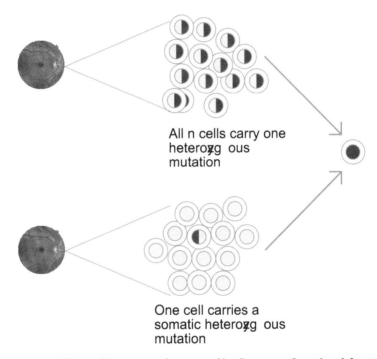

All n cells carry one heterozyg ous mutation

One cell carries a somatic heterozyg ous mutation

Figure 29.2. Hereditary and sporadic forms of retinoblastoma. Top: in hereditary retinoblastoma, all cells in the retina carry a heterozygous *RB1* mutation, and a second *RB1* mutation in any retinal cell will lead to the development of a retinoblastoma (a malignant tumor of the retina). Bottom: In sporadic forms of retinoblastoma, tumor development occurs if there are two independent *RB1* mutations in the same cell.

NGS gene panel, WES, or WGS analysis for hereditary mutations in cancer susceptibility genes is performed in the same way that was de-

---

tumor genetics, it is far from being applicable to tumor suppressors or cancer in general.

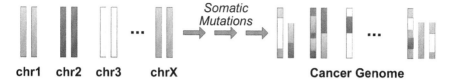

chr1   chr2   chr3   chrX   Cancer Genome

Figure 29.3. Genomic instability. Schematic representation of genomic instability leading to somatic structural variants in the cancer genome.

scribed in the previous chapters of this book for germline mutations in Mendelian or other genetic disease. The analysis of somatic mutations is usually performed by sequencing a tumor sample and a germline sample independently and searching for variants that are only present in the tumor sample. This procedure will be described in the following chapters.

29.3.1 Structural variation in cancer

Genomic instability is an important feature of many tumors (see Figure 29.3). Indeed, most tumors have at least a few somatic structural variants such as copy number alterations or inter- and intra-chromosomal translocations. Figure 29.4 shows the Circos plot of a tumor with several intra-chromosomal translocations on chromosomes 3 and 5, an inter-chromosomal translocation between chromosomes 2 and 19, and multiple copy number alterations including a loss of one copy of chromosome 6 and a very high copy number (amplification) of a small portion of chromosome 9.[4] These are typical somatic structural variants which can be observed in many cancer samples.

In some extreme cases, tens to hundreds of locally clustered rearrangements can affect parts of chromosomes or even entire chromosomes, caused presumably by a catastrophic event termed "chromothripsis" [375, 410]. It is likely that a cell can survive such extensive DNA damage only if it co-occurs with a loss of *TP53* [357] which is an important tumor suppressor and mediates processes such as cell-cycle arrest, apoptosis and DNA repair [233].

---

[4]Circos [229], which can be obtained at `http://circos.ca/`, is a highly configurable and versatile tool that is used to generate circular representations of genomic and other data. It is frequently used for genome-wide visualization of structural variants.

The impact of structural variants on tumor cells may vary from the loss of tumor suppressors due to deletions, duplications or amplifications of oncogenes, to the creation of fusion genes through translocations. Even in cases where these events do not directly affect oncogenes or tumor suppressors, they may still drive tumorigenesis by altering gene expression, as is illustrated by structural variants that bring activating enhancers close to the proto-oncogenes *GFI1* and *GFI1B* in some medulloblastomas [312].

29.4 USEFUL DATABASES FOR CANCER BIOINFORMATICS

There are many useful resources for cancer bioinformatics which can help in analyzing DNA sequencing data from tumor samples and interpreting the results. The following short list reviews only a few of the more prominent databases.

- **COSMIC** (Catalogue Of Somatic Mutations In Cancer)[5] [28, 122] is a large collection of somatic mutations found in cancer. It can be browsed and searched for specific mutations, genes and samples (i.e., collections of mutations from individual patients). The grouping of mutations per patient can be especially of interest to study, for example, combinatorics like frequent co-occurrences and mutual exclusiveness. It should however be kept in mind that this database collects all somatic mutations, independent of their status as drivers or passengers.

- The **Cancer Gene Census**[6] [128] is a subset of COSMIC which is specifically dedicated to genes and mutations that are known to be causally implicated in human cancer.

- The **International Cancer Genome Consortium (ICGC)**[7] [483] and **The Cancer Genome Atlas (TCGA)**[8] are two large-scale projects that provide different types of cancer-related data including mutation databases as well as raw sequencing data (controlled access).

- Raw sequencing data from cancer samples can also be found

---

[5]http://cancer.sanger.ac.uk/cosmic/
[6]http://cancer.sanger.ac.uk/census/
[7]http://icgc.org/ and https://dcc.icgc.org/
[8]https://cancergenome.nih.gov/

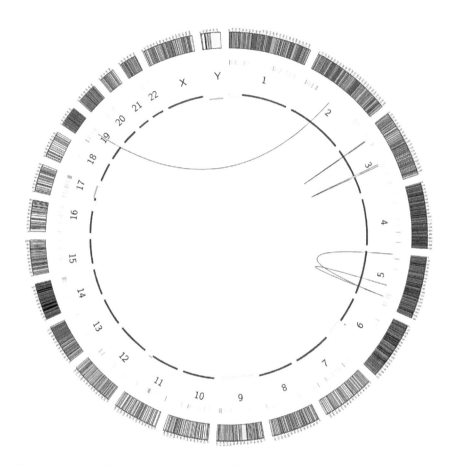

Figure 29.4. Circos plot showing the somatic structural variation in a cancer sample. Inter- and intra-chromosomal translocations are indicated by red lines. Large copy number alterations are shown in the inner circle and highlighted by cold colors (blue for copy number loss) and warm colors (orange and red for copy number gain), with one exception: the lack of chromosome Y merely hints at the patient being female.

in sequence data archives like the **Sequence Read Archive (SRA)**[9] [245] and the **European Genome-phenome Archive (EGA)**[10] [237] (see also Section 4.3).

- The **dbSNP**[11] [393] database is mostly dedicated to germline mutations rather than somatic mutations, but it can be used for filtering called somatic variants. Indeed, somatic mutations which are also frequently observed as common SNPs throughout the population are unlikely to have any tumor-promoting effects. However, such a filter is not straightforward since dbSNP also contains many hereditary mutations of cancer-relevant genes that are implicated in familial cancer syndromes.

- The **cBioPortal for Cancer Genomics**[12] is a Web-based resource for exploring, visualizing, and analyzing multidimensional cancer genomics data in multiple and flexible ways [63].

- **Therapeutically Applicable Research to Generate Effective Treatments (TARGET)**,[13] which is part of the Cancer Genomics Hub (CGHub), offers many datasets and tools for the analysis of genomic changes that drive childhood cancers [460].

Many additional useful resources have been described in a recent review of databases for cancer genomics [472].

FURTHER READING

Hanahan, D. and Weinberg, R.A. (2000). The hallmarks of cancer. *Cell* **100**(1):57–70.

Hanahan, D. and Weinberg, R.A. (2011). Hallmarks of cancer: the next generation. *Cell* **144**(5):646–674.

Stratton, M.R., Campbell, P.J. and Futreal, P.A. (2009). The cancer genome. *Nature* **458**(7239):719–724.

---

[9]https://www.ncbi.nlm.nih.gov/sra/
[10]https://www.ebi.ac.uk/ega/home/
[11]https://www.ncbi.nlm.nih.gov/SNP/
[12]http://www.cbioportal.org/
[13]https://ocg.cancer.gov/programs/target

CHAPTER 30

Somatic Variants in Cancer

S OMATIC mutations occur in all body tissues throughout life and are not present in the germline from which the individual developed.[1] The majority of somatic mutations have no phenotypic effect. However, occasionally they may have functional consequences that confer a selective advantage on the cell owing to preferential growth or survival. These so-called driver mutations may ultimately lead to oncogenesis.[2] On the other hand, passenger mutations do not contribute to cancer development, but occur at random during cell division and are carried along like passengers during the clonal expansion that characterizes the development of a tumor [411]. Indeed, many tumors are associated with elevated mutation rates that lead to an overall increase in the number of somatic mutations.

A major goal of cancer genomics is to reliably identify those somatic mutations that cause or contribute to the initial occurrence or the later development of the tumor, and are thus under positive selection in the microenvironment of the tissue in which the cancer arises. For this purpose, however, first one needs to identify those variants which are somatic and distinguish them from those which were already present in the germline.[3]

---

[1]The Greek word "*soma*" means body.

[2]A process in which normal cells are transformed into malignant cancer cells.

[3]We note that the identification of germline variants is essential for the diagnosis of familial (hereditary) cancer syndromes (see Section 29.3), and the general analysis procedure is largely the same as has been discussed for other inherited genetic disorders in previous chapters.

30.1 TUMOR SAMPLE VERSUS MATCHED CONTROL

The most important strategy for characterizing somatic mutations in cancer is the comparison of DNA from a cancer sample (e.g., from the biopsy of a tumor) and a matched normal sample (e.g., blood) from the same individual. This allows acquired/somatic mutations (which usually make up <0.01% of variants that are observed in exome or genome data from a typical tumor) and inherited germline variation (>99.99% of variants) to be distinguished.[4]

As we will see, though, there are many sources of noise and bias and it is not sufficient to merely subtract variants identified in a normal sample from the variants found in a matched tumor sample. There are many approaches to this, but we will illustrate the topic using an application called VarScan2 [204, 205], which employs a simple but effective heuristic strategy.

First, however, we need to consider some particular quality issues that can arise when comparing a tumor sample to a matched control.

30.1.1 Quality control issues

As for all DNA sequencing analyses, quality control is essential to ensure a meaningful interpretation of the results. In general, there is not much difference between quality control of tumor-related sequencing data and quality control of germline data. There are, however, some additional considerations which should not be ignored.

First, since tumor samples are often compared to matched normal samples, it is imperative to ascertain that the two samples actually come from the same individual. While sample swapping (e.g., due to erroneous sample labeling before sequencing) is not very frequent, it can easily happen in larger datasets. One approach to verifying the correct sample identities of the tumor and the control sample is by DNA fingerprinting, i.e., by obtaining information on their genotypes regarding a set of known, common polymorphisms [321, 327].

Second, in order to identify somatic mutations, which are present in the tumor sample but not in the control sample, it is important to make sure that the control sample is not contaminated by infiltrating

---

[4]An important issue, which we will not address here, is the fact that many of the somatic mutations identified in a tumor sample may actually have occurred prior to tumor initiation [432]. Still, these somatic mutations contribute to the genetic background in which the tumor develops.

malignant cells.[5] Still, in many cases a tiny fraction of tumor cells in the control sample cannot be avoided. Even control samples taken from the blood of the patient, instead of adjacent tissue, can contain circulating tumor cells. In most cases, however, the fraction of tumor cells in a control sample should be so low that reads from somatic, tumor-related mutations will hardly be distinguishable from the general noise of the sample. Still, since a higher contamination is possible, the distribution of variant allele fractions in the control sample should be checked for unusual peaks that occur in addition to the expected peaks at frequencies of 0.5 and 1.

30.2   SOMATIC SNVS AND INDELS

As with DNA sequencing of germline variants, the most important somatic alterations are single nucleotide variants (SNVs), short insertions and deletions (indels), and larger structural variants. We will first demonstrate variant calling for somatic SNVs and indels based on a tumor-normal sample pair using VarScan2 [204, 205].

30.3   PILEUP FORMAT

VarScan2 makes use of pileup files produced by SAMtools [254] for its analysis, so we will begin with an explanation of this particular format.[6] There is no standard file extension for a pileup file, but ".pileup" is commonly used.

The pileup format provides base-pair information at each chromosomal position in form of a summary that enables both manual checking of alignment and base-calling results and can also be used by downstream programs or scripts for variant calling.

The default pileup format used by SAMtools has six tab-separated columns:[7]

1. The reference sequence (in this case the chromosome).

2. The 1-based position (coordinate) on the chromosome.

---

[5]For instance, contamination can occur if the "normal" sample was obtained from tissue that lies adjacent to the tumor in a biopsy specimen.

[6]At the time of this writing, the current version of SAMtools was 1.3.1. See http://www.htslib.org/doc/.

[7]There are alternative formats depending on the options used for SAMtools but we will exclusively concentrate on the default here.

3. The reference base.

4. The number of reads covering the site.

5. The read bases, i.e., the bases that each read reports at this position (encoded as explained below).

6. The base qualities corresponding to the read bases. This column is optional.

For instance, the following shows a typical line in a pileup file:

```
chr3    315877    C    13    ,$,....,A,,,a.    >A@@A@A?AA@=@
```

The pileup format uses conventions that indicate whether a matching or mismatching base in the alignment column is located on the forward or reverse strand. A **dot** (.) indicates a match to the reference base on the forward strand. A **comma** (,) indicates a match on the reverse strand. An **upper-case character** (ACGTN) stands for a mismatch on the forward strand, and a **lower-case character** (acgtn) stands for a mismatch on the reverse strand.

Thus, in the example above, the nucleotide at position 315,877 of chromosome 3 is covered by a total of 13 reads. Eleven reads (five on the forward strand and six on the reverse strand) favor the reference base C, and two reads favor an alternate base, A (one read on the forward strand, one on the reverse strand).

There is one such line for each position that is covered by at least one read. Here is a an excerpt of the same pileup file that shows seven contiguous lines.

```
chr1    13807    C    5    ,,T,,     BB00<
chr1    13808    C    5    ,,.,,     BB<7<
chr1    13809    C    6    ,,,.,,    BBBB0B
chr1    13810    A    6    ,,,.,,    B<B<0B
chr1    13811    C    3    ,,.       BB7
chr1    13812    C    6    ,,,.,,    BB7B7<
chr1    13813    T    6    ,,g.,,    BB<B0B
```

If these were the only symbols used in the pileup format, important information about the alignment would be lost and the original read alignments could not be reconstructed. Therefore, the pileup format defines additional symbols to mark the beginning and the end of a read, as well as insertions and deletions on single reads:

- `^`: marks the start of a read and is followed by an additional ASCII character reporting its associated mapping quality (ASCII character minus 33).

- `$`: marks the end of a read.

- `+`: indicates an insertion between this reference base and the following reference base. The `+` is followed by a string satisfying the regular expression "`[0-9]+[ACGTNacgtn]+`" to specify the inserted nucleotides (e.g., `+4AGTG`)

- `-`: indicates a deletion from the reference sequence. It is followed by a string satisfying the same regular expression to specify the deleted nucleotides (e.g., `-4AGTG`)

- `*`: indicates a deleted base in a multiple basepair deletion that was previously specified using the `-` notation.

Let us consider a further example that will serve three purposes: (i) to understand how these marks are used in pileups, (ii) to illustrate the connection between read alignments and the corresponding pileup information, (iii) to show how these pileups can be used for variant calling. We will visualize a portion of an alignment on chromosome 10 using the SAMtools text alignment viewer `tview`. In order to generate Figure 30.1, we caused `tview` to output the visualization as HTML with the `-d H` argument. If this is left out, the alignment can be visualized directly in the terminal with various possibilities for navigation and coloring the output.[8]

```
$ samtools tview -p chr10:69023352 NA12892.bam hs38DH.fa
```

This following excerpt of the mpileup file produced by the command shown above corresponds to part of the alignment shown in Figure 30.1. For legibility, we have omitted the base quality strings. The files produced by the mpileup command can be very large, and users may want to restrict the output to a certain region of the genome using the `-r` option (e.g., `-r chr10:1-1000`).

---

[8] See online documentation for details. We do not show a screenshot here, because `tview` uses a black background that does not reproduce well.

```
$ samtools mpileup -f hs38DH.fa \
                   -r chr10:69023370-69023374 \
                   -Q 0 \
                   NA12892.bam
```

```
chr10   69023370   A   23   .$,,,,,,,,.,.,,,,,..,.,,.
chr10   69023371   G   22   ,,,,,,,,,.,.,,,,,..,.,,,.
chr10   69023372   C   24   ,,,,a,a,.,A,,,aAA,.,a.^]A^],
chr10   69023373   T   24   ,,,,,,,,.,.,,,,,..,.,,,..,
chr10   69023374   C   24   ,,,,,,,,.,.,,,,,..,.,,,..,
```

In the first of these pileup lines (optional base qualities omitted), the '$' symbol states that there is a read whose last base is mapped to position 69,023,370 on chromosome 10, with the last base being 'A'. (If the read was on the reverse strand, there would be a ',' instead of a '.' before '$'). The "^]A" and "^]," in the third line mean that there is a forward-strand read whose mapped first base is 'A' instead of 'C' and a reverse-strand read whose mapped first base is 'C', both with mapping quality (MAPQ) ']', or 93 minus 33, i.e., 60. It is important to note that the '^+MAPQ' is specified *before* the symbol for the first base of the corresponding read and the '$' symbol is specified *after* the symbol for the last base of the read.

The third line reports that 8 out of 24 reads favor an alternate base A (4 reads on the forward strand, 4 on the reverse), suggesting the existence of a single nucleotide variant at position 69,023,372. How this is related to the corresponding read alignments can be seen in Figure 30.1.[9]

There are additional features of the SAMtools pileup format that we will not cover here, such as the **angle brackets >** and **<** that are used for a reference skip on a forward and a reverse strand read.[10]

---

[9]We included the -Q 0 argument in the samtools command used to generate the pileup file, because otherwise samtools would remove any base with base quality under 13 from the pileup. Since the samtools tview command shows all bases regardless of quality, the pileup and the tview output would not matched unless we adjusted the base quality threshold in this way.

[10]See the documentation at http://www.htslib.org/doc/.

chr10:69023352

```
   69023361   69023371   69023381   69023391   69023401   69023411   69023421
TACCACCCGCAGAGGGAGAGCTCCTGTGTGACATCATCAAGGAGAAGCTGTGCTAAGTCGCCCTGGACTTTGAGCAGGAG
.........................M......................................................
.... ,,,,,,,,,,,,,,a,,,,,,,,,,,,,,,,,,,,,,,,,,,,,,,,,,,,,,,,,,,,,,,,,,,,,,,,,,,,,
,,,,,,,,,, ....................................................................
..................A.............................................................
,,,,,,,,,,,,,,,,,,,,,,,,,                ........................................
,,,,,,,,,,,,,,,,,,,,,,,,,,,,,,,          ........................................
,,,,,,,,,,,,,,,,,,,,,,,,,,,,,,,,,,,,          ...................................
,,,,,,,,,,,,,,,,,,,a,,,,,,,,,,,,,             ...................................
,,,,,,,,,,,,,,,,,,,,,,,,,,,,,,,,,             ...................................
,,,,,,,,,,,,,,,,,,a,,,,,,,,,,,,,,,,,,,,,          ...............................
,,,,,,,,,,,,,,,,,,,,,,,,,,,,,,,,,,,,,,,,,,,.A..... ..............................
..............................................................A..... ...........
................G...A.....................................           ...........
,,,,,,,,,,,,,,,,,,,,,,,,,,,,,,,,,,,,,,,,,,,,,,,,                       .....
,,,,,,,,,,,,,,,,,,,,,,,,,,,,,,,,,,,,,,,,,,,,,,,,,,,,,,,,,,,               ,,,,
,,,,,,,,,,,,,,,,,,,,,,,,,,,,,,,,,,,,,,,,,,,,,,,,,,,,,,,,,,,,,,,,,,,         ,,,
,,,,,,,,,,,,,,,,,,a,,,,,,,,,,,,,,,,,,,,,,,,,,,,,,,,,,,,,,,,,,,,,,,,,,,,,,,,,,,,,,
..............................A.............................................C.
..............................A.................................................
,,,,,,,,,,,,,,,,,,,,,,,,,,,,,,,,,,,,,,,,,,,,,,,,,,,,,,,,,,,,,,,,,,,,,,,,,,,,,,,,,,
................................................................................
,,,,,,,,,,,,,,,,,,,,,,,,,,,,,,,,,,,,,,,,,,,,,,,,,,,,,,,,,,,,,,,,,,,,,,,,,,,,,,,,,,
             ,,,,,,,,,,,,,,,,,,,,,,,,,,,,,,,,,,,,,,,,,,,,,,,,,,,,,,,,,,,,,,,,,,,,
             ...................................................................
                 ...............................................................
                     ...........................................................
                                 ,,,,,,,,,,,,,,,,,,,,,,,,,,,,,,,,,,,,,,,,,,,,,,,,
                                 ,,,,,,,,,,,,,,,,,,,,,,,,,,,,,,,,,,,,,,,,,,,,,,,,
                                   .............................................
                                   .........................
```

Figure 30.1. SAMtools tview. SAMtools alignment view showing linked variants on chromosome 10. The meaning of the period, comma, and capitalization of alternate bases is identical as described for the pileup format. With `tview`, the first line shows the genome coordinates, the second line shows the reference sequence, and the third line shows the consensus sequence determined from the aligned reads (the "M" stands for A/C). A period indicates a match to the reference genome, and a comma indicates a match on the reverse strand.

30.3.1 Filtering of SAMtools pileup

The pileups produced by SAMtools can be filtered in various ways according to attributes of specific reads or bases. Some frequently used options are[11]

- A minimum mapping quality below which reads are omitted (`--min-MQ`).

- A minimum base quality below which bases are not included in the output (`--min-BQ`).

- Specific read bitflags that need to be set or unset (see Table 9.3 in Chapter 9). By default, reads that are unmapped (0x4), secondary reads (0x100), reads not passing quality controls. (0x200), and duplicate reads (0x400) are ignored (`--incl-flags` indicates required flags and will skip reads whose mask bits are unset; and `--excl-flags` indicates filter flags and will skip reads whose mask bits are set).

30.3.2 VarScan2

VarScan2 employs a heuristic, pileup-based approach towards variant calling that shows good performance in practice.[12] VarScan2 is designed to perform variant calling in several different scenarios, but we will present only its functionality for somatic mutation detection in tumor-normal pairs.

The following steps are performed for each position of genome in parallel for the tumor sample and the matched normal sample:

1. Determine if both samples meet the minimum coverage requirement (by default, eight reads with base quality ≥ 15 in the normal sample and six such reads in the tumor)

2. Determine a genotype for each sample individually based upon the read bases observed. By default, a variant allele must be supported by at least two independent reads.

---

[11]For a complete reference, see the official SAMtools website at `http://www.htslib.org/doc/samtools.html`.

[12]Here, we use the newest version available in March 2017 (v2.4.3). VarScan2 can be obtained from `http://dkoboldt.github.io/varscan/`.

3. Variants are called homozygous if supported by 75% or more of all reads at a position; otherwise they are called heterozygous. However, the purity of the samples will be considered for evaluating the genotype (see the next section).

Following this, VarScan2 compares the results for each sample to make the final call. If the genotype determined for the normal and the tumor sample match and both differ from the reference sequence, then VarScan2 calls a germline variant.

If the genotypes do not match, then their read counts are evaluated by one-tailed Fisher's exact test in a two-by-two contingency table:

	Reference	Alternate
Tumor reads	Ref. bases in tumor	Alt. bases in tumor
Normal reads	Ref. bases in normal	Alt. bases in normal

The variant is called **somatic** if the p-value of the Fisher's exact test is significant and the normal sample has a reference base and a variant base is called for the tumor sample.

30.3.3 Running VarScan2 on glioblastoma data

For calling somatic variants, Varscan2 requires an mpileup file (multiway pileup) generated for a tumor sample and its matched normal control. We will demonstrate the tool for the analysis of open-access *IDH1*-mutated glioblastoma cancer data from Seoul National University Hospital [322]. The first step is to create an mpileup file for the tumor-normal sample pair. We use the SAMtools mpileup command and pass the names of the input files with the normal sample first and the tumor sample second. The BAM files have to be generated from the FASTQ files downloaded from SRA (see Section 4.3) and postprocessed as described in previous chapters (duplicate removal, local realignment around known indel sites, etc.). It is recommended to use the option -q 1 for samtools, which limits the pileup to reads with mapping quality greater than zero (for more stringency even higher minimum mapping qualities can be used).

```
$ samtools mpileup -q 1 -f hs38DH.fa normal.bam \
    tumor.bam > normal_tumor.mpileup
```

This will result in a two-sample mpileup file representing the base calls at each position. For example, here is a base with 5 reads in the normal sample and 8 in the tumor. We are looking at position 13,042 of chromosome 1, and the normal sample has four reads on the forward strand and one on the reverse strand, the tumor has four reads on the forward strand and four on the reverse strand.

```
chr1   13042   C   5   ..,..   FIAFF   8   ..,,,,..   DDGIDDHF
```

VarScan2 is then used to perform somatic mutation calling. The following command will report germline, somatic, and loss-of-heterozygosity events at positions where both the normal and the tumor samples have sufficient coverage (the default threshold is a coverage of ≥ 8 reads for the normal sample and ≥ 6 reads for the tumor).

```
$ java -jar VarScan.jar somatic [normal_tumor.mpileup] \
    [out] --mpileup 1 [OPTIONS]
```

The prefix for the output files is indicated as [out], and causes VarScan2 to output files named out.snp and out.indel. There are many options that control the minimum coverage and variant frequency required to call variants in different constellations. Important for us are the options for normal and tumor purity (--normal-purity and --tumor-purity). These values should be numbers between 0 and 1; the default value 1 implies that the normal sample is 100% "normal" with no contaminating tumor cells and that the tumor is composed of 100% tumor cells with no contaminating normal stromal cells.

The control sample is often obtained from the patient's blood. The blood sample conceivably can contain circulating tumor cells, but the proportion is usually very small, such that in most cases a normal purity of (very close to) 100% can be assumed. Tumor samples, however, do in most cases contain a significant proportion of normal cells, as we have discussed in Chapter 29. Here, a good estimate for the tumor purity of the sample will help to get reliable results. We will describe in Chapter 31 how such an estimate can be obtained.

For the malignant glioblastoma sample we use here, as can be seen in Supplementary Table 4 of the original publication [322], the amount of tumor tissue could not be reliably estimated by the authors. Since they obtained a tumor purity between 61% and 92% for all other tumor samples in their study (74% and 92% for the other malignant samples), we will assume a value of 80% for the tumor purity.

Other options that may be useful include --strand-filter. If this is set to 1, variants with >90% strand bias are removed. This may be helpful because variant read counts that are highly imbalanced towards the forward or the reverse strand may be signs that the variant call is artefactual. Setting the option --output-vcf to 1 will produce the output in VCF format (with an additional filename suffix ".vcf", e.g., out.snp.vcf), which can be useful because many downstream analysis tools work with VCF files as input. The command for calling somatic variants for the glioblastoma tumor-normal pair is thus:

```
$ java -jar VarScan.jar somatic \
    normal_tumor.mpileup  tumor  --mpileup 1 \
    --tumor-purity 0.80 --strand-filter 1 \
    --output-vcf 1
```

Unfortunately, the mpileup file consumes a lot of disk space (in this case 56G if uncompressed). But if it is not required as input for other tools, it can be directly sent from SAMtools to VarScan2 using a named first-in-first-out (FIFO) pipe. This not only saves disk space but also speeds up the entire process because it avoids slow file I/O operations. In addition, if you have multiple cores or processors, SAMtools and VarScan2 can be run in parallel, with VarScan2 "consuming" the pileup data while samtools is "producing" it. This is how it works:

```
$ mkfifo /tmp/mpileup-pipe

$ samtools mpileup -q 1 -f hs38DH.fa normal.bam \
                   tumor.bam > /tmp/mpileup-pipe &

$ java -jar VarScan.jar somatic \
    /tmp/mpileup-pipe  tumor  --mpileup 1 \
    --tumor-purity 0.80 --strand-filter 1 \
    --output-vcf 1
```

In this workflow, SAMtools needs to run as a background process, which is achieved by the ampersand (&) at the end of the SAMtools command line.

The named pipe is initialized by mkfifo which creates a FIFO file on the file system. Instead of being written to this file, though, the data is passed in memory from the producing process to the consuming process. The FIFO file can be deleted after VarScan2 has finished.

```
$ ls -l /tmp/mpileup-pipe
prw-r--r-- 1 piro 10345 0 Mar  8 00:30 /tmp/mpileup-pipe

$ rm /tmp/mpileup-pipe
```

30.3.4   Filtering somatic variants

Like most other variant callers, VarScan2 tries to minimize the false
negative calls, i.e., not to miss variants that are truly present. This,
however, means that the list of called variants generally also contains
a high proportion of false positive calls.

Therefore, it is necessary to filter the results. Of course, this also
means a trade-off between reducing the number of false positives and
overlooking true somatic mutations. Somatic variants obtained from
VarScan2 cannot be easily filtered using techniques like variant-quality
score recalibration (VQSR, see Chapter 12) because VarScan2 does not
determine Phred-scaled quality scores for the variant calls. Instead,
filtering can be done in the following steps [204]:

Step 1: Determine high-confidence somatic mutations

The first step is to separate the called variants based on somatic status—germline, somatic and loss-of-heterozygosity (LOH)—
and above all based on confidence according to empirically-derived criteria, writing high-confidence calls into separate files:

```
$ java -jar VarScan.jar processSomatic tumor.snp.vcf \
    --min-tumor-freq=0.1 --max-normal-freq=0.05 \
    --p-value=0.05

[...]
90247 VarScan calls processed
2467 were Somatic (531 high confidence)
77964 were Germline (71732 high confidence)
9760 were LOH (9741 high confidence)
```

where `min-tumor-freq` is the minimum variant allele fraction (VAF),
i.e., minimum fraction of reads, that a somatic variant should have in
the tumor, `--max-normal-freq` is the maximum VAF it should have
in the normal control sample, and `p-value` is the *p*-value threshold for
high-confidence calling.

Even if it might seem reasonable, the maximum VAF in the normal control should not be set to zero because the control sample might contain a small fraction of migrating or infiltrating tumor cells. The minimum VAF of 10% in the tumor sample represents a trade-off between overlooking true somatic mutations present only in a subclone of the tumor and retaining noise due to, for example, sequencing errors. Depending on the goals of analysis, it may be appropriate to use a different value for the minimum VAF.

Running `processSomatic` on the SNV calls will produce the following files ("hc" = high-confidence):

```
tumor.snp.Germline.hc.vcf
tumor.snp.Germline.vcf
tumor.snp.LOH.hc.vcf
tumor.snp.LOH.vcf
tumor.snp.Somatic.hc.vcf
tumor.snp.Somatic.vcf
```

The same should be done for indels:

```
$ java -jar VarScan.jar processSomatic tumor.indel.vcf \
    --min-tumor-freq=0.1 --max-normal-freq=0.05 \
    --p-value=0.05

[...]
10695 VarScan calls processed
483 were Somatic (129 high confidence)
9553 were Germline (8594 high confidence)
657 were LOH (655 high confidence)
```

In the following, we will only look at high-confidence somatic mutations. For clinical analysis, it is important to assess germline mutations in cancer susceptibility genes, and loss of heterozygosity (LOH). We leave this to the reader as an exercise.

Step 2: Filter false positives due to systematic artifacts

As we have seen in previous chapter, short-read NGS data is prone to various systematic errors such as GC bias and ambiguous mapping of sequences from repetitive regions of the genome. Such artifacts can often be recognized by analyzing read position bias, strand representation and other criteria [204]. In order to generate all of the data required to create a false-positive filter, the BAM file needs to be processed using the tool `bam-readcount`, which computes read statistics for the bases at the called variants:[13]

```
$ cat tumor.snp.Somatic.hc.vcf | \
    awk '{OFS="\t"; if (!/^#/){ print $1,$2,$2 } }' \
    > snp.regions

$ bam-readcount \
    --min-mapping-quality=1 --min-base-quality=20 \
    --reference-fasta=hs38DH.fa --site-list=snp.regions \
    tumor_aln_sorted.rmdup.realigned.bam > snp.readcounts
```

The first command extracts the variant's coordinates from the VCF file and transforms them into a BED-like file format (`chrom`, `chromStart` and `chromEnd`) to be passed to `bam-readcount`.

For indels, especially for deletions, the regions to be specified for `bam-readcount` are ranges of bases involved in the indel. This is a little more complex:

```
$ cat tumor.indel.Somatic.hc.vcf | \
    perl -e 'while(<STDIN>){
      next if (/^\#/);
      @tok=split(/\t/);
      printf "@tok[0]\t$tok[1]\t%d\n",
        ($tok[1]+length($tok[3]));
    }' > indel.regions

$ bam-readcount \
    --min-mapping-quality=1 --min-base-quality=20 \
    --reference-fasta=hs38DH.fa \
    --site-list=indel.regions \
    tumor_aln_sorted.rmdup.realigned.bam \
    > indel.readcounts
```

The read statistics determined in this way are now used to filter false positive variant calls:[14]

---

[13]`https://github.com/genome/bam-readcount/releases`

[14]We show the command for SNVs; an analogous command applies for indels.

```
$ java -jar VarScan.jar fpfilter \
    tumor.snp.Somatic.hc.vcf snp.readcounts \
    --output-file tumor.snp.Somatic.hc.fpfilt.vcf

[...]
531 variants in input file
529 had a bam-readcount result
521 had reads1>=2
190 passed filters
341 failed filters
2 failed because no readcounts were returned
168 failed minimim variant count < 4
0 failed minimum variant freq < 0.05
69 failed minimum strandedness < 0.01
1 failed minimum reference readpos < 0.1
3 failed minimum variant readpos < 0.1
1 failed minimum reference dist3 < 0.1
13 failed minimum variant dist3 < 0.1
94 failed maximum reference MMQS > 100
141 failed maximum variant MMQS > 100
123 failed maximum MMQS diff (var - ref) > 50
0 failed maximum mapqual diff (ref - var) > 50
8 failed minimim ref mapqual < 15
28 failed minimim var mapqual < 15
0 failed minimim ref basequal < 15
0 failed minimim var basequal < 15
3 failed maximum RL diff (ref - var) > 0.25
```

Analogous steps are done for high-confidence germline and LOH calls.

VarScan's fpfilter allows several important parameters to be adjusted, only some of which we mention here. Running the command $ java -jar VarScan.jar fpfilter --help displays additional filter options and their default values.

- --min-var-count: Minimum number of required variant-supporting reads (default: 4)

- --min-var-freq: Minimum variant allele fraction (default: 0.05)

- --min-ref-readpos: Minimum relative read position of reference-supporting bases with respect to the ends of the reads (default: 0.1)

- `--min-var-readpos`: Minimum relative read position of variant-supporting bases with respect to ends of the reads (default: 0.1)

- `--min-strandedness`: Minimum fraction of variant reads from each strand (default: 0.01)

- `--min-strand-reads`: Minimum allele depth required to perform the strandness test (default: 5)

- `--min-ref-basequal`: Minimum average base quality for the reference allele (default: 15)

- `--min-var-basequal`: Minimum average base quality for the variant allele (default: 15)

- `--min-ref-mapqual`: Minimum average mapping quality for the reference allele (default: 15)

- `--min-var-mapqual`: Minimum average mapping quality for the variant allele (default: 15)

The default values are reasonable for most SNV calls. Since indels are much more difficult to call than SNVs, for samples with comparably low coverage it may be advisable to lower the minimum variant count to 2 (`--min-var-count 2`) and ignore the strandness (`--min-strandedness 0`).

The options `--min-var-readpos` and `--min-ref-readpos` make sure that bases at the very ends of reads are not considered for variant and reference base counts. A value of 0.1, for example, means that only bases at read positions between 10% and 90% of the read length will be taken into account. This filter is especially important for SNV calls close to indels (see Chapter 10 for more details).[15]

Finally, using the option `--keep-failures 1` produces a single VCF file containing both variants which passed all filters (`FILTER` field set to `PASS`) and variants that were excluded by one or more of the filters.

---

[15]In previous versions of VarScan2, this issue was addressed by the `somaticFilter` command. See `https://github.com/dkoboldt/varscan/blob/master/VarScan.v2.4.3.description.txt`

The optimal settings for these filters depends on the data itself. The tumor purity, for example, influences the maximum variant allele fractions at which somatic mutations can be observed. The coverage or depth of sequencing influences the minimum read count that can be expected for variants. Generally, it is a good idea to explore the effects of the various filters and verify called variants but also excluded variants in the IGV browser.[16]

30.3.5   Other somatic variant callers

Variant calling is still a topic of active research. Current callers and pipelines produce widely divergent results and there is no one-fits-all solution. Several articles have been published with evaluations of some of the competing methods (see [227, 368, 451, 467] and Further Reading at the end of this chapter). In general, it is advisable not to rely on just a single method for somatic variant screening.

30.3.6   Annotation of somatic SNVs and indels

The biological and medical evaluation of somatic mutations usually starts from transcript annotations. Tools such as Jannovar (see Chapters 14 and 15) can be used to produce annotated VCF files with basic information for each called variant such as the affected gene (if any) and the potential effect of a coding SNV on the encoded protein. Additionally, specialized tools have been developed to provide oncology-specific information about variants. For example, it is often of use to know whether somatic mutations are present in the COSMIC database, which would indicate that an identical variant was previously called in other tumor samples. Even more important is the question whether a gene affected by a somatic mutation is a known cancer gene. Here, we will present one such tool. Oncotator[17] can annotate variant calls in different formats including VCF and MAF (Mutation Annotation Format),[18] and provides information from a rich set of data sources [351], including:

---

[16]Optimal filter settings for a tumor of 100% purity containing subclones (variant allele fractions of 50%, 33%, and 20%) are suggested at: https://github.com/dkoboldt/varscan/blob/master/VarScan.v2.4.3.description.txt

[17]Web tool or download: http://portals.broadinstitute.org/oncotator/

[18]https://wiki.nci.nih.gov/display/TCGA/Mutation+Annotation+Format+(MAF)+Specification

- The reference sequence and GC content around the variant.

- Annotations about affected genes and transcripts.

- Annotations and predictions of functional consequences of variants.

- Site-specific protein annotations from UniProt.

- Cancer mutation frequencies taken from COSMIC.

- Cancer gene annotations from the Cancer Gene Census (CGC) and the Familial Cancer Database.[19]

- Cancer variant annotations from ClinVar and the Cancer Cell Line Encyclopedia.[20]

- Non-cancer variant and common SNP annotations from dbSNP, 1000 Genomes and the NHLBI GO Exome Sequencing Project (ESP).[21]

Once Oncotator and the associated databases have been downloaded, annotating variants is straightforward:

```
$ oncotator --input_format=VCF --output_format=VCF \
    --db-dir /path/to/oncotator_v1_ds_April052016/ \
    --log_name <out.log> <in.vcf> <out.vcf> hg19
```

At the time of writing, the latest data bundle[22] was available only for the hg19 reference genome. Therefore, variant calls that have been determined using hg38 need to be remapped to the hg19 chromosomal coordinates[23] (see Section 8.1.3).

For a specific tumor-relevant variant at chrX:77,634,628 (equivalent to chrX:76,890,119 on hg19) the original INFO field in the VarScan2 VCF file is:

```
DP=408;SOMATIC;SS=2;SSC=255;GPV=1E0;SPV=1.5924E-45
```

---

[19]http://www.familialcancerdatabase.nl/

[20]https://portals.broadinstitute.org/ccle/home

[21]https://esp.gs.washington.edu/drupal/

[22]oncotator_v1_ds_April052016 from April 2016

[23]Or the entire analysis needs to be performed using the GRCh37 genome assembly.

Jannovar provides information about the location and possible functional effect of the somatic mutation, and adds the annotations to the INFO field:

```
ANN=C|missense_variant|MODERATE|ATRX|546|transcript|
NM_000489.4|Coding|17/35|c.4775T>G|p.(Leu1592Arg)|
5043/281392|4775/7479|1592/2493||;
DP=408;GPV=1E0;SOMATIC;SPV=1.5924E-45;SS=2;SSC=255
```

Hence, the somatic variant is a missense mutation (`c.4775T>G`, `p.Leu1592Arg`) in the *ATRX* gene (NCBI/Entrez gene ID 546, RefSeq transcript ID `NM_000489.4`). Oncotator aggregates genomic annotations, site-specific protein annotations and functional impact predictions from dbNSFP, as well as cancer-specific annotations from COSMIC, the Cancer Gene Census, the cancer cell line encyclopedia, the familial cancer database, and ClinVar [351]:[24]

```
transcript_exon=17;
gc_content=0.363;
ref_context=ACAGTGGGCAAGAATGCATCC;
CGC_Cancer_Somatic_Mut=yes;
CGC_Tumour_Types__(Somatic_Mutations)=
    "Pancreatic_neuroendocrine_tumors|_paediatric_GBM";
CGC_Other_Germline_Mut=yes;
CGC_Other_Syndrome/Disease=
    ATR-X_(alpha_thalassemia/mental_retardation)_syndrome;
COSMIC_n_overlapping_mutations=3;
COSMIC_Tissue_total_alterations_in_gene=2745;
COSMIC_Tissue_tissue_types_affected=
    NS(9)|adrenal_gland(118)|autonomic_ganglia(39)|
    biliary_tract(25)|bone(3)|breast(191)|
    central_nervous_system(710)|endometrium(19)|
    haematopoietic_and_lymphoid_tissue(852)|kidney(200)|
    large_intestine(77)|lung(36)|oesophagus(3)|ovary(42)|
    pancreas(154)|prostate(8)|skin(35)|soft_tissue(22)|
    stomach(58)|thymus(24)|upper_aerodigestive_tract(120);
```

Therefore, the gene has somatic mutations in many tumors present

---

[24]Oncotator additionally reports gene, transcript and protein IDs from Ensembl and UniProt, and GO annotations.

in COSMIC and is listed in the Cancer Gene Census as a known cancer gene. Germline mutations of the gene can cause the alpha thalassemia/mental retardation syndrome.

30.3.7 SNVs and indels in the glioblastoma

Calling somatic variants in the glioblastoma sample applying VarScan2 and the above described filters, we obtain the following non-synonymous variants in coding regions:[25]

```
 1 disruptive_inframe_deletion
 4 frameshift_variant
54 missense_variant
 1 missense_variant&splice_region_variant
 2 splice_acceptor_variant&coding_transcript_intron_variant
 1 splice_donor_variant&coding_transcript_intron_variant
 6 stop_gained
```

Additionally, some variants are synonymous or lie in intronic regions but are very close to splice sites. These variants conceivably could lead to splice defects:

```
 7 splice_region_variant&coding_transcript_intron_variant
 1 splice_region_variant&non_coding_transcript_exon_variant
 1 splice_region_variant&synonymous_variant
```

Here, we will discuss only a few of these mutations. The most important somatic mutation we can identify is the missense mutation in isocitrate dehydrogenase 1 (*IDH1*-R132H) that is commonly found in a subset of glioblastoma [83, 470]:

```
ANN=T|missense_variant|MODERATE|IDH1|3417|transcript|
NM_001282386.1|Coding|4/10|c.395G>A|
p.(Arg132His)|601/18232|395/1245|132/415||;
```

IDH1-R132H mutations occur early in the development of a subset of low-grade astrocytomas which later progress to become glioblastomas [83]. The mutated IDH1 enzyme, instead of producing α-ketoglutarate (α-KG), catalyzes a metabolic reaction that converts α-KG to the presumed oncometabolite 2-hydroxyglutarate (2-HG) [96] which acts as an antagonist of α-KG. The exact mechanism by which this changed metabolic function drives oncogenesis is not known, but the mutation is associated with an altered cellular metabolism and

---

[25]Jannovar annotations are shown in this section.

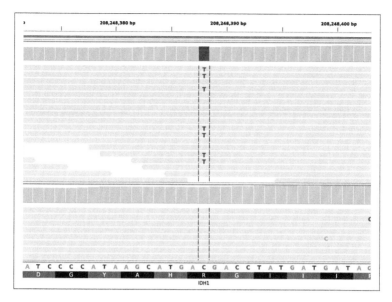

Figure 30.2. Somatic *IDH1*-missense mutation (c.395G>A, p.R132H) on chromosome 2. Upper panel: tumor; lower panel: normal control.

increased histone and DNA methylation [83, 470]. This missense mutation is present in the tumor sample but not in the matched control (Figure 30.2).

Another missense variant in the *ATRX* gene, which is frequently altered in *IDH*-mutated glioblastoma [50] causes the substitution of leucine by an arginine residue (p.L1592R):

```
ANN=C|missense_variant|MODERATE|ATRX|546|transcript|
NM_000489.4|Coding|17/35|c.4775T>G|
p.(Leu1592Arg)|5043/281392|4775/7479|1592/2493||;
```

Loss of *ATRX* in glioblastoma multiforme has been shown to increase mutation rates at the SNV level, as well as genomic instability by impairing nonhomologous end joining and increasing sensitivity to DNA-damaging agents which induce double-stranded DNA breaks [218]. COSMIC reports a somatic missense mutation to a histidine at this position (c.4775T>A, p.L1592H) in a hematopoietic neoplasm, and predicts it to be pathogenic (COSMIC mutation ID COSM4385419).

Figure 30.3 illustrates that the decision whether a candidate variant is a true somatic event or not is anything but straightforward. The

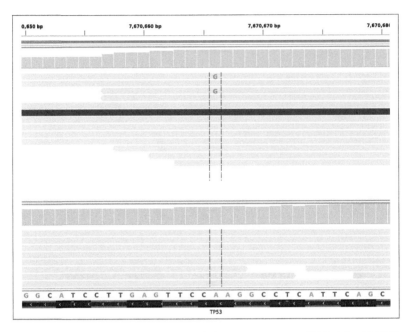

Figure 30.3. Low-confidence candidate variant in *TP53* (c.1043T>C, p.L348S) on chromosome 17. Upper panel: tumor; lower panel: normal control.

figure shows a candidate missense mutation (p.L348S) in the important tumor suppressor gene *TP53*:

```
ANN=G|missense_variant|MODERATE|TP53|7157|transcript|
NM_000546.5|Coding|10/11|c.1043T>C|
p.(Leu348Ser)|1245/19149|1043/1182|348/394||;
```

This event was not called as a variant by the original authors [322] and also in our case it is excluded in step 1 of the filtering procedure. That is, it is not considered a high-confidence variant. Still, two out of twelve high-quality reads (~17%) would be consistent with a heterozygous variant in a tumor subclone that constitutes around 30–35% of cells in the sample, especially given the fact that the alternate bases have high base qualities (35 and 41) and both reads have a high mapping quality (60). Strandness, too, appears to be favorable: one read maps to the forward strand, the other to the reverse strand. The region affected is highly conserved in mammals and this leucine in particular is conserved also in more distant species like zebrafish,[26] suggesting that a mutation at this position might be deleterious. COSMIC reports two somatic mutations for this base:

- COSM46015: a nonsense/stop gain mutation (c.1043T>A, p.L348*) in an esophageal squamous cell carcinoma, predicted to be pathogenic.

- COSM5013841: a missense mutation (c.1043T>G, p.L348W) in a renal cell carcinoma, predicted to be pathogenic.

However, taking a closer look into the BAM file or the IGV browser we find that the two reads hinting at a variant belong to the same read pair, i.e., they are actually overlapping reads from the same, short DNA fragment (insert size of 137 bases). It is thus only a single fragment that points to a variant (one out of eleven, ∼9%). This could be either a true heterozygous variant in about 15–20% of the tumor cells or, at least as likely, an error during library preparation or sequencing. It is therefore justified to exclude this candidate from the high-confidence variants, although we cannot rule out that *TP53* is actually mutated in some of the tumor cells.

30.4 SOMATIC STRUCTURAL VARIANTS

Structural variants like duplications, deletions, intra- and interchromosomal translocations, and inversions are frequent events in many cancer types and can result in tumor-promoting effects like the loss of tumor suppressors, oncogenic gene fusions or an amplification of otherwise unaltered oncogenes.

While the identification of structural variants in tumors is generally performed as outlined in Chapter 19, there are some particular aspects related to the distinction between somatic events and variants already present in the germline.

30.4.1 Somatic CNVs: CopywriteR

Somatic copy number variants (CNVs) can in principle also be called using the VarScan2 `copynumber` and `copyCaller` commands:

```
$ java -jar VarScan.jar copynumber normal_tumor.mpileup \
```

---

[26]This can be easily verified in the UCSC Genome Browser.

```
    tumor  --mpileup 1
```

```
$ java -jar VarScan.jar copyCaller tumor.copynumber \
    --output-file tumor.copynumber.called \
    --output-homdel-file tumor.copynumber.homozygdel
```

This requires, however, external software such as the R/Biocon-ductor package DNAcopy for smoothing and segmentation of the raw output from VarScan2 `copynumber`. Also the final merging of adjacent segments of similar copy number needs to be done with an external VarScan2 script.[27]

Here, we will show how to call somatic CNVs in WES data with the R/Bioconductor package CopywriteR [231]. In contrast to other CNV detection tools for WES, CopywriteR does not use on-target reads, i.e., reads that fall within the exonic regions targeted by the extraction kit. Instead, it uses off-target reads which are largely distributed at random across introns and intergenic regions. This random distribution entails that off-target regions with copy number changes will accumulate more (or less) random reads than regions with a neutral copy number [231]. This has two advantages. First, it avoids the problem of widely differing target capture efficiency across the exons. Second, it covers a much larger fraction of the genome than the approaches based on on-target, exonic reads. In practice, CopywriteR treats off-target reads as if they were derived from low-coverage WGS, which is sometimes used for CNV detection.

Application of CopywriteR in R is fairly easy and requires only the BAM files of the tumor and normal control samples. First, the working directory (or any other desired folder) is prepared as a workspace:

```
> library(CopywriteR)
> data.folder <-
    tools::file_path_as_absolute(file.path(getwd()))
> preCopywriteR(output.folder = file.path(data.folder),
    bin.size = 20000, ref.genome = "hg38", prefix = "chr")
```

This creates a folder in the workspace named "hg38_20kb_chr" containing some auxiliary files with genome characteristics for the hg38

---

[27]`mergeSegements.pl` available at: `https://sourceforge.net/projects/varscan/files/scripts/`

reference genome,[28] computed over 20kb windows or bins (`bin.size`) and using "chr" as a prefix for chromosome labels (e.g., "chr12" or "chrX"). The prefix can also be omitted if plain chromosome labels are required (e.g., "12" or "X"). The auxiliary files need to be generated only once using `preCopywriteR` if later runs of CopywriteR use the same configuration.

Then, the auxiliary files produced in the preparation step are loaded:

```
> load(file = file.path(data.folder, "hg38_20kb_chr",
    "blacklist.rda"))
> load(file = file.path(data.folder, "hg38_20kb_chr",
    "GC_mappability.rda"))
```

The first auxiliary file ("`blacklist.rda`") contains a list of blacklisted genomic regions of known common CNVs. The second ("`GC_mappability.rda`"), specifies the mappability and GC content computed over the desired window size or bin width. This is particularly important because the local GC content and the genomic mappability (uniqueness and complexity of the sequence) are the major sources of bias that affect the obtainable read counts (Section 19.4).

CopywriteR supports parallel processing via the R/Bioconductor package BiocParallel[29] and hence requires information on the type and characteristics of the parallel processing environment. The simplest possible specification is:

```
> bp.param <- SnowParam(workers = 1, type = "SOCK")
```

This corresponds to a simple network of workstations (SNOW) with only 1 processor core (`workers`), i.e., no parallel processing at all. By default, if the number of workers is not specified, all available cores will be used for parallel processing of different samples. The `type` of SNOW you will most likely have is `SOCK` (default) which indicates that data communication is done via Unix domain sockets between processes executing on the same host. This is the case for a normal multi-core PC. If you're working on a true cluster, BiocParallel `SnowParam` also supports the Message Passing Interface (`type` = `"MPI"`).[30]

---

[28] Other reference genomes accepted by CopywriteR are hg18 and hg19 for human, and mm9 and mm10 for mouse.

[29] http://bioconductor.org/packages/release/bioc/html/BiocParallel.html

[30] See the BiocParallel documentation for further details.

An R `data.frame` is used to specify the samples to be analyzed. The first column of the `data.frame` should list all samples, and the second column contains the corresponding normal control samples. CopywriteR requires that the normal control samples are also indicated in the first row. This is required for plotting of the results (see below). Since our example data set has only one tumor-control sample pair, the `data.frame` has only two rows.

```
> sample.control <- data.frame(
  samples=c("/path/to/tumor.bam", "/path/to/normal.bam"),
  controls=c("/path/to/normal.bam", "/path/to/normal.bam")
)
```

Finally, CNVs are called and plotted by CopywriteR:

```
> CopywriteR(sample.control = sample.control,
    destination.folder = file.path(data.folder),
    reference.folder = file.path(data.folder,
                               "hg38_20kb_chr"),
    bp.param = bp.param)
```

All results will be saved to a subfolder `CNAprofiles` located in the workspace (or wherever `destination.folder` points to):

- `CopywriteR.log`: A log file for the CopywriteR run.

- `input.Rdata`: An R object that contains information necessary for plotting the results with `plotCNA`, such as the bin size and chromosome names.

- `log2_read_counts.igv`: A tab-separated table containing the log2-transformed, normalized read count ratios for the tumor and normal sample.

- `qc`: A folder with additional QC plots.

- `read_counts.txt`: A tab-separated table containing read counts for every 20kb window and every sample. For every sample, it also reports the fraction of the window that effectively contributes to the read count ignoring the overlaps with "peaks" called at on-target regions in the matched reference sample (the sample in column 2 of the `sample.control` table). Read counts are specified both uncompensated and compensated (i.e., normalized to the remaining effective window size).

We obtain the following `log2_read_counts.igv` table:[31]

```
$ head -3 CNAprofiles/log2_read_counts.igv
#track viewLimits=-3:3 graphType=heatmap color=255,0,0
Chr.  Start   End     Feature              log2.tumor  log2.normal
chr1  820001  840000  chr1:820001-840000   -0.1320047  -0.01060641
chr1  840001  860000  chr1:840001-860000   -1.1150983  -0.77120495
chr1  860001  880000  chr1:860001-880000   -0.8370817   0.50211847
```

This file can be used both for downstream analysis and for visualization in IGV (see below).

The final step in the CopywriteR workflow identifies segments of similar copy number and plots the resulting copy number profiles for all samples and chromosomes:

```
> plotCNA(destination.folder = file.path(data.folder))
```

This will add the following files to the previously created subfolder (`<destination.folder>/CNAprofiles`):

- plots: A folder with comparison plots[32] for the copy number profiles of all individual chromosomes as well as a plot for the entire genome.

- segment.Rdata: An R data object containing all log2 ratios from individual 20kb windows (`segment.CNA.object$data`), used to compute the profiles, and the mean log2 ratios for the identified segments of similar copy number (`segment.CNA.object$output`).

The sample comparisons in `segment.CNA.object$output` can be filtered for the particular comparison of interest as indicated in the ID column. The table has the following format:

```
> head(segment.CNA.object$output)
                        ID chrom  loc.start  loc.end num.mark seg.mean
1 log2.tumor.vs.log2.normal     1    830000.5  2430000       78  -0.8708
2 log2.tumor.vs.log2.normal     1   2450000.5  5970000      168  -0.5656
3 log2.tumor.vs.log2.normal     1   5990000.5  7830000       93  -0.8322
4 log2.tumor.vs.log2.normal     1   7850000.5 11390000      178  -1.1642
5 log2.tumor.vs.log2.normal     1  11410000.5 11590000       10  -0.6929
6 log2.tumor.vs.log2.normal     1  11610000.5 21690000      467  -0.0292
```

---

[31]The version of the table shown here contains some abbreviations and rounded values.

[32]In our case tumor versus normal, tumor versus none (without reference), normal versus none (without reference), normal versus normal.

Row four, for instance, reports a segment with a mean log2 ratio of -1.16 between tumor and normal, i.e., with the tumor having about half the copy number of the normal. This segment lies on chromosome 1, begins with the 20kb window that starts at position 7,850,000, and ends with the 20kb that starts at position 11,390,000.

For further use, the data contained in `segment.CNA.object` can be written to comma- or tab-separated text files using the standard R command `write.table`.

Chapter 19 has additional information on structural variation.

30.4.2   Copy number alterations in the glioblastoma

The copy number profiles generated by CopywriteR for the glioblastoma sample suggest that several chromosomes harbor structural variants. Here, we illustrate only two examples.

Figure 30.4 shows extensive copy number changes on chromosome 9. The entire chromosome arm 9q and a significant part of arm 9p have a third copy. This is especially well visible when loading `log2_read_counts.igv` into IGV (Figure 30.5, upper panel). Additionally a smaller somatic duplication is observed on chromosome 9p22 as well as a somatic homozygous deletion on 9p21.3 (see also Figure 30.5, lower panel).

Homozygous deletions of 9q21.3 are known to be recurrent events in glioblastomas and usually involve the gene cluster encoding the cyclin-dependent kinase inhibitors CDKN2A and CDKN2B [120] (see the lower panel of Figure 30.5). The inactivation of these tumor suppressors plays an important role in many malignant gliomas [398].

Interestingly, chromosome 9p21.2 harbors a small duplication involving only a few genes that is already present in the germline (see Figure 30.4, upper panel, and Figure 30.5, central panel). One of these genes is the tyrosine kinase receptor TEK/Tie2 which has been found to promote glioma progression from a lower to higher grade [243] as well as an invasive phenotype by influencing the interaction between endothelial cells and the brain tumor's stem cells [262]. Whether a germline duplication of this gene increases the risk of developing brain tumors has not yet been studied.

Another especially interesting somatic copy number change is the deletion on chromosome 1p36 shown in Figures 30.6 and 30.7. This region is frequently deleted in glioblastomas and other astrocytic tumors [177]. The region deleted in this patient contains several known or

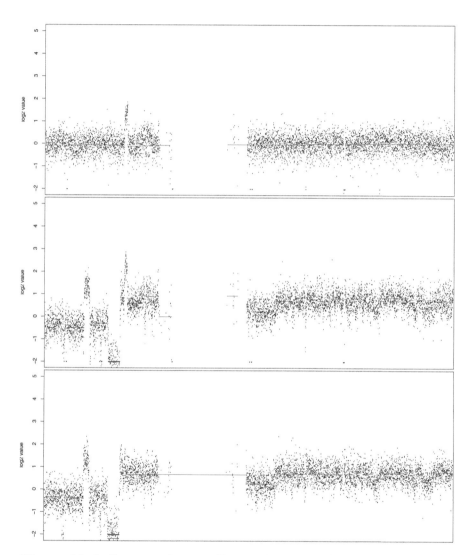

Figure 30.4. Copy number profiles as plotted by CopywriteR for chromosome 9 of the glioblastoma. Panels from top to bottom: (i) normal control sample; (ii) tumor sample; (iii) comparison tumor versus normal.

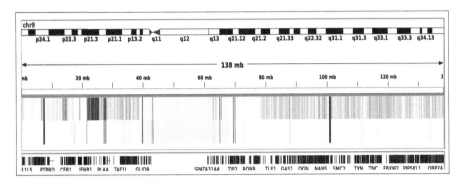

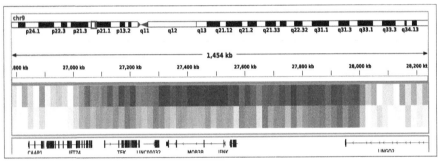

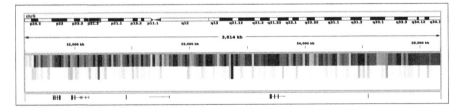

Figure 30.5. IGV plot of `log2_read_counts.igv` for chromosome 9 of the glioblastoma. The middle section of each panel shows copy numbers in both the tumor (top layer) and normal sample (bottom layer), using red for increased copy number and blue for decreased copy number. Upper panel: chromosome 9; compare this to Figure 30.4. Central panel: germline duplication on chromosome 9p22. Lower panel: somatic homozygous deletion on chromosome 9p21 (including the CDKN2A/B locus).

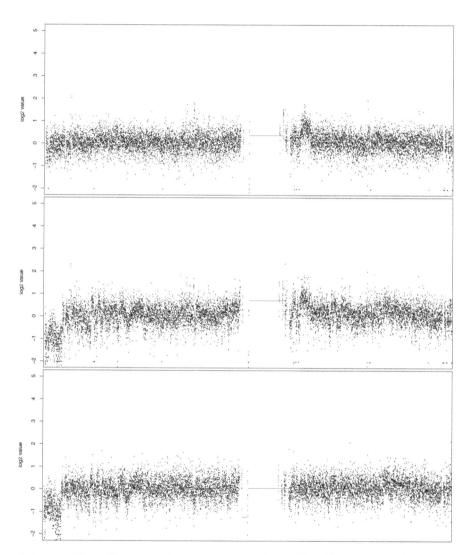

Figure 30.6. Copy number profiles as plotted by CopywriteR for chromosome 1 of the glioblastoma. Panels from top to bottom: (i) normal control sample; (ii) tumor sample; (iii) comparison tumor versus normal.

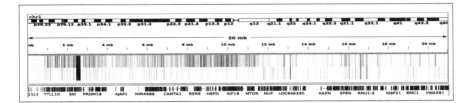

Figure 30.7. IGV plot of `log2_read_counts.igv` showing the somatic deletion on chromosome 1p36 of the glioblastoma. The middle section shows copy numbers in both the tumor (top layer) and normal sample (bottom layer), using red for copy number increases and blue for copy number decreases. Compare this to Figure 30.6

suspected tumor suppressor genes including the protein-coding genes *TP73* [348], *CAMTA1* [388], *ERRFI1* [110], and *CHD5* [23], and the microRNAs miR-200b [257] and miR-34a [355].

30.5 MEDICAL INTERPRETATION AND DIAGNOSTICS

Our understanding of cancer biology and genomics, while still far from complete, has progressed so rapidly in the last decades that NGS-based diagnostics is now becoming increasingly important for the clinical care of individuals with cancer. We will not attempt to provide a comprehensive description of the role of genomic diagnostics in oncology, which would require a book on its own. Instead, this section will review several areas in which bioinformaticians and computational analysts are likely to interact with the medical team.

Traditional chemotherapy employs agents that attack quickly dividing cells, so-called cytotoxic drugs. Clinical research in oncology was previously focused on investigations that aimed to determine the best combinations of (non-specific) cytotoxic drugs for particular types and stages of cancer. Targeted therapy relies instead upon the identification of characteristic molecular abnormalities of an individual tumor that are both responsible for carcinogenesis or tumor growth and can be specifically targeted by agents such as monoclonal antibodies or small-molecule medications. In many cases, the "target" is a specific mutation in a cancer-related gene. Correspondingly, NGS-based diagnostics including WGS, WES or gene-panel assays of sets of cancer-related genes is playing an ever growing role in oncological diagnostics.

Therefore, clinical genomic diagnostics in oncology is required to generate accurate reports with informative interpretations of the genomic alterations. The reports list the variants and structural alterations found in the clinical sample and provide a summary of relevant clinical information, often depending on the tumor type and anatomical site [173].

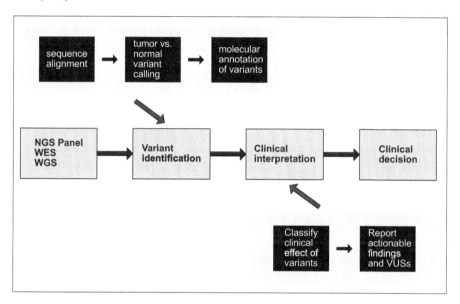

Figure 30.8. Clinical interpretation of somatic variants. Figure adapted and redrawn from [437].

The workflow, illustrated in Figure 30.8, for interpretation of genomic results begins with Q/C steps, alignment, variant calling and tumor-normal comparison. After an initial list of variants has been assembled and the molecular effects have been annotated, clinical interpretation aims to determine the variant effects and whether there is evidence supporting the use of a targeted therapy. Variants of unknown significance (VUS) are reported as such. Finally, the clinician uses this information as a part of his or her decision process [437].

Consequently, one very important role of computational analysis of tumor samples is to support medical interpretation by collecting all information about the somatic mutations identified in the sample for which a targeted therapy might be available. Databases such as COSMIC [122] and The Cancer Genome Atlas (TCGA) [57] are not intended for direct clinical use and contain little information about specific drug responses and clinical trials. However, there are multiple publicly available databases for this purpose as well as many commercial offerings such as IBM Watson for Oncology.[33] Table 30.1 provides an overview of several publicly available databases.

Table 30.1. Knowledge bases for precision genomic medicine in oncology.

Database	URL
Precision Medicine Knowledge Base [173]	`https://pmkb.weill.cornell.edu/`
JAX-Clinical Knowledgebase [188]	`https://ckb.jax.org/`
CIViC [145]	`https://civic.genome.wustl.edu`
My Cancer Genome [425]	`https://www.mycancergenome.org/`
Database of Curated Mutations [8]	`http://docm.genome.wustl.edu/`
DGIdb [446]	`http://dgidb.genome.wustl.edu/`
CanDL [94]	`https://candl.osu.edu/`

Still, the existing knowledge on cancer therapy, biomarkers (such as specific mutations), and pharmacogenomics is not well structured [131]. Obviously, a great deal of caution is necessary for the interpretation of genomic findings and their potential therapeutic relevance, and this requires substantial clinical expertise and experience. Our ability to interpret genomic findings is still very incomplete. For instance, knowledge about targeted therapies that may be indicated in the presence of a certain mutation is often restricted to only a few cancer types, and the fact that a mutation predicts response to a targeted therapy in one type of cancer does not necessarily mean that another type of cancer with the same mutation will respond to the same targeted therapy.

In practice, the final analysis of the relevance of the mutations and of which targeted therapies (if any) are indicated for a given patient are dependent on the clinical situation. However, we will explore the kind of information that is available for some of the mutations found in our example glioblastoma exome. A clinical WES report would summarize this and other information in order to help the clinician make final decisions about the indicated treatment modalities.

Several interpretations are available for the somatic variants identified in the glioblastoma sample analyzed above (shown in excerpts here):

---

[33]`https://www.ibm.com/watson/health/oncology/`

- The *IDH1* missense variant NM_001282386.1:c.395G>A, p.R132H, has been found in multiple tumors including glioblastoma, breast and liver cancer, and leukemia. The mutation is commonly found in malignant gliomas and acts as a gain of function, promoting the production of the onco-metabolite 2-hydroxyglutarate (2-HG) that may contribute to the formation and malignant progression of gliomas [96]. A phase I clinical trial of the AG-120 IDH1 inhibitor in *IDH1*-mutant positive solid tumors is currently underway.[34]

- The mutated leucine at position 1592 of the *ATRX*-encoded protein lies within its helicase ATP-binding domain. While the effect of the mutation on the protein is unclear, it is possible that its function is severely impaired. In IDH-mutant astrocytomas which were treated either with temozolomide or with a combination of procarbazine, lomustine and vincristine, the time to treatment failure and progression-free survival were greater in cases with a loss of *ATRX* expression [458].

- In a single case study of a patient with metastatic, recurrent/refractory cholangiocarcinoma, harboring a nonsense mutation (E384X) in *ERRFI1* in absence of any mutations or amplifications in other EGFR signaling members, the treatment with the EGFR kinase inhibitor erlotinib led to a partial response after 3 months [44]. Admittedly, this is a different cancer type and the glioblastoma shows a deletion of the gene (on chromosome 1p36), not a truncation of the protein. Nevertheless, this finding might be of interest because ERRFI1 is a negative regulator of EGFR and response to erlotinib treatment of glioma patients was found to be associated with EGFR expression or amplification [152].

Variant prioritization approaches similar to those described in Part VI of the book have also been developed for personalized cancer genomics. For instance, HitWalker uses a protein–protein interaction (PPI) network (similar to the algorithm in Chapter 25) to prioritize variants detected in cancer patients or cohorts [46, 47].

The task of the clinician is to integrate the available information about the mutations identified in a cancer and decide upon the indicated treatment regime and clinical management. Therefore, the mere

---

[34]See https://www.clinicaltrials.gov, NCT02746081.

identification of somatic mutations is just the initial step of the clinical analysis.

FURTHER READING

Alioto TS, Buchhalter I, Derdak S, Hutter B, et al. (2015) A comprehensive assessment of somatic mutation detection in cancer using whole-genome sequencing. *Nature Communications* **6**:10001.

Krøigård AB, Thomassen M, Lænkholm A-V, Kruse TA, Larsen MJ (2016) Evaluation of nine somatic variant callers for detection of somatic mutations in exome and targeted deep sequencing data. *PLoS One* **11**(3):e0151664.

Macintyre G, Ylstra B, Brenton JD (2016) Sequencing structural variants in cancer for precision therapeutics. *Trends in Genetics* **32**(9):530–542.

Tubio JMC (2015) Somatic structural variation and cancer. *Briefings in Functional Genomics* **14**(5):339–351.

Xu H, DiCarlo J, Satya RV, Peng Q, Wang Y (2014) Comparison of somatic mutation calling methods in amplicon and whole exome sequence data. *BMC Genomics* **15**:244.

30.6   EXERCISES

Exercise 1

The glioblastoma tumor-normal pair described in this chapter comes from a set of three samples that also includes an initial low grade astrocytoma that only later progressed into the glioblastoma [322].

Download the astrocytoma from SRA using the accession/run number SRR1823085 (see Section 4.3). The tumor purity of this sample was estimated by the original authors to be 63% (see Supplementary Table 4 of [322]).

Use VarScan2 to call SNVs and indels, and CopywriteR to produce copy number profiles. The normal control sample is, of course, the same as the one used for the glioblastoma (it is the same patient!).

Analyze the results and answer the following questions:

1. Is the *CDKN2A-CDKN2B* gene cluster on 9q21.3 deleted also in the astrocytoma or did this deletion occur only during the progression of the tumor to a glioblastoma?

2. Does chromosome 1p36 of the astrocytoma show the same typical deletion?

3. Are all somatic non-synonymous mutations (missense, nonsense, indels, and splice site mutations) of the low grade astrocytoma still present after progression to a glioblastoma? Reflect about your observation and provide an explanation.

Exercise 2

Explore some of the knowledge bases offering mutation-specific clinical information listed in Table 30.1. Try to gain insight about effects of deletions of *CDKN2A* and *CDKN2B* in glioblastoma. Can you identify possible treatment options? Do not forget to evaluate the status of the studies suggesting the treatment options (preclinical study, case study, clinical trial).

Exercise 3

Compare the pileup data for positions 69,023,372 on page 434 with the corresponding alignment in Figure 30.1 and verify that they reflect the information specified in the pileup.

CHAPTER 31

Tumor Evolution and Sample Purity

T UMOR sequencing data are often characterized by a high degree of heterogeneity because in most cases tumor samples are composed not only of cancerous tissue but also non-cancerous blood vessels, stromal cells, and adjacent tissues. These non-cancerous parts of a tumor are important for cancer biology, should be taken into account in the computational analysis of cancer genomics data, and in many cases are important for the biological and medical interpretation of the results of computational analysis [20]. Additionally, even the cancerous components of a tumor can be heterogeneous. Different subclones of the tumor harbor partially overlapping sets of somatic alterations that may be responsible for the distinct behavior of individual subclones (see Chapter 29).

31.1   ESTIMATION OF TUMOR PURITY AND CLONALITY

Contamination of clinical tumor samples by non-cancerous cells can make it more difficult to reliably perform mutation testing because the observed allele fractions are reduced. Light microscopic (histopathological) estimation of tumor purity was commonly used to assess the proportion of cancerous tissue in samples prior to molecular testing, although histopathological estimates are not always reliable and there may be a low correlation between histopathological and molecular estimates of tumor purity [186, 408]. A number of computational methods have been developed to estimate tumor purity from genomic data.

Knowledge about tumor purity and clonal composition may help in assessing what approximate allele fractions are to be expected for variants that are shared among all tumor clones, or those that are specific to only a subset of the clones. Shared variants are likely to have arisen early in the tumor history or may have given one of the tumor clones a selective advantage which led to the clone supplanting other clones. On the other hand, clone-specific mutations may also be of clinical importance. For instance, if a mutation that confers resistance to a therapeutic agent is present in a subclone that makes up only a small proportion of tumor tissue, this subclone could be selected for and expand if the patient receives the therapeutic agent.

While the estimation of sample purity and the determination of the subclonal composition of a tumor can be seen as distinct problems, some approaches can address them jointly by treating the normal cells in the sample as an additional (non-mutated or wildtype) subclone of the tumor.

Table 31.1 lists several tools for tumor sample purity estimation and inferring tumor heterogeneity or clonality from whole genome or whole exome sequencing data. A recent review presented a comparative assessment of five tools for computational sample purity estimation [468].

Table 31.1. Tools for estimating sample purity (P) and/or tumor heterogeneity or clonality (H).

Tool	P	H	URL
ABSOLUTE [61]	✓		archive.broadinstitute.org/cancer/cga/absolute
CLONET [343]	✓	✓	demichelislab.unitn.it/doku.php?id=public:clonet
ExPANdS [18]	✓	✓	cran.r-project.org/package=expands
PurBayes [238]	✓	✓	cran.r-project.org/package=PurBayes
PurityEst [414]	✓		odin.mdacc.tmc.edu/~xsu1/PurityEst.html
SciClone [293]		✓	tvap.genome.wustl.edu/tools/sciclone/
SubcloneSeeker [346]	✓	✓	github.com/yiq/SubcloneSeeker
THetA2 [315]	✓	✓	compbio.cs.brown.edu/projects/theta/
TrAp [412]	✓	✓	sourceforge.net/projects/klugerlab/files/TrAp/

In the following, we will review the approaches of two of these tools, PurityEst and PurBayes. Readers will be asked to apply PurBayes to the low-grade astrocytoma data we investigated in Exercise 1 of Chapter 30.

31.1.1 Tumor purity: PurityEst

The PurityEst algorithm[1] [414] infers the level of tumor purity of a given tumor sample with a matched control using a simple heuristic. In a pure sample without normal cell content, heterozygous somatic mutations should ideally have a variant allele fraction of 50%. If consistently lower allele fractions are observed, then one can infer the presence of non-cancerous cells in the sample (which do not carry the somatic variants). The percentage of non-cancerous cells in the sample can be estimated according to the observed allele fractions.

This simple approach ignores two important factors. First, tumor samples are usually heterogeneous mixtures of multiple clones. This means that even if no normal cells are present, individual somatic mutations can exhibit much lower variant allele fractions, depending on what proportion of the tumor clones or cells they affect. Second, observed fractions can also be affected by copy number changes. If, for example, the mutant allele is duplicated, the fraction of reads displaying the variant increases. If, on the contrary, the wild-type allele is duplicated, the variant allele fraction is reduced. Loss of heterozygosity (LOH), which is commonly observed in many cancer types, refers to the loss of wildtype alleles due to copy-number losses or other genomic alterations. LOH thus leads to a change in polymorphic markers from a heterozygous state in the germline to an apparently homozygous state in the tumor DNA [382].

PurityEst partly addresses the second factor by computing purity estimates for each chromosome (except the sex chromosomes X and Y) and then taking a robust average as an estimate for the entire sample, such that the final estimate possesses some robustness with respect to individual increases or reductions of variant allele fractions due to copy number changes.

The first factor, however, is not addressed by PurityEst. Instead, samples are assumed to be composed of normal cells and only one tumor clone. This approximation ignores tumor heterogeneity but may be justified in cases in which most of the tumor cells stem from one dominating tumor clone. In practice, many clones share sets of genetic alterations that arose before their evolutionary divergence. If the proportion of subclonal[2] alterations is comparably small, a simple approach like this may still provide satisfactory estimates.

---

[1] Available at: http://odin.mdacc.tmc.edu/~xsu1/PurityEst.html
[2] Specific to a subclone of the tumor.

The purity estimate γ for a given chromosome is taken as

$$\gamma = \frac{\mu_t}{\mu_n} \tag{31.1}$$

where $\mu_t = \frac{\sum_i V_i}{\sum_i (V_i + W_i)}$ is the variant allele fraction for tumor t, given by the variant read counts V_i and the wild-type read counts W_i summed over all i called heterozygous single-nucleotide variants.

The variant allele fraction of heterozygous SNPs in the normal control sample (μ_n) would be 0.5 if sequencing, alignment, and variant calling were without biases, but in practice, μ_n is often slightly lower than the expected value.[3] Therefore, μ_n is explicitly computed in the same way as μ_t, by summing over all heterozygous SNPs from the normal sample, rather than just assuming a value of 0.5.

31.1.2 Tumor clonality: PurBayes

PurBayes [238] is an R package available at CRAN[4] that implements a Bayesian approach to sample purity estimation combined with the detection of tumor heterogeneity. It models the variant allele read counts V_i as a binomial–binomial hierarchical mixture model with marginal distribution $Bin(N_i, \gamma/2)$ that depends on the total read counts $N_i = V_i + W_i$ and the tumor purity γ.

To determine intratumor heterogeneity, PurBayes assumes the existence of $1, \ldots, J - 1$ subclonal populations with cellularities $\gamma_1 < \cdots < \gamma_{J-1} < \gamma_J$ and $\gamma_J \cong \gamma$.

The set $\mathbf{Y} = (Y_1, \ldots, Y_S)$ of variant read counts from S heterozygous somatic mutations[5] are modeled under a Bayesian finite mixture model where $\kappa_{i,j}$ indicates the probability that mutation i belongs to the clonal population j for $j = 1, \ldots, J$, with corresponding cellularity values of γ_j, where $\sum_j \kappa_{i,j} = 1$.

PurBayes determines both J and the set of cellularities γ_j by starting with a single tumor clone, i.e., assuming tumor homogeneity, and iteratively adding individual subclonal populations until a sufficiently accurate model fit is obtained.

PurBayes requires as input the set $\mathbf{Y}$ of variant read counts of

---

[3] Reads that carry the non-reference allele can have a lower probability of mapping correctly to the reference genome, leading to a slight reference bias [100].

[4] https://cran.r-project.org/package=PurBayes

[5] PurBayes uses only single-nucleotide variants (SNVs) because their read counts tend to be more reliable than with indels or more complicated variants.

heterozygous somatic mutations and the associated set **N** of total read counts. Optionally, the variant read counts **M** and total read counts **Z** for heterozygous germline variants from the control sample can be specified in order to correct for the reference bias, as with PurityEst.

```
library(PurBayes)
result <- PurBayes(N, Y, M, Z)
```

Results can be explored by summarizing and plotting them:

```
summary(result)
plot(result)
```

PurBayes and PurityEst were shown to provide nearly identical results for homogeneous tumor data but PurBayes was more accurate with simulated heterogeneous tumor samples [238]. The accuracy of sample purity estimation and of the identification and quantification of its subclonal structure is dependent on the amount of data available for inferring proportions of cellular populations. Thus, WGS data will often yield more accurate results than WES data.

31.2   ISSUES WITH PURITY AND CLONALITY ESTIMATION

PurBayes can be used to estimate both the tumor purity of a sample and the number of subclonal populations present in a tumor. However, it does not trace the chronology of the clonal evolution or identify which of the somatic mutations are present in the individual subclones. Various tools and approaches have been devised for this purpose, such as SciClone [293], SubcloneSeeker [346], and TrAp [412], but it remains very challenging to accurately reconstruct the development of tumor clones. The accuracy of purity and clonality prediction strongly depends on the number of somatic mutations that can be evaluated. WGS data is therefore preferable to WES data, and substantially higher coverage is required than for many of the other applications that have been presented in this book. It is obviously difficult to obtain an accurate measure of the variant allele fraction of an SNV if the region is not covered well. Can we be sure that the true VAF is 33% if we observe two alternate bases out of six? How about 20 alternate bases out of 60? The VAF computed from the higher number of covering bases is much more likely to be accurate. The higher the sequencing depth, the easier is the identification of different subclones, especially if they make up only a minor fraction of the tumor cells.

31.3   INTERPRETING VARIANT ALLELE FRACTIONS

Even if we have a sufficient coverage to compute comparably accurate VAFs, we still face the difficulty of correctly interpreting them. For germline SNVs this is usually easy: we expect heterozygous SNVs to have a VAF of about 50% and homozygous SNVs to have a VAF of about 100% (with deviations related to sequencing errors and random chance). It is substantially more difficult to interpret the variant allele fraction for somatic SNVs identified in tumors. For instance, a somatic VAF of 30% might be observed with

- A homozygous variant in 30% of the sequenced cells

- A heterozygous variant in 60% of the sequenced cells

- A heterozygous variant in 86% of the sequenced cells in which the other, non-mutated allele was duplicated (i.e., the mutated allele resides on one out of three copies)[6]

- A heterozygous variant in 35% of the cells in which the mutated allele was duplicated (two out of three copies)[7]

Thus, the VAF alone is not sufficient to fully understand the fraction of cells affected by an individual mutation.

FURTHER READING

Ding L, Raphael BJ, Chen F, Wendl MC (2013) Advances for studying clonal evolution in cancer. *Cancer Letters* **340**:212–219.

Marusyk A, Almendro V, Polyak K (2012) Intra-tumour heterogeneity: a looking glass for cancer? *Nature Reviews Cancer* **12**:323–334.

Yadav VK, De S (2015) An assessment of computational methods for estimating purity and clonality using genomic data derived from heterogeneous tumor tissue samples. *Briefings in Bioinformatics* **16**:232–241.

31.4   EXERCISES

Exercise 1

In this exercise, we will apply PurBayes to the read counts of heterozygous somatic mutations identified in the low-grade astrocytoma

---

[6]$0.3 \approx 86/(86 \cdot 3 + 14 \cdot 2)$
[7]$0.3 \approx (35 \cdot 2)/(35 \cdot 3 + 65 \cdot 2)$

that we analyzed in the first exercise of Chapter 30. For this tumor, the original authors estimated a tumor purity of 63% using SNP array data (see Supplementary Table 4 of the original publication [322]).

For purity estimation based on sequencing data, it is important to use heterozygous somatic mutations independently of their functional annotations, that is, including synonymous and non-coding mutations. The functional effect of variants is irrelevant for the investigation of tumor purity. One should also take the high-confidence heterozygous germline variants into account as a background distribution in order to correct for reference bias, as explained above.

Deciding which somatic mutations are heterozygous is difficult. In principle, VarScan2 produces also genotype calls but these are based on the specification of the tumor purity for variant calling (option `--tumor-purity`, see Chapter 30). Consequently, using these genotype calls for the present exercise would lead to biased results and it is preferable to use an ad-hoc filtering to separate heterozygous variants from homozygous ones.

As rule of thumb, we could exclude all somatic mutations with a variant allele fraction (VAF) of >75% as likely homozygous mutations. Indeed, even if the tumor purity was 100%, many of these mutations would likely be homozygous. For germline mutations, we will use the rule of thumb that heterozygous variants in the germline have a VAF between 25% and 75%. In order to be able to compute reasonably accurate VAFs, one can limit the analysis to variants that have a coverage (total read count) of at least ten reads.

Additionally, it may be wise to use only somatic variants for which there is at most one variant base in the normal control sample. A single read should be permitted because it can stem from circulating or infiltrating tumor cells.

For obtaining the read counts for reference and variant bases in both the tumor and the normal sample, we can use the DP4 information in the VCF files. "DP4" stands roughly for "high-quality base allelic read DePth, specified in 4 values". These four values are comma-separated and have the following meaning:

1. Number of high-quality reference bases on the forward strand (ref-forward).

2. Number of high-quality reference bases on the reverse strand (ref-reverse).

3. Number of high-quality altered (variant) bases on the forward strand (alt-forward).

4. Number of high-quality altered (variant) bases on the reverse strand (alt-reverse).

In the latest versions of the VCF standard, the DP4 field is deprecated[8] and will be eventually replaced by ADF ("Allelic Depths on the Forward strand") and ADR ("Allelic Depths on the Reverse strand"), but many variant callers still provide DP4 rather than ADF and ADR.

If you used VarScan2 for calling somatic variants (see Chapter 30), for both somatic and germline mutations you will find the read counts from the normal sample in the DP4 field of the 10th column and the read counts from the tumor sample in the DP4 field of the 11th column. The exact location of the DP4 field within these columns is indicated in the legend in the FORMAT field of the VCF file:[9]

```
GT:GQ:DP:RD:AD:FREQ:DP4
```

That is, the DP4 field is the last colon-separated entry in the 10th and 11th column. For example, the DP4 contained in

```
0/1:.:14:8:6:42.86%:2,6,2,4
```

is 2,6,2,4, reporting 2+6=8 reference bases and 2+4=6 variant bases, and thus a total read count or coverage of 14. Please note that the Oncotator (see Chapter 30) removes the DP4 fields. Jannovar, instead, preserves them.

Use the read counts from the DP4 fields of both somatic and germline variants and filter them according to the suggestions above. Compute the required input to run **PurBayes** and answer the following questions:

1. What purity estimate do you obtain? Compare your result to the estimate of 63% of the original authors.

2. How many populations of variants, i.e. subclones, are detected by PurBayes? Is this realistic? If not, what might be the reason?

---

[8]See http://www.htslib.org/doc/vcf.html

[9]See Chapter 13. Unfortunately, the placement of the DP4 field is not standard, and other variant callers output the DP4 field in the INFO field or elsewhere. Sometimes only the tumor DP4 is recorded and the normal DP4 is omitted or specified in another format.

3. How does the purity estimate change if you do not use the germline variants for background correction? Can you explain your observation?

Exercise 2

Estimate the tumor purity of the glioblastoma using the approach of PurityEst. Instead of running the tool, you can simply use the read counts that you already obtained from the DP4 information in Exercise 1 to estimate the purity (see Equation 31.1). Since there are only few somatic, heterozygous high-confidence variants per chromosome, it is easiest to compute a single purity estimate using the read counts for the entire genome. In this exercise we're less interested in accuracy than in conveying the general idea by means of a practical example.

Did you obtain the same estimate as in Exercise 1?

CHAPTER 32

Driver Mutations and Mutational Signatures

T HE computational analysis of cancer genomics data as presented in Chapter 30 had the goal of supporting precision cancer medicine by identifying the mutations associated with an individual cancer specimen and searching for specific mutations that are amenable to targeted therapies. Another field of cancer bioinformatics aims to use cancer genomics data to better understand cancer biology by discovering new driver genes and characterizing mutational signatures associated with specific types of cancer. This type of analysis generally investigates large cohorts of patients rather than individual samples.

32.1 DRIVER OR PASSENGER?

One of the most pressing issues in cancer research is the identification of driver mutations that are causally liked to tumorigenesis and/or therapy resistance. Driver mutations, which were introduced on page 422, are mutations that confer a selective growth advantage to tumor cells. Most somatic mutations identified in cancer genomes are likely to be mere "passengers" [144]. Passenger mutations confer no selective advantage to tumor cells. One might assume that passenger mutations occur at random throughout the genome, but this is not entirely true because some mutational processes take place preferentially at certain sequence motifs, leading to recognizable mutational signatures.

32.1.1 Recurrent Mutations

Tumors, even those considered to be of the same cancer type (e.g., breast cancer or melanoma), typically show highly diverse genetic and genomic landscapes with largely different sets of mutations and mutated genes.

Recurrent mutations, i.e., identical mutations that arise independently in different tumors and affect a substantial proportion of patients with a given tumor type, are the exception rather than the rule. Some notable examples are:

- Virtually all secretory meningiomas harbor the same missense mutation that changes a lysine to a glutamine (p.K409Q) within the DNA-binding domain of the transcription factor *KLF4* [80, 363].

- The *BRAF* p.V600E missense mutation occurs frequently in different cancer types, such as melanomas, colorectal and thyroid cancers [58].

- Point mutations of arginine 132 (R132) in *IDH1* and the analogous residue in *IDH2* (R172) are associated with multiple cancer types, in particular with the majority of low-grade gliomas and secondary glioblastomas [83, 470].

- The *BCR-ABL* fusion gene, generated by the so-called "Philadelphia chromosome translocation" and resulting in an aberrant ABL tyrosine kinase activity, is a hallmark of most chronic myelogenous leukemias [385].

- *NAB2-STAT6* fusion genes have been identified in both meningeal and non-central nervous system hemangiopericytoma (HPCs) and solitary fibrous tumors (SFTs) [69, 370, 391].

The specificity of these recurrent mutations suggests that they are gain-of-function mutations with an oncogenic effect. Such specific, recurrent mutations, however, constitute only a small minority of driver mutations.

32.1.2 Recurrently mutated genes

Tumor suppressors genes are deactivated by loss-of-function mutations in many cancer types. There are multiple ways in which the function of a tumor suppressor can be eliminated, including deletions, nonsense mutations, frameshift mutations and missense mutations at one of the critical residues of the affected protein, and thus one does not expect each tumor to have an identical mutation. Therefore, it also makes sense to search for recurrently mutated genes that show a significant enrichment of mutations within a patient group.[1] Such recurrently mutated genes are likely to be causally linked to tumor development, suggesting that many of the mutations affecting them will be driver mutations (although some of them might in principle also be passengers).

Some notable examples are:

- Loss of function of the tumor suppressor *TP53* is one of the most frequent genetic alterations in human cancer, with more than half of all tumors exhibiting a *TP53* mutation [124, 153, 316]. Many different mutations affect *TP53*, including deletions of the entire gene and various mutations of different sizes within its coding sequence. Another prominent example for a frequently mutated tumor suppressor is *PTEN*, whose level of expression appears to be critical [59].

- Although loss of function mutations in *NF1* are associated with the hereditary tumor predisposition syndrome neurofibromatosis type 1 (hence the gene's name), it has also been found to be somatically altered in sporadic malignancies such as colon cancer and anaplastic astrocytoma [324].

- *RB1* is frequently mutated or deleted in tumors, especially in retinoblastomas [105].

Although the notion that tumor suppressor genes are inactivated by arbitrary loss-of-function mutations seems intuitive, biological reality can be more complicated. TP53 is a classic tumor suppressor gene, and TP53 mutations can abrogate TP53 tumor suppression functions. However, the observed mutational spectrum in TP53 is far from uniform, and it is thought that some TP53 missense mutations confer novel activities that may drive tumor progression [318]. In fact, the

---

[1]The notion of a significant enrichment is very important here. Very long genes, for example, can be affected by random passenger mutations more frequently than short genes. Hence, the frequency alone by which a gene is affected by somatic mutations is not enough to reliably identify drivers. We need a gene to be mutated *more frequently than expected by chance.*

spectrum of somatic TP53 mutations in human cancers shows several hotspots, with the three most commonly mutated residues being located in the DNA-binding domain (residues 101–306). p.Arg175Ala, three mutations of the arginine residue at position 248 (p.Arg248Glu, p.Arg248Gly, and p.Arg248Leu) and three mutations of the arginine residue at position 275 (p.Arg273Cys, p. Arg275His, and p.Arg275Leu) have been shown experimentally to affect DNA binding or transactivation of TP53 target genes (Figure 32.1), and in general the clustering of mutations in the DNA binding domain suggests an essential role for this domain.

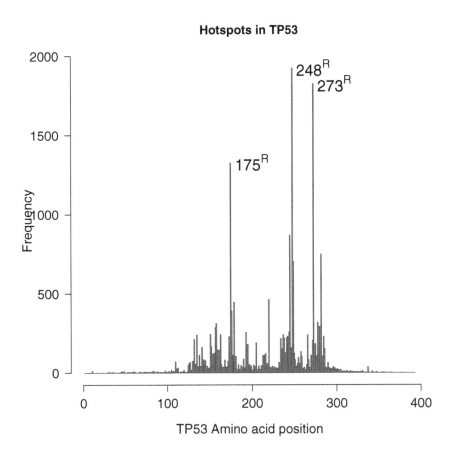

Figure 32.1. Spectrum of somatic mutations in the *TP53* gene. The locations of the three most commonly mutated residues are shown. The chart was generated with data from the IARC TP53 Database [48].

32.1.3 Recurrently mutated pathways

The search for recurrently mutated genes relies mostly on the idea that driver mutations accumulate in tumor-relevant genes due to positive selection. While this is true for some driver genes, many others are only rarely mutated.[2] Driver mutations can, for example, show functional redundancy with mutations of other genes in the same biological process or pathway. Also driver mutations in genes which play a role only in later stages of tumor development (e.g., for the formation of metastasis) might be less frequently observed in a multi-stage dataset. Identifying less frequently mutated driver genes by their mutational frequency alone requires increasingly large cohorts.

Consequently, an alternative way of identifying driver mutations is to search for candidate driver pathways that collectively contain more mutations than expected from the background mutation frequency, i.e., that are affected in more patients than expected by chance.

32.1.4 Tools for identifying drivers

Several effective approaches to the identification of recurrently mutated genes or pathways have been developed in recent years. Here, we will discuss only DrGaP because it identifies both driver genes and driver pathways and requires minimal input. We will mention some other frequently used tools without going into detail.[3]

32.1.4.1 DrGaP

DrGaP (driver genes and pathways)[4] [172] is a tool which incorporates knowledge of biological pathways (in form of user-defined gene sets) to identify both individual driver genes and entire driver pathways. The latter is achieved by treating each pathway as one "big" gene (combining the individual genes and their mutations) and applying the same statistical analysis that is used for individual genes.

---

[2]The mutation frequency of a driver gene may vary largely across different cancer types. The oncogene *KRAS*, for example, is mutated in 20–45% of uterine corpus endometrial carcinomas, lung adenocarcinomas as well as colon and rectal carcinomas, and in 4% of acute myeloid leukemias, but in only 0–1.2% of samples from seven other cancer types [190].

[3]Further tools were recently reviewed by Cheng et al. [67], and a more comprehensive list can be found at: https://omictools.com/driver-mutations-category

[4]DrGaP is available at: https://code.google.com/archive/p/drgap/

As can be explored in an exercise at the end of the chapter, using DrGaP is fairly simple:

```
drgap -i <input.mt> -g <exp.mt> -f <genome.fasta> \
    -b <pathway_file>
```

- <in.mt>: A tab-separated input file specifying the called mutations in the following columns:

  1. sample_id: The ID of the sample or patient in which the mutation was called

  2. gene: The ID or symbol of the gene affected by the mutation

  3. chr: The chromosome on which the mutation lies

  4. pos: The position of the mutated nucleotide

  5. ref: The reference base according to the reference genome

  6. var: The variant base produced by the mutation

  7. mutation_type: The type of mutation (silent, missense, nonsense, splicing, FS_indel or nFS_indel)

- <exp.mt>: A predefined gene mutation table that can be downloaded from the DrGaP website.

- <genome.fasta>: The reference genome in FASTA format.

- <pathway_file>: A user-defined list of gene sets/pathways. KEGG pathways and several other predefined gene sets can be downloaded from the DrGaP website. The tab-separated file should have the following columns:

  1. pathway_id: An ID for the pathway or gene set (e.g., hsa00061)

  2. pathway_name: The name of the pathway or gene set (e.g., Fatty_acid_biosynthesis)

  3. number_of_genes: The number of genes in the pathway or gene set (e.g., 6)

  4. gene_list: A space-separated list of gene symbols or IDs (e.g., ACACA ACACB MCAT FASN OXSM OLAH)

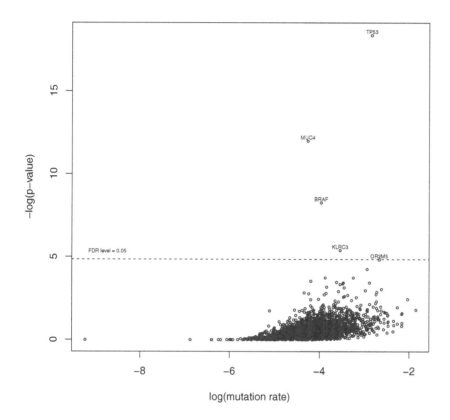

Figure 32.2. Recurrently mutated genes. Example of recurrently mutated genes identified by DrGaP. Data were obtained from a subset of cell lines of the COSMIC Cell Lines Project at `http://cancer.sanger.ac.uk/cell_lines` (see exercises at the end of the chapter for more information).

If the pathway file (option -b) is not specified, DrGaP will only evaluate individual genes.

Figures 32.2 and 32.3 show examples of recurrently mutated genes and pathways inferred by DrGaP using a set of variants from cancer cell lines, including mostly Ewing sarcomas, neuroblastomas, colon carcinomas and skin melanomas (see exercises). Notably, among the recurrently mutated genes identified by DrGaP we find the tumor-suppressor gene *TP53* and the oncogene *BRAF*. DrGaP identified

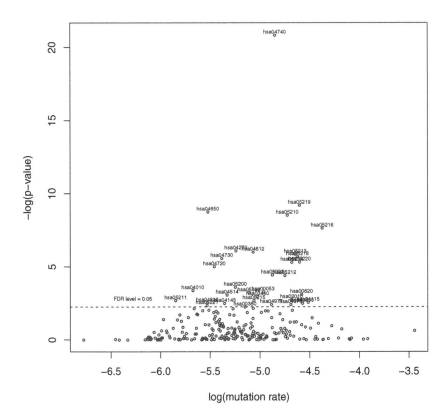

Figure 32.3. Recurrently mutated pathways. Example of recurrently mutated pathways identified by DrGaP. Data were obtained from a subset of cell lines of the COSMIC Cell Lines Project at http://cancer.sanger.ac.uk/cell_lines (see exercises at the end of the chapter for more information).

several recurrently mutated pathways including the MAPK signaling pathway (KEGG hsa04010), metabolic pathways like the pyruvate metabolism (hsa00620), and disease-related pathways like colorectal cancer (hsa05210).

32.1.4.2 MutSig

MutSig (Mutational Significance) [239] is a frequency-based approach to identifying genes that are mutated more often than expected by chance. MutSig analyzes mutations (single point mutations and indels) identified in a set of patients, taking into account gene characteristics such as the length of the coding sequence to build a background model from the spectrum of mutations, and thereby reflecting the mutational processes at work in the dataset.

MutSigCV,[5] the most recent version of MutSig, can be either run in Matlab or using the compiled version with a free Matlab runtime environment.

32.1.4.3 MuSiC

Much like MutSig, MuSiC (Mutational Significance in Cancer) [99][6] provides a set of tools helping to determine the tumor relevance of somatic mutations from a cohort of cancer samples.

As inputs, MuSiC requires the BAM files containing the mapped reads for all tumor/normal sample pairs, the reference sequence, the called SNVs and indels, a specification of the regions of interest (e.g., the boundaries of the coding exons), and clinically relevant data (numeric and/or categorical) for the set of patients.

32.1.4.4 OncodriveFM

OncodriveFM [138] can be used either as a standalone tool or within the IntOGen pipeline.[7] In contrast to most other tools, OncodriveFM does not rely on mutation frequencies in order to identify driver genes. Instead, it assumes that the impact on protein function of mutations in driver genes in a set of tumor samples will deviate from a null distribution which is determined from all variants observed across the tumor samples. In other words, the positive selection of mutations in driver genes will be observable as an accumulation of mutations with high functional impact.

---

[5]http://archive.broadinstitute.org/cancer/cga/mutsig

[6]MuSiC version 0.4 is available at http://gmt.genome.wustl.edu/packages/genome-music/ for Ubuntu 10.04. MuSiC2 version 0.2 is available at https://github.com/ding-lab/MuSiC2. For recent Debian Linux distributions, MuSiC can be installed using the following command:

```
$ sudo apt-get install libgenome-model-tools-music-perl
```

[7]https://www.intogen.org/search

For this purpose, OncodriveFM evaluates the called variants based on functional impact scores assigned by three other tools: SIFT, PolyPhen2 and MutationAssessor which need to be installed and run independently to assemble the necessary input data for OncodriveFM.

Like DrGaP, OncodriveFM can additionally evaluate pathways or gene sets for an enrichment in mutations of high functional impact.

32.2 MUTATIONAL SIGNATURES

In Section 32.1.2 we mentioned that background mutation frequencies may differ from dataset to dataset, depending on the underlying mutational processes. It is important to take this into account when deciding whether a gene or pathway is recurrently mutated. The diverse mutagenic processes that are associated with tumors can lead to different characteristic nucleotide changes [15, 164]. Prominent examples are

- Ultraviolet (UV) light, exposure to which can cause skin cancer, is predominantly associated with changes of cytosine to thymine (C>T) or cytosine–cytosine to thymine–thymine (CC>TT) [168, 392].

- Age-related spontaneous deamination of 5-methyl-cytosine also causes characteristic C>T transitions which in this case mostly occur at CpG dinucleotides [330].

- Lung cancers from smokers have been found to harbor more C>A transitions than those from nonsmokers, which occur less frequently at CpG dinucleotides [12].

This illustrates that the proportions of point mutations which represent specific transitions (C>T and T>C) and specific transversions (C>A, C>G, T>A, and T>G) can vary significantly from dataset to dataset and even from sample to sample. Transitions and transversions of adenines (A) and guanines (G) to other nucleotides do not need to be included because G>A, for example, is equivalent to C>T on the reverse strand. That is, each point mutation always changes one purine (A or G) on one strand *and* one pyrimidine (C or T) on the other. Thus, for the purpose of specifying mutational signature it is conventional to count the mutated pyrimidine (C or T).

Nucleotide changes induced by mutational processes depend in many cases not only on the changed nucleotides themselves but also on their sequence context. Mutational signatures are therefore often determined not with respect to the six possible changes of the affected nucleotide (C>A, C>G, C>T, T>A, T>C, T>G), but for the 96 possible changes of the central base of a trinucleotide:

$$A[C>A]A, A[C>A]C, \ldots, C[C>A]A, \ldots, T[T>G]T$$

The difference is illustrated in Figure 32.4 for the signature of the age-related endogenous mutational process associated with the spontaneous deamination of 5-methylcytosine. While the mutational process is clearly associated with C>T transitions (upper panel), the analysis based on trinucleotides shows that these transitions mostly occur where the cytosine is followed by a guanine (lower panel). Indeed, the four highest mutation fractions are observed at A[C>T]G, C[C>T]G, G[C>T]G, and T[C>T]G.

32.2.1 Basic concept and mathematical description

A catalog of 30 mutational signatures and their associated mutation frequencies[8] was derived by a non-negative matrix factorization (NMF) applied to thousands of tumors of various types [13, 14]. In brief, the method analyzed the somatic point mutations observed in a dataset of G tumor genomes as a $96 \times G$-matrix $\mathcal{M}$ where $\mathcal{M}_{ig}$ is the count of mutations of the i-th trinucleotide change (e.g., A[C>T]G) in the tumor genome $g \in G$. NMF is applied to determine a $96 \times N$-matrix $\mathcal{P}$ of signatures for N mutational processes (with $\mathcal{P}_{in}$ being the fraction of mutations in signature $n \in N$ that correspond to the i-th trinucleotide change), and an $N \times G$-matrix $\mathcal{E}$ of "exposures" (with $\mathcal{E}_{ng}$ being the number of mutations in $g \in G$ that are caused by process/signature $n \in N$), such that:

$$\mathcal{M} \approx \mathcal{P} \times \mathcal{E} \tag{32.1}$$

$$\mathcal{M}_{ig} = \sum_{n=1}^{N} \mathcal{P}_{in} \mathcal{E}_{ng} + \epsilon_{ig} \tag{32.2}$$

where ϵ_{ig} are error terms to be minimized.

This allows both the mutational signatures of the underlying processes and the counts of mutations caused by these processes to be inferred in the individual cancer genomes of the dataset. For further details, we refer to the original work by Alexandrov et al. [13, 14].[9]

---

[8]http://cancer.sanger.ac.uk/cosmic/signatures and http://cancer.sanger.ac.uk/cancergenome/assets/signatures_probabilities.txt.

[9]Shiraishi et al. [396] have recently presented a different notion of mutational

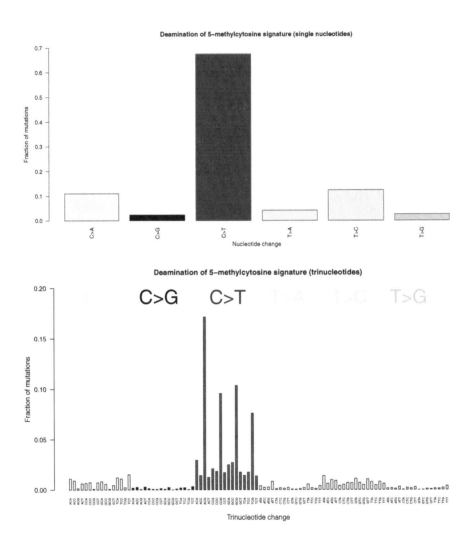

Figure 32.4. Mutational signature associated with the spontaneous deamination of 5-methylcytosine. Fractions of the possible substitutions with respect to the single base affected (upper panel) and the corresponding trinucleotide analysis (lower panel). Data was obtained from: http://cancer.sanger.ac.uk/cosmic/signatures.

32.2.2 SomaticSignatures

SomaticSignatures [130], one of the few available methods for the discovery of mutational signatures, is an R/Bioconductor package that implements the NMF-based framework of Alexandrov et al. and requires only a VCF file and a reference genome as input.[10] These are loaded as `VRanges` and `BSgenome` objects, respectively.

In contrast to the above described mathematical framework, however, which uses individual tumor genomes to infer mutational signatures, SomaticSignatures is intended to work with tumor classes, i.e., it groups tumor samples according their annotated tumor types and applies the NMF-based framework to these groups rather than individual samples.

Since deriving mutational signatures requires the mutational information of many samples, we illustrate the procedure using a subset of the mutation data of the COSMIC Cell Lines Project (see Exercises).[11]

The mutation data obtained from the COSMIC Cell Lines Project can be reduced to the most essential information required for applying SomaticSignatures, and organized into a VCF file:

```
##fileformat=VCFv4.1
##source=VarScan2
##INFO=<ID=CL,Number=1,Type=String,Description="cell line">
##INFO=<ID=TM,Number=1,Type=String,Description="tumor type">
#CHROM  POS       ID  REF  ALT  QUAL  FILTER  INFO
10   1000733    .  G    A    .    PASS    CL=Mewo;TM=skin_melanoma
10   100174241  .  G    A    .    PASS    CL=MZ2-MEL;TM=skin_melanoma
10   100185624  .  C    T    .    PASS    CL=HCC-44;TM=lung_adenocarcinoma
10   100194497  .  T    G    .    PASS    CL=MZ7-mel;TM=skin_melanoma
10   100209764  .  G    A    .    PASS    CL=A2058;TM=skin_melanoma
10   100262000  .  C    T    .    PASS    CL=MZ7-mel;TM=skin_melanoma
[...]
```

Only SNVs are required by SomaticSignatures, indels can be ig-

---

signature that is related to position weight matrices and requires less parameters. Here, however, we focus on the model originally proposed by Alexandrov et al.

[10]Alternatively, SomaticSignatures allows to use principal component analysis (PCA) or user-defined decomposition functions instead. Here, we discuss only the use of NMF.

[11]Mutational signature analysis should generally be performed with WGS data in order to have the greatest possible number of somatic mutations available for inferring or detecting signatures. For this demonstration, we chose the COSMIC Cell Lines dataset, which includes only coding variants, for convenience and ease of access.

nored. The nucleotide change of the SNVs needs to be referred to the genomic sequence, not the coding sequence. For example, in the above VCF file, the second mutation at position 100,174,241 on chromosome 10 is not specified as C>T (c.471C>T) but as REF=G and ALT=A because the corresponding gene lies on the reverse strand.

Apart from the position and nucleotide change, only information about the sample and the sample group, in which the mutation occurred, is required. Here, these correspond to the cell line (specified as "CL") and tumor type (specified as "TM").

Table 32.1. Tumor types and cell lines of the COSMIC Cell Lines Project used for deriving mutational signatures.

Tumor Type	Cell Lines	SNVs
Neuroblastoma	34	12481
Astrocytoma	28	8768
Acute myeloid leukemia (AML)	27	10936
Burkitt lymphoma	14	16995
Lung adenocarcinoma	35	20370
Skin melanoma	55	41641

We use only a subset of the tumor types for which mutation data is available (see Table 32.1). After having transformed the mutation information to a VCF file, we can load it as a VRanges object in R:

```
> library(SomaticSignatures)
> clmutlist <- readVcfAsVRanges("selected_mutations.vcf")
> sampleNames(clmutlist) <- clmutlist$CL
> genome(clmutlist) <- "GRCh38"
```

The last two commands serve to set the sample names according to the cell line information we kept in the INFO field (CL=...) and to set the reference genome (GRCh38) for correctly interpreting the positions specified in the VCF file.

The mutation data of the COSMIC Cell Lines Project does not use the standard chromosome names "X" and "Y" but uses "23" and "24" instead, so we have to change these in the loaded VRanges object:

```
> newSeqnames <- gsub("23","X", seqlevels(clmutlist))
> newSeqnames <- gsub("24","Y", newSeqnames)
> seqlevels(clmutlist) <- newSeqnames
```

Now, we can determine the sequence context (neighboring nucleotides) for all variants and count the occurrences of different mutational motifs (i.e., changes of trinucleotides):[12]

```
> library(BSgenome.Hsapiens.NCBI.GRCh38)
> mutmotifs <- mutationContext(clmutlist,
                        BSgenome.Hsapiens.NCBI.GRCh38)
```

The identified sequence motifs and nucleotide changes can be grouped according to tumor type ("TM=..." in the VCF file's INFO field), normalized as fractions of mutation counts, and organized as a motif matrix. This matrix corresponds to the mutational catalog $\mathcal{M}$ described above, except that the individual samples are grouped according to tumor type.

```
> mmat <- motifMatrix(mutmotifs, group="TM",
                        normalize=TRUE)
> head(round(mmat,4))
          AML astroc. Burkitt lung ad. neurobl. skin mel.
CA A.A 0.0090  0.0072  0.0044   0.0165   0.0113    0.0029
CA A.C 0.0069  0.0092  0.0039   0.0141   0.0090    0.0030
CA A.G 0.0037  0.0033  0.0039   0.0116   0.0047    0.0012
CA A.T 0.0045  0.0043  0.0022   0.0098   0.0062    0.0017
CA C.A 0.0111  0.0089  0.0073   0.0324   0.0207    0.0062
CA C.C 0.0080  0.0075  0.0071   0.0288   0.0101    0.0033
```

The command plotMutationSpectrum(mutmotifs, "TM") allows the mutated sequence motifs, i.e., the trinucleotides affected by the SNVs, and their occurrences in the tumor types to be plotted (Figure 32.5).

After the mutation data has been preprocessed, we can proceed with the identification of mutational signatures. The first step is to determine how many signatures are actually required to explain the variance of the data:

```
> n_sign = 2:6
> gof_nmf <- assessNumberSignatures(mmat, n_sign,
                        nReplicates = 5)
> plotNumberSignatures(gof_nmf)
```

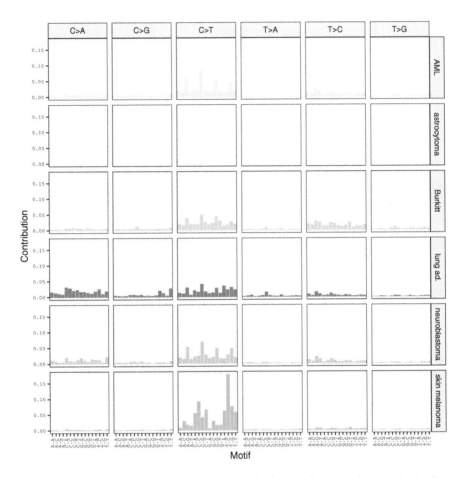

Figure 32.5. Mutated sequence motifs (trinucleotides) and their frequencies in the cancer cell lines.

This produces a plot that helps to decide what number of signatures should be inferred (see Figure 32.6). In this example, up to six signatures can be tested because the data in the motif matrix consists of six tumor types, i.e., it has six columns. If the decomposition method (here: NMF) uses random seeding, multiple runs should be used for assessing the number of signatures (nReplicates>1). As the figure shows, five signatures are sufficient to explain close to 100% of

---

[12]The reference genome BSgenome.Hsapiens.NCBI.GRCh38 can be installed as a standard R/Bioconductor package.

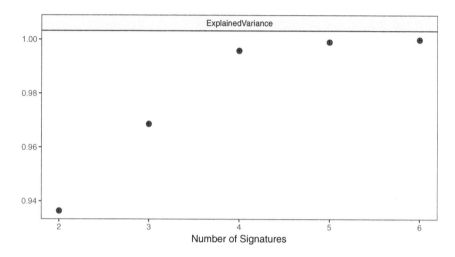

Figure 32.6. Explained variance of the mutation data as a function of the number of inferred signatures.

the variance in this data. The benefit of a sixth signature would be negligible.

Now we have all we need to finally infer the mutational signatures present in the data:

```
> n_sign = 5
> sign_nmf <- identifySignatures(mmat, n_sign,
                                  nmfDecomposition)
```

The results can be visualized in different plots using the graphical R library ggplot2:[13]

```
> library(ggplot2)

> plotSignatures(sign_nmf)
> plotSignatureMap(sign_nmf)

> plotSamples(sign_nmf)
> plotSampleMap(sign_nmf)
```

Both signatures and sample exposures can be plotted either as barplots or heatmaps (Figures 32.7 and 32.8). The results show that

---
[13]http://ggplot2.org/

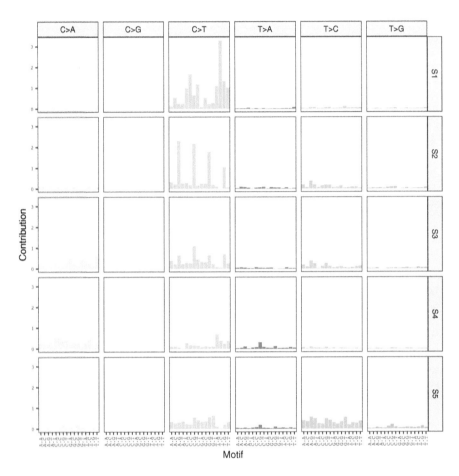

Figure 32.7. Mutational signatures identified in the cell lines of six selected tumor types.

skin melanoma is strongly associated with mutations described by signature S1 (compare Figures 32.5, 32.7, and 32.8) which is characterized by a high frequency of T[T>C]C and C[T>C]C nucleotide changes. Indeed, this signature is consistent with the T>C mutations ascribed to ultraviolet (UV) light [168, 392] and highly similar to the UV-related signature 7 identified by [13].[14] In contrast, astrocytomas appear more associated with signature S2 which very much resembles the age-related signature shown in Figure 32.4.

---

[14]http://cancer.sanger.ac.uk/signatures/Signature-7.png

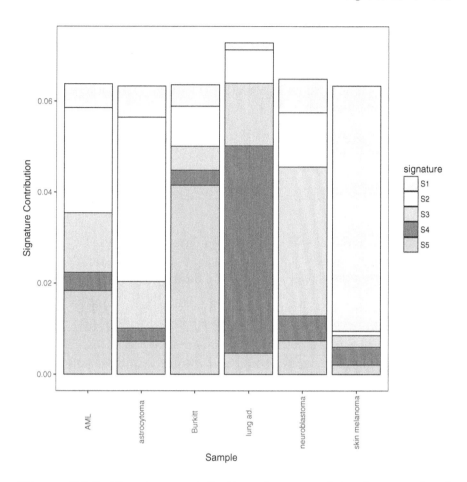

Figure 32.8. Signature contributions (exposures) to the mutational catalogs of the cell lines of six selected tumor types.

FURTHER READING

Alexandrov LB, Nik-Zainal S, Wedge DC, et al. (2013) Deciphering Signatures of Mutational Processes Operative in Human Cancer, *Cell Reports* **3**(1):246–259.

Alexandrov LB, Nik-Zainal S, Wedge DC, et al. (2013) Signatures of mutational processes in human cancer. *Nature* **500**(7463):415–421.

Alexandrov LB, Stratton MR (2014) Mutational signatures: the patterns of somatic mutations hidden in cancer genomes. *Current Opinion in Genetics & Development* **24**(100):52–60.

Cheng F, Zhao J, Zhao Z (2016) Advances in computational approaches for prioritizing driver mutations and significantly mutated genes in cancer genomes. *Briefings in Bioinformatics* **17**(4):642–656.

Greenman C, Stephens P, Smith R, et al. (2007) Patterns of somatic mutation in human cancer genomes. *Nature* **446**:153–158.

Helleday T, Saeed E, Nik-Zainal S (2014) Mechanisms underlying mutational signatures in human cancers. *Nature Reviews Genetics* **15**:585–598.

32.3   EXERCISES

Exercise 1

Download the complete (coding) mutation data file from the COSMIC Cell Lines Project, `CosmicCLP_MutantExport.tsv.gz`,[15] and filter the file to use only the following subset of the data:

1. Use only the 37 cell lines whose IDs start with "COLO" (e.g., COLO-205), "NB" (e.g., NB10) and "EW-" (e.g., EW-16). These are mostly neuroblastomas, Ewing sarcomas, colon carcinomas and skin melanomas, but contain also some cell lines from other carcinomas and an acute myeloid leukemia.

2. Use only coding SNVs, that is, variants specified as "Substitution - Nonsense", "Substitution - Missense" and "Substitution - coding silent" (these need to be renamed to "nonsense", "missense" and "silent").

3. Exclude variants without well defined chromosomal locations (chromosome and position).

4. Since some variants are listed multiple times (e.g., per gene and transcript), make sure that each combination of cell line, chromosome, position, reference base and variant base is used only once (this is, in fact, one variant).

You should obtain a list of roughly 16,000 variants which can be handled by DrGaP using slightly over 4 GB of RAM.

Transform this list to obtain the tab-separated input file `<in.mt>` specifying the filtered variants. The file should look like this (see Section 32.1.4.1 for more details):

---

[15]`http://cancer.sanger.ac.uk/cell_lines/download`

```
COLO-205    CRB1      1   197427545   C   T   silent
COLO-205    DNAJB4    1   78013597    C   T   missense
COLO-205    FCRLA     1   161713185   G   A   silent
COLO-205    GJA9      1   38875018    G   T   nonsense
COLO-205    GSTM2     1   109708567   G   A   silent
COLO-205    HFM1      1   91315950    C   T   missense
COLO-205    KANK4     1   62268427    G   T   missense
COLO-205    LCK       1   32275027    C   T   silent
COLO-205    MTOR      1   11210890    C   T   missense
COLO-205    OR2M7     1   248323665   A   G   missense
[...]
```

Now, run DrGaP to identify recurrently mutated genes and pathways within this set of cell lines:

```
$ drgap -i <in.mt> -g hg38_refGene.exp -f hs38DH.fa \
        -b KEGG_plus_4.txt
```

This will produce an output directory called `drgap_out`. Explore the results and compare them to Figures 32.2 and 32.3.

Note: the preprocessed reference gene file which uses the human genome hg38 (`hg38_refGene.exp`) is not provided with DrGaP. It can be produced from the version provided for hg19 (`hg19_refGene.exp`) using `liftOver` or one of the other conversion tools discussed in Section 8.1.3. Alternatively, we provide the converted version for hg38 on the book's website.

Exercise 2

Take alls SNVs from the 37 cell lines used for Exercise 1 and produce a VCF file with the SNVs which reports the cell line and tumor type as described in this chapter. For simplicity, you may assign the different carcinoma types to a single tumor type "carcinoma". You should have 5 tumor types (carcinoma, neuroblastoma, Ewing sarcoma, acute myeloid leukemia, and skin melanoma).

Use the SomaticSignatures R package to discover mutational signatures in these tumor types.

- How many signatures do you need to explain about 99% of the variance?

- Are the tumor types associated with different signatures or do they show similar mutational profiles?

Hints and Answers

Hints are provided for selected exercises.

A.1 CHAPTER 3: HINTS AND ANSWERS

Exercise 1

If you cannot answer this question, review the difference between read length and insert size.

Exercise 2

Consider that there is some probability that two independent DNA fragments that start at the same genomic position will both be sequenced. Now consider the probability that such fragments additionally have the same length.

A.2 CHAPTER 6: HINTS AND ANSWERS

Exercise 1

$$Q_{PHRED} = -10 \log_{10} p \tag{A.1}$$

$Q = 0$ corresponds to an error probability of 1, and $Q = 41$ corresponds to an error probability of $10^{-41/10} = 7.943282 \times 10^{-05}$.

A.2.1 Chapter 9: Hints and Answers

Exercise 3

(a) This corresponds to the CIGAR string 3M1I3M1D5M starting at POS 5, i.e., the read aligns starting at position 5 on the reference.

The CIGAR says that the first 3 bases of the read sequence align with the reference. The next base of the read does not exist in the reference (insertion). Then 3 bases align with the reference. The next reference base does not exist in the read sequence (deletion), then 5 more bases align with the reference. Note that at position 14, the base in the read is different than the reference nucleotide, but it still counts as an M since it aligns to that position.

The sequence mismatch at this position affects both the NM and MD tags:

The read and reference have an edit distance of three (hence, NM:i:3) counting the inserted base, the deleted base and the mismatched base.

The reference-centered description of the alignment is MD:Z:6^C2A2. From the starting position of the alignment (fifth base in the reference), the first 6 reference bases are matched by the read's bases, then a C of the reference is missing in the read (^C), another two reference bases are matched, followed by a mismatch (A in the reference) and two final matches. As explained in Chapter 9, insertions are not recorded in the MD tag.

A.2.2 Chapter 13: Hints and Answers

Exercise 3

Combine the Unix commands **grep**, **cut**, and **wc** using pipes (|). Alternatively, use **bcftools** (see Section 12.3).

Exercise 4

Write an **awk** script that skips comment lines and that iterates over the variant lines, incrementing a variable if the QUAL column is less than 30.

Exercise 6

Recall that the values of the PL field are Phred-scaled genotype likelihoods rounded to the closest integer. The most likely genotype (given in the GT field) is scaled so that its value is $P = 1.0$ (0 when Phred-scaled), and the other likelihoods are relative to this. The following function can be used to calculate the likelihood of a singe Phred-scaled value.

```
convert.phred <- function(p) { 10^{-p/10}}
```

A.2.3 Chapter 15: Hints and Answers

Exercise 1

Can you tell from the description what position of which transcript is affected? [440].

A.2.4 Chapter 22: Hints and Answers

Exercise 1

To find gene-based information about a SNP, navigate on the db-SNP webpage until you find the section entitled SNP Details are organized in the following sections. Click on GeneView.

Exercise 7

Consider that the disease in question is a rare disease.

References

[1] 1000 Genomes Project Consortium, et al. (2010) A map of human genome variation from population-scale sequencing. *Nature* **467**:1061–1073.

[2] Abel H. J., et al. (2014) Detection of gene rearrangements in targeted clinical next-generation sequencing. *The Journal of Molecular Diagnostics* **16**:405–417.

[3] Abyzov A., et al. (2011) CNVnator: An approach to discover, genotype, and characterize typical and atypical CNVs from family and population genome sequencing. *Genome Research* **21**:974–984.

[4] Acuna-Hidalgo R., et al. (2015) Post-zygotic point mutations are an underrecognized source of de novo genomic variation. *American Journal of Human Genetics* **97**:67–74.

[5] Acuna-Hidalgo R., Veltman J. A., and Hoischen A. (2016) New insights into the generation and role of de novo mutations in health and disease. *Genome Biology* **17**:241.

[6] Adzhubei I. A., et al. (2010) A method and server for predicting damaging missense mutations. *Nature Methods* **7**:248–249.

[7] Aerts S., et al. (2006) Gene prioritization through genomic data fusion. *Nature Biotechnology* **24**:537–544.

[8] Ainscough B. J., et al. (2016) DoCM: A database of curated mutations in cancer. *Nature Methods* **13**:806–807.

[9] Akawi N., et al. (2015) Discovery of four recessive developmental disorders using probabilistic genotype and phenotype matching among 4,125 families. *Nature Genetics* **47**:1363–1369.

[10] Aken B. L., et al. (2017) Ensembl 2017. *Nucleic Acids Research* **45**:D635–D642.

[11] Albers C. A., et al. (2012) Compound inheritance of a low-frequency regulatory SNP and a rare null mutation in exon-junction complex subunit RBM8A causes TAR syndrome. *Nature Genetics* **44**:435–439.

[12] Alexandrov L. B., et al. (2016) Mutational signatures associated with tobacco smoking in human cancer. *Science* **354**:618–622.

[13] Alexandrov L. B., et al. (2013) Signatures of mutational processes in human cancer. *Nature* **500**:415–421.

[14] Alexandrov L. B., et al. (2013) Deciphering signatures of mutational processes operative in human cancer. *Cell Reports* **3**:246–259.

[15] Alexandrov L. B. and Stratton M. R. (2014) Mutational signatures: The patterns of somatic mutations hidden in cancer genomes. *Current Opinion in Genetics & Development* **24**:52–60.

[16] Alkodsi A., Louhimo R., and Hautaniemi S. (2015) Comparative analysis of methods for identifying somatic copy number alterations from deep sequencing data. *Briefings in Bioinformatics* **16**:242–254.

[17] Amberger J. S., et al. (2015) OMIM.org: Online Mendelian Inheritance in Man (OMIM), an online catalog of human genes and genetic disorders. *Nucleic Acids Research* **43**:D789–D798.

[18] Andor N., et al. (2014) EXPANDS: Expanding ploidy and allele frequency on nested subpopulations. *Bioinformatics* **30**:50–60.

[19] Antonarakis S. E. (1998) Recommendations for a nomenclature system for human gene mutations. nomenclature working group. *Human Mutation* **11**:1–3.

[20] Aran D., Sirota M., and Butte A. J. (2015) Systematic pan-cancer analysis of tumour purity. *Nature Communications* **6**:8971.

[21] Arnaud P., et al. (2017) Homozygous and compound heterozygous mutations in the FBN1 gene: Unexpected findings in molecular diagnosis of Marfan syndrome. *Journal of Medical Genetics* **54**:100–103.

[22] Auffray C., et al. (2016) From genomic medicine to precision medicine: Highlights of 2015. *Genome Medicine* **8**:12.

[23] Bagchi A., et al. (2007) CHD5 is a tumor suppressor at human 1p36. *Cell* **128**:459–475.

[24] Bai H., et al. (2014) The genome of a Mongolian individual reveals the genetic imprints of Mongolians on modern human populations. *Genome Biology and Evolution* **6**:3122–3136.

[25] Bailey J. A., et al. (2002) Recent segmental duplications in the human genome. *Science* **297**:1003–1007.

[26] Baird P. A., et al. (1988) Genetic disorders in children and young adults: A population study. *American Journal of Human Genetics* **42**:677–693.

[27] Baker M. (2012) Structural variation: The genome's hidden architecture. *Nature Methods* **9**:133–137.

[28] Bamford S., et al. (2004) The COSMIC (Catalogue of Somatic Mutations in Cancer) database and website. *British Journal of Cancer* **91**:355–358.

[29] Barabasi and Albert. (1999) Emergence of scaling in random networks. *Science (New York, N.Y.)* **286**:509–512.

[30] Baralle D. and Baralle M. (2005) Splicing in action: Assessing disease causing sequence changes. *Journal of Medical Genetics* **42**:737–748.

[31] Barba M., Czosnek H., and Hadidi A. (2014) Historical perspective, development and applications of next-generation sequencing in plant virology. *Viruses* **6**:106–136.

[32] Barber L. J., Davies M. N., and Gerlinger M. (2015) Dissecting cancer evolution at the macro-heterogeneity and micro-heterogeneity scale. *Current Opinion in Genetics & Development* **30**:1–6.

[33] Bayat A., et al. (2017) Improved VCF normalization for accurate VCF comparison. *Bioinformatics* **33**:964–970.

[34] Beaulieu C. L., et al. (2014) FORGE Canada Consortium: Outcomes of a 2-year national rare-disease gene-discovery project. *American Journal of Human Genetics* **94**:809–817.

[35] Bell C. J., et al. (2011) Carrier testing for severe childhood recessive diseases by next-generation sequencing. *Science Translational Medicine* **3**:65ra4.

[36] Bellos E. and Coin L. J. M. (2014) cnvOffSeq: Detecting intergenic copy number variation using off-target exome sequencing data. *Bioinformatics* **30**:i639–i645.

[37] Benjamini Y. and Speed T. P. (2012) Summarizing and correcting the GC content bias in high-throughput sequencing. *Nucleic Acids Research* **40**:e72.

[38] Beroukhim R., et al. (2006) Inferring loss-of-heterozygosity from unpaired tumors using high-density oligonucleotide SNP arrays. *PLOS Computational Biology* **2**(5):e41.

[39] Besenbacher S., et al. (2015) Novel variation and de novo mutation rates in population-wide de novo assembled Danish trios. *Nature Communications* **6**:5969.

[40] Bolger A. M., Lohse M., and Usadel B. (2014) Trimmomatic: A flexible trimmer for Illumina sequence data. *Bioinformatics* **30**:2114–2120.

[41] Bone W. P., et al. (2016) Computational evaluation of exome sequence data using human and model organism phenotypes improves diagnostic efficiency. *Genetics in Medicine* **18**:608–617.

[42] Bonfield J. K. (2014) The Scramble conversion tool. *Bioinformatics* **30**:2818–2819.

[43] Booms P., et al. (1997) A novel de novo mutation in exon 14 of the fibrillin-1 gene associated with delayed secretion of fibrillin in a patient with a mild marfan phenotype. *Human Genetics* **100**:195–200.

[44] Borad M. J., et al. (2014) Integrated genomic characterization reveals novel, therapeutically relevant drug targets in FGFR and EGFR pathways in sporadic intrahepatic cholangiocarcinoma. *PLOS Genetics* **10**:e1004135.

[45] Botstein D. and Risch N. (2003) Discovering genotypes underlying human phenotypes: Past successes for Mendelian disease, future approaches for complex disease. *Nature Genetics* **33 Suppl**:228–237.

[46] Bottomly D., McWeeney S. K., and Wilmot B. (2016) Hit-Walker2: Visual analytics for precision medicine and beyond. *Bioinformatics* **32**:1253–1255.

[47] Bottomly D., et al. (2013) HitWalker: Variant prioritization for personalized functional cancer genomics. *Bioinformatics* **29**:509–510.

[48] Bouaoun L., et al. (2016) TP53 variations in human cancers: New lessons from the IARC TP53 Database and Genomics Data. *Human Mutation* **37**:865–876.

[49] Bragin E., et al. (2014) DECIPHER: Database for the interpretation of phenotype-linked plausibly pathogenic sequence and copy-number variation. *Nucleic Acids Research* **42**:D993–D1000.

[50] Brennan C., et al. (2013) The somatic genomic landscape of glioblastoma. *Cell* **155**:462–477.

[51] Brookes A. J. and Robinson P. N. (2015) Human genotype-phenotype databases: Aims, challenges and opportunities. *Nature Reviews Genetics* **16**:702–715.

[52] Bulmer M. (1979) *Principles of statistics.* Dover Publications.

[53] Buntru A., et al. (2016) Current approaches toward quantitative mapping of the interactome. *Frontiers in Genetics* **7**:74.

[54] Burrows M. and Wheeler D. J. (1994) A block-sorting lossless data compression algorithm. Technical Report Digital Systems Research Center.

[55] Buske O. J., et al. (2015) PhenomeCentral: A portal for phenotypic and genotypic matchmaking of patients with rare genetic diseases. *Human Mutation* **36**:931–940.

[56] Caminsky N., Mucaki E. J., and Rogan P. K. (2014) Interpretation of mRNA splicing mutations in genetic disease: Review of the literature and guidelines for information-theoretical analysis. *F1000Research* **3**:282.

[57] Cancer Genome Atlas Research Network, et al. (2013) The cancer genome atlas pan-cancer analysis project. *Nature Genetics* **45**:1113–1120.

[58] Cantwell-Dorris E. R., O'Leary J. J., and Sheils O. M. (2011) BRAFV600E: Implications for carcinogenesis and molecular therapy. *Molecular Cancer Therapeutics* **10**(3):385–394.

[59] Carracedo A., Alimonti A., and Pandolfi P. P. (2011) PTEN level in tumor suppression: How much is too little? *Cancer Research* **71**:629–633.

[60] Carson A. R., et al. (2014) Effective filtering strategies to improve data quality from population-based whole exome sequencing studies. *BMC Bioinformatics* **15**:125.

[61] Carter S. L., et al. (2012) Absolute quantification of somatic DNA alterations in human cancer. *Nature* **30**:413–421.

[62] Carvalho C. M. B. and Lupski J. R. (2016) Mechanisms underlying structural variant formation in genomic disorders. *Nature Reviews Genetics* **17**:224–238.

[63] Cerami E., et al. (2012) The cBio cancer genomics portal: An open platform for exploring multidimensional cancer genomics data. *Cancer Discovery* **2**:401–404.

[64] Chan L. F., et al. (2015) Whole-exome sequencing in the differential diagnosis of primary adrenal insufficiency in children. *Frontiers in Endocrinology* **6**:113.

[65] Chen C., et al. (2014) Software for pre-processing Illumina next-generation sequencing short read sequences. *Source Code for Biology and Medicine* **9**:8.

[66] Chen K., et al. (2009) BreakDancer: An algorithm for high-resolution mapping of genomic structural variation. *Nature Methods* **6**:677–681.

[67] Cheng F., Zhao J., and Zhao Z. (2016) Advances in computational approaches for prioritizing driver mutations and significantly mutated genes in cancer genomes. *Briefings in Bioinformatics* **17**:642–656.

[68] Chilamakuri C. S. R., et al. (2014) Performance comparison of four exome capture systems for deep sequencing. *BMC Genomics* **15**:449.

[69] Chmielecki J., et al. (2013) Whole-exome sequencing identifies a recurrent NAB2-STAT6 fusion in solitary fibrous tumors. *Nature Genetics* **45**:131–132.

[70] Choi J., et al. (2016) Hemangiopericytomas in the central nervous system: A multicenter study of Korean cases with validation of the usage of STAT6 immunohistochemistry for diagnosis of disease. *Annals of Surgical Oncology* **23**:954–961.

[71] Choi Y. and Chan A. P. (2015) PROVEAN web server: A tool to predict the functional effect of amino acid substitutions and indels. *Bioinformatics* **31**:2745–2747.

[72] Chong J. X., et al. (2015) The genetic basis of Mendelian phenotypes: Discoveries, challenges, and opportunities. *American Journal of Human Genetics* **97**:199–215.

[73] Chong J. X., et al. (2012) A population-based study of autosomal-recessive disease-causing mutations in a founder population. *American Journal of Human Genetics* **91**:608–620.

[74] Church D. M., et al. (2011) Modernizing reference genome assemblies. *PLoS Biology* **9**:e1001091.

[75] Church D. M., et al. (2015) Extending reference assembly models. *Genome Biology* **16**:13.

[76] Cibulskis K., et al. (2013) Sensitive detection of somatic point mutations in impure and heterogeneous cancer samples. *Nature Biotechnology* **31**:213–219.

[77] Cingolani P., et al. (2012) A program for annotating and predicting the effects of single nucleotide polymorphisms, SnpEff: SNPs in the genome of Drosophila melanogaster strain w1118; iso-2; iso-3. *Fly* **6**:80–92.

[78] Cirstea I. C., et al. (2010) A restricted spectrum of NRAS mutations causes Noonan syndrome. *Nature Genetics* **42**:27–29.

[79] Clark M. J., Chen R., and Snyder M. (2013) Exome sequencing by targeted enrichment. *Current Protocols in Molecular Biology* **102**:7.12.1–7.12.21.

[80] Clark V. E., et al. (2013) Genomic analysis of non-NF2 meningiomas reveals mutations in TRAF7, KLF4, AKT1, and SMO. *Science* **339**:1077–1080.

[81] Cock P. J. A., et al. (2010) The Sanger FASTQ file format for sequences with quality scores, and the Solexa/Illumina FASTQ variants. *Nucleic Acids Research* **38**:1767–1771.

[82] Cogill S. and Wang L. (2016) Support vector machine model of developmental brain gene expression data for prioritization of autism risk gene candidates. *Bioinformatics* **32**:3611–3618.

[83] Cohen A. L., Holmen S. L., and Colman H. (2013) IDH1 and IDH2 mutations in gliomas. *Current Neurology and Neuroscience Reports* **13**:345.

[84] Coin L. J. M., et al. (2012) An exome sequencing pipeline for identifying and genotyping common CNVs associated with disease with application to psoriasis. *Bioinformatics* **28**:i370–i374.

[85] Collins F. S., et al. (2003) A vision for the future of genomics research. *Nature* **422**:835–847.

[86] Collins F. S. and Varmus H. (2015) A new initiative on precision medicine. *The New England Journal of Medicine* **372**:793–795.

[87] Cook C. (2009) Is clinical gestalt good enough? *The Journal of Manual & Manipulative Therapy* **17**:6–7.

[88] Cooper G. M. and Shendure J. (2011) Needles in stacks of needles: Finding disease-causal variants in a wealth of genomic data. *Nature Reviews Genetics* **12**:628–640.

[89] Cormen T. H., et al. (2009) *Introduction to Algorithms, Third Edition.* The MIT Press.

[90] Corpas M. (2012) A family experience of personal genomics. *Journal of Genetic Counseling* **21**:386–391.

[91] Corpas M. (2013) Crowdsourcing the corpasome. *Source Code for Biology and Medicine* **8**:13.

[92] Corpas M., et al. (2015) Crowdsourced direct-to-consumer genomic analysis of a family quartet. *BMC Genomics* **16**:910.

[93] Dahlqvist J., et al. (2010) A single-nucleotide deletion in the POMP 5' UTR causes a transcriptional switch and altered epidermal proteasome distribution in KLICK genodermatosis. *American Journal of Human Genetics* **86**:596–603.

[94] Damodaran S., et al. (2015) Cancer Driver Log (CanDL): Catalog of potentially actionable cancer mutations. *The Journal of Molecular Diagnostics* **17**:554–559.

[95] Danecek P., et al. (2011) The variant call format and VCFtools. *Bioinformatics* **27**:2156–2158.

[96] Dang L., et al. (2009) Cancer-associated IDH1 mutations produce 2-hydroxyglutarate. *Nature* **462**:739–744.

[97] Davydov E. V., et al. (2010) Identifying a high fraction of the human genome to be under selective constraint using GERP++. *PLoS Computational Biology* **6**:e1001025.

[98] de Ligt J., et al. (2012) Diagnostic exome sequencing in persons with severe intellectual disability. *The New England Journal of Medicine* **367**:1921–1929.

[99] Dees N. D., et al. (2012) MuSiC: Identifying mutational significance in cancer genomes. *Genome Research* **22**:1589–1598.

[100] Degner J. F., et al. (2009) Effect of read-mapping biases on detecting allele-specific expression from RNA-sequencing data. *Bioinformatics* **25**:3207–3212.

[101] Deininger P. (1999) Genetic instability in cancer: Caretaker and gatekeeper genes. *The Ochsner Journal* **1**:206–209.

[102] Del Fabbro C., et al. (2013) An extensive evaluation of read trimming effects on Illumina NGS data analysis. *PLoS One* **8**:e85024.

[103] den Dunnen J. T., et al. (2016) HGVS recommendations for the description of sequence variants: 2016 update. *Human Mutation* **37**:564–569.

[104] DePristo M. A., et al. (2011) A framework for variation discovery and genotyping using next-generation DNA sequencing data. *Nature Genetics* **43**:491–498.

[105] Di Fiore R., et al. (2013) RB1 in cancer: Different mechanisms of RB1 inactivation and alterations of pRb pathway in tumorigenesis. *Journal of Cellular Physiology* **228**:1676–1687.

[106] Ding L., et al. (2012) Clonal evolution in relapsed acute myeloid leukaemia revealed by whole-genome sequencing. *Nature* **481**:506–510.

[107] Ding L., et al. (2013) Advances for studying clonal evolution in cancer. *Cancer Letters* **340**:212–219.

[108] Doelken S. C., et al. (2013) Phenotypic overlap in the contribution of individual genes to CNV pathogenicity revealed by cross-species computational analysis of single-gene mutations in humans, mice and zebrafish. *Disease Models & Mechanisms* **6**:358–372.

[109] Doyle A. J., et al. (2012) Mutations in the TGF-β repressor SKI cause Shprintzen–Goldberg syndrome with aortic aneurysm. *Nature Genetics* **44**:1249–1254.

[110] Duncan C., et al. (2010) Integrated genomic analyses identify ERRFI1 and TACC3 as glioblastoma-targeted genes. *Oncotarget* **1**:265–277.

[111] Eberle M. A., et al. (2017) A reference data set of 5.4 million phased human variants validated by genetic inheritance from sequencing a three-generation 17-member pedigree. *Genome Research* **27**:157–164.

[112] Ehmke N., et al. (2014) First description of a patient with Vici syndrome due to a mutation affecting the penultimate exon of EPG5 and review of the literature. *American Journal of Medical Genetics. Part A* **164A**:3170–3175.

[113] Eilbeck K., et al. (2005) The Sequence Ontology: A tool for the unification of genome annotations. *Genome Biology* **6**:R44.

[114] Emmert-Buck M. R., et al. (1996) Laser capture microdissection. *Science* **274**:998–1001.

[115] ENCODE Project Consortium. (2012) An integrated encyclopedia of DNA elements in the human genome. *Nature* **489**:57–74.

[116] EURORDIS. What is a rare disease? `http://www.eurordis.org/sites/default/files/publications/Fact_Sheet_Eurordiscare2.pdf`. Accessed: 2016-09-17.

[117] Ewing B., et al. (1998) Base-calling of automated sequencer traces using phred. I. accuracy assessment. *Genome Research* **8**:175–185.

[118] Fares F., et al. (2008) Carrier frequency of autosomal-recessive disorders in the Ashkenazi Jewish population: Should the rationale for mutation choice for screening be reevaluated? *Prenatal Diagnosis* **28**:236–241.

[119] Faust G. G. and Hall I. M. (2014) SAMBLASTER: Fast duplicate marking and structural variant read extraction. *Bioinformatics* **30**:2503–2505.

[120] Feng J., et al. (2012) An integrated analysis of germline and somatic, genetic and epigenetic alterations at 9p21.3 in glioblastoma. *Cancer* **118**:232–240.

[121] Ferragina P. and Manzini G. (2000) Opportunistic data structures with applications. In *Proceedings of the 41st Annual Symposium on Foundations of Computer Science (FOCS '00)* pages 390–398 Washington, DC. IEEE Computer Society.

[122] Forbes S. A., et al. (2017) COSMIC: Somatic cancer genetics at high-resolution. *Nucleic Acids Research* **45**:D777–D783.

[123] Francioli L. C., et al. (2015) Genome-wide patterns and properties of de novo mutations in humans. *Nature Genetics* **47**:822–826.

[124] Freed-Pastor W. A. and Prives C. (2012) Mutant p53: One name, many proteins. *Genes & Development* **26**:1268–1286.

[125] Freeman J. L., et al. (2006) Copy number variation: New insights in genome diversity. *Genome Research* **16**:949–961.

[126] Fromer M., et al. (2012) Discovery and statistical genotyping of copy-number variation from whole-exome sequencing depth. *American Journal of Human Genetics* **91**:597–607.

[127] Fu W., et al. (2013) Analysis of 6,515 exomes reveals the recent origin of most human protein-coding variants. *Nature* **493**:216–220.

[128] Futreal P., et al. (2004) A census of human cancer genes. *Nature Reviews Cancer* **4**:177–183.

[129] Garrison E. and Marth G. (2012) Haplotype-based variant detection from short-read sequencing. *ArXiv* **1207**:1207.3907.

[130] Gehring J. S., et al. (2015) SomaticSignatures: Inferring mutational signatures from single-nucleotide variants. *Bioinformatics* **31**:3673.

[131] Geifman N., et al. (2014) Promoting precision cancer medicine through a community-driven knowledgebase. *Journal of Personalized Medicine* **4**:475–488.

[132] George R. A., et al. (2006) Analysis of protein sequence and interaction data for candidate disease gene prediction. *Nucleic Acids Research* **34**:e130.

[133] Gilissen C., et al. (2014) Genome sequencing identifies major causes of severe intellectual disability. *Nature* **511**:344–347.

[134] Gilissen C., et al. (2012) Disease gene identification strategies for exome sequencing. *European Journal of Human Genetics* **20**:490–497.

[135] Gilkes J. A. and Heldermon C. D. (2014) Mucopolysaccharidosis III (sanfilippo syndrome)- disease presentation and experimental therapies. *Pediatric Endocrinology Reviews* **12 Suppl** 1:133–140.

[136] Glusman G., et al. (2011) Kaviar: An accessible system for testing SNV novelty. *Bioinformatics* **27**:3216–3217.

[137] González-Pérez A. and López-Bigas N. (2011) Improving the assessment of the outcome of nonsynonymous SNVs with a consensus deleteriousness score, Condel. *American Journal of Human Genetics* **88**:440–449.

[138] González-Pérez A. and López-Bigas N. (2012) Functional impact bias reveals cancer drivers. *Nucleic Acids Research* **40**:e169.

[139] Gout A. M., et al. (2007) Analysis of published PKD1 gene sequence variants. *Nature Genetics* **39**:427–428.

[140] Goyal S., et al. (2016) Confirmation of TTC8 as a disease gene for nonsyndromic autosomal recessive retinitis pigmentosa (RP51). *Clinical Genetics* **89**:454–460.

[141] Grantham R. (1974) Amino acid difference formula to help explain protein evolution. *Science* **185**:862–864.

[142] Greaves M. and Maley C. C. (2012) Clonal evolution in cancer. *Nature* **481**:306–313.

[143] Greene D., et al. (2016) Phenotype similarity regression for identifying the genetic determinants of rare diseases. *American Journal of Human Genetics* **98**:490–499.

[144] Greenman C., et al. (2007) Patterns of somatic mutation in human cancer genomes. *Nature* **446**:153–158.

[145] Griffith M., et al. (2017) CIViC is a community knowledgebase for expert crowdsourcing the clinical interpretation of variants in cancer. *Nature Genetics* **49**:170–174.

[146] Groza T., et al. (2015) The Human Phenotype Ontology: Semantic unification of common and rare disease. *American Journal of Human Genetics* **97**:111–124.

[147] GTEx Consortium. (2013) The genotype-tissue expression (GTEx) project. *Nature Genetics* **45**:580–585.

[148] Guo Y., et al. (2013) MitoSeek: Extracting mitochondria information and performing high-throughput mitochondria sequencing analysis. *Bioinformatics* **29**:1210–1211.

[149] Guo Y., et al. (2014) Three-stage quality control strategies for DNA re-sequencing data. *Briefings in Bioinformatics* **15**:879–889.

[150] Guo Y., et al. (2014) Multi-perspective quality control of Illumina exome sequencing data using QC3. *Genomics* **103**:323–328.

[151] Gussow A. B., et al. (2016) The intolerance to functional genetic variation of protein domains predicts the localization of pathogenic mutations within genes. *Genome Biology* **17**:9.

[152] Haas-Kogan D. A., et al. (2005) Epidermal growth factor receptor, protein kinase B/Akt, and glioma response to erlotinib. *Journal of the National Cancer Institute* **97**:880.

[153] Hainaut P. and Pfeifer G. P. (2016) Somatic TP53 mutations in the era of genome sequencing. *Cold Spring Harbor Perspectives in Medicine* **6**:a026179.

[154] Hajirasouliha I., et al. (2010) Detection and characterization of novel sequence insertions using paired-end next-generation sequencing. *Bioinformatics* **26**:1277–1283.

[155] Hanahan D. and Weinberg R. A. (2000) The hallmarks of cancer. *Cell* **100**:57–70.

[156] Hanahan D. and Weinberg R. A. (2011) Hallmarks of cancer: The next generation. *Cell* **144**:646–674.

[157] Hansen K. D., Brenner S. E., and Dudoit S. (2010) Biases in Illumina transcriptome sequencing caused by random hexamer priming. *Nucleic Acids Research* **38**:e131.

[158] Hansen P., et al. (2016) Q-nexus: A comprehensive and efficient analysis pipeline designed for ChIP-nexus. *BMC Genomics* **17**:873.

[159] Hansen P., et al. (2015) Saturation analysis of ChIP-seq data for reproducible identification of binding peaks. *Genome Research* **25**:1391–1400.

[160] Head S. R., et al. (2014) Library construction for next-generation sequencing: Overviews and challenges. *BioTechniques* **56**:61–77.

[161] Heinrich V., et al. (2017) A likelihood ratio-based method to predict exact pedigrees for complex families from next-generation sequencing data. *Bioinformatics* **33**:72–78.

[162] Heinrich V., et al. (2013) Estimating exome genotyping accuracy by comparing to data from large scale sequencing projects. *Genome Medicine* **5**:69.

[163] Heinrich V., et al. (2012) The allele distribution in next-generation sequencing data sets is accurately described as the result of a stochastic branching process. *Nucleic Acids Research* **40**:2426–2431.

[164] Helleday T., Eshtad S., and Nik-Zainal S. (2014) Mechanisms underlying mutational signatures in human cancers. *Nature Reviews Genetics* **15**:585–598.

[165] Hill R. E. and Lettice L. A. (2013) Alterations to the remote control of Shh gene expression cause congenital abnormalities. *Philosophical Transactions of the Royal Society of London. Series B, Biological Sciences* **368**:20120357.

[166] Hogeweg P. (2011) The roots of bioinformatics in theoretical biology. *PLoS Computational Biology* **7**:e1002021.

[167] Hoischen A., et al. (2010) De novo mutations of SETBP1 cause Schinzel–Giedion syndrome. *Nature Genetics* **42**:483–485.

[168] Howard B. D. and Tessman I. (1964) Identification of the altered bases in mutated single-stranded DNA. II. in vivo mutagenesis by 5-bromodeoxyuridine and 2-aminopurine. *Journal of Molecular Biology* **9**:364–371.

[169] Howard M. F., et al. (2014) Mutations in PGAP3 impair GPI-anchor maturation, causing a subtype of hyperphosphatasia with mental retardation. *American Journal of Human Genetics* **94**:278–287.

[170] Howe D. G., et al. (2017) The Zebrafish Model Organism Database: New support for human disease models, mutation details, gene expression phenotypes and searching. *Nucleic Acids Research* **45**:D758–D768.

[171] Hsi-Yang Fritz M., et al. (2011) Efficient storage of high throughput DNA sequencing data using reference-based compression. *Genome Research* **21**:734–740.

[172] Hua X., et al. (2013) DrGaP: A powerful tool for identifying driver genes and pathways in cancer sequencing studies. *American Journal of Human Genetics* **93**:439–451.

[173] Huang L., et al. (2017) The cancer precision medicine knowledge base for structured clinical-grade mutations and interpretations. *Journal of the American Medical Informatics Association* **24**:513–519.

[174] Huddleston J., et al. (2017) Discovery and genotyping of structural variation from long-read haploid genome sequence data. *Genome Research* **27**:677–685.

[175] Hwang S., et al. (2015) Systematic comparison of variant calling pipelines using gold standard personal exome variants. *Scientific Reports* **5**:17875.

[176] Ibn-Salem J., et al. (2014) Deletions of chromosomal regulatory boundaries are associated with congenital disease. *Genome Biology* **15**:423.

[177] Ichimura K., et al. (2008) 1p36 is a preferential target of chromosome 1 deletions in astrocytic tumours and homozygously deleted in a subset of glioblastomas. *Oncogene* **27**:2097–2108.

[178] Ingram V. M. (1956) A specific chemical difference between the globins of normal human and sickle-cell anaemia haemoglobin. *Nature* **178**:792–794.

[179] International HapMap Consortium. (2003) The International HapMap Project. *Nature* **426**:789–796.

[180] International Human Genome Sequencing Consortium. (2004) Finishing the euchromatic sequence of the human genome. *Nature* **431**:931–945.

[181] Jacob H. J., et al. (2013) Genomics in clinical practice: Lessons from the front lines. *Science Translational Medicine* **5**:194cm5.

[182] Jäger M., et al. (2011) Composite transcriptome assembly of RNA-seq data in a sheep model for delayed bone healing. *BMC Genomics* **12**:158.

[183] Jäger M., et al. (2016) Alternate-locus aware variant calling in whole genome sequencing. *Genome Medicine* **8**:130.

[184] Jäger M., et al. (2014) Jannovar: A java library for exome annotation. *Human Mutation* **35**:548–555.

[185] Jensen S. A., et al. (2015) A microfibril assembly assay identifies different mechanisms of dominance underlying Marfan syndrome, stiff skin syndrome and acromelic dysplasias. *Human Molecular Genetics* **24**:4454–4463.

[186] Jo H., et al. (2014) Correlating molecular and histopathologic tumor purity: An analysis of 816 patients. *Cancer Research* **74**:Abstract nr 5598.

[187] Jones S. J., et al. (2010) Evolution of an adenocarcinoma in response to selection by targeted kinase inhibitors. *Genome Biology* **11**:R82.

[188] Kaisaki P. J., et al. (2016) Targeted next-generation sequencing of plasma DNA from cancer patients: Factors influencing consistency with tumour DNA and prospective investigation of its utility for diagnosis. *PLoS One* **11**:e0162809.

[189] Kamihara J., Rana H. Q., and Garber J. E. (2014) Germline TP53 mutations and the changing landscape of Li–Fraumeni syndrome. *Human Mutation* **35**:654–662.

[190] Kandoth C., et al. (2013) Mutational landscape and significance across 12 major cancer types. *Nature* **502**:333–339.

[191] Karakoc E., et al. (2011) Detection of structural variants and indels within exome data. *Nature Methods* **9**:176–178.

[192] Kaye J. (2012) The tension between data sharing and the protection of privacy in genomics research. *Annual Review of Genomics and Human Genetics* **13**:415–431.

[193] Keller I., Bensasson D., and Nichols R. A. (2007) Transition-transversion bias is not universal: A counter example from grasshopper pseudogenes. *PLoS Genetics* **3**:e22.

[194] Kim K., et al. (2015) Effect of next-generation exome sequencing depth for discovery of diagnostic variants. *Genomics & Informatics* **13**:31–39.

[195] Kinzler K. W. and Vogelstein B. (1997) Gatekeepers and caretakers. *Nature* **386**:761–763.

[196] Kircher M., Heyn P., and Kelso J. (2011) Addressing challenges in the production and analysis of Illumina sequencing data. *BMC Genomics* **12**:382.

[197] Kircher M., et al. (2014) A general framework for estimating the relative pathogenicity of human genetic variants. *Nature Genetics* **46**:310–315.

[198] Kirkpatrick B. E., et al. (2015) GenomeConnect: Matchmaking between patients, clinical laboratories, and researchers to improve genomic knowledge. *Human Mutation* **36**:974–978.

[199] Kishnani P. S., et al. (2014) Diagnosis and management of glycogen storage disease type I: A practice guideline of the American College of Medical Genetics and Genomics. *Genetics in Medicine* **16**:e1.

[200] Klambauer G., et al. (2012) cn.MOPS: Mixture of Poissons for discovering copy number variations in next-generation sequencing data with a low false discovery rate. *Nucleic Acids Research* **40**:e69.

[201] Klockow B., et al. (2000) Oncogenic insertional mutations in the P-loop of Ras are overactive in MAP kinase signaling. *Oncogene* **19**:5367–5376.

[202] Knecht C., et al. (2017) IMHOTEP-a composite score integrating popular tools for predicting the functional consequences of nonsynonymous sequence variants. *Nucleic Acids Research* **45**:e13.

[203] Knudson A. G. (1971) Mutation and cancer: Statistical study of retinoblastoma. *Proceedings of the National Academy of Sciences of the United States of America* **68**:820–823.

[204] Koboldt D. C., Larson D. E., and Wilson R. K. (2013) Using VarScan 2 for germline variant calling and somatic mutation detection. *Current Protocols in Bioinformatics* **44**:15.4.1–15.4.17.

[205] Koboldt D. C., et al. (2012) VarScan 2: Somatic mutation and copy number alteration discovery in cancer by exome sequencing. *Genome Research* **22**:568–576.

[206] Kodama Y., et al. (2012) The Sequence Read Archive: Explosive growth of sequencing data. *Nucleic Acids Research* **40**:D54–D56.

[207] Köhler S., et al. (2008) Walking the interactome for prioritization of candidate disease genes. *American Journal of Human Genetics* **82**:949–958.

[208] Köhler S., et al. (2011) Improving ontologies by automatic reasoning and evaluation of logical definitions. *BMC Bioinformatics* **12**:418.

[209] Köhler S., et al. (2014) The Human Phenotype Ontology project: Linking molecular biology and disease through phenotype data. *Nucleic Acids Research* **42**:D966–D974.

[210] Köhler S., et al. (2013) Construction and accessibility of a cross-species phenotype ontology along with gene annotations for biomedical research. *F1000Research* **2**:30.

[211] Köhler S. and Robinson P. N. (2017) Diagnostics in human genetics: Integration of phenotypic and genomic data. *Bundesgesundheitsblatt, Gesundheitsforschung, Gesundheitsschutz* **60**:542–549.

[212] Köhler S., et al. (2014) Clinical interpretation of CNVs with cross-species phenotype data. *Journal of Medical Genetics* **51**:766–772.

[213] Köhler S., et al. (2009) Clinical diagnostics in human genetics with semantic similarity searches in ontologies. *American Journal of Human Genetics* **85**:457–464.

[214] Köhler S., et al. (2017) The Human Phenotype Ontology in 2017. *Nucleic Acids Research* **45**:D865–D876.

[215] Kolanczyk M., et al. (2015) Missense variant in CCDC22 causes X-linked recessive intellectual disability with features of Ritscher–Schinzel/3C syndrome. *European Journal of Human Genetics* **23**:633–638.

[216] Kong A., et al. (2012) Rate of de novo mutations and the importance of father's age to disease risk. *Nature* **488**:471–475.

[217] Korbel J. O., et al. (2007) Paired-end mapping reveals extensive structural variation in the human genome. *Science* **318**:420–426.

[218] Koschmann C., et al. (2016) ATRX loss promotes tumor growth and impairs nonhomologous end joining DNA repair in glioma. *Science Translational Medicine* **8**:328ra28.

[219] Kotlarz D., et al. (2013) Loss-of-function mutations in the IL-21 receptor gene cause a primary immunodeficiency syndrome. *The Journal of Experimental Medicine* **210**:433–443.

[220] Kou T., et al. (2016) The possibility of clinical sequencing in the management of cancer. *Japanese Journal of Clinical Oncology* **46**:399–406.

[221] Kraulis P. J. (1992) MOLSCRIPT: A program to produce both detailed and schematic plots of protein structures. *Journal of Applied Crystallography* **24**:946–950.

[222] Krawczak M., Reiss J., and Cooper D. N. (1992) The mutational spectrum of single base-pair substitutions in mRNA splice junctions of human genes: Causes and consequences. *Human Genetics* **90**:41–54.

[223] Krawitz P., et al. (2015) The genomic birthday paradox: How much is enough? *Human Mutation* **36**:989–997.

[224] Krawitz P. M., et al. (2012) Mutations in PIGO, a member of the GPI-anchor-synthesis pathway, cause hyperphosphatasia with mental retardation. *American Journal of Human Genetics* **91**:146–151.

[225] Krawitz P. M., et al. (2013) PGAP2 mutations, affecting the GPI-anchor-synthesis pathway, cause hyperphosphatasia with mental retardation syndrome. *American Journal of Human Genetics* **92**:584–589.

[226] Krawitz P. M., et al. (2010) Identity-by-descent filtering of exome sequence data identifies PIGV mutations in hyperphosphatasia mental retardation syndrome. *Nature Genetics* **42**:827–829.

[227] Krøigård A. B., et al. (2016) Evaluation of nine somatic variant callers for detection of somatic mutations in exome and targeted deep sequencing data. *PLoS One* **11**:e0151664.

[228] Krumm N., et al. (2012) Copy number variation detection and genotyping from exome sequence data. *Genome Research* **22**:1525–1532.

[229] Krzywinski M., et al. (2009) Circos: An information aesthetic for comparative genomics. *Genome Research* **19**:1639–1645.

[230] Kuchenbecker L., et al. (2015) IMSEQ–a fast and error aware approach to immunogenetic sequence analysis. *Bioinformatics* **31**:2963–2971.

[231] Kuilman T., et al. (2015) CopywriteR: DNA copy number detection from off-target sequence data. *Genome Biology* **16**:49.

[232] Lage K., et al. (2007) A human phenome-interactome network of protein complexes implicated in genetic disorders. *Nature Biotechnology* **25**:309–316.

[233] Lakin N. D. and Jackson S. P. (1999) Regulation of p53 in response to DNA damage. *Oncogene* **18**:7644–7655.

[234] Lander E. S., et al. (2001) Initial sequencing and analysis of the human genome. *Nature* **409**:860–921.

[235] Landrum M. J., et al. (2016) ClinVar: Public archive of interpretations of clinically relevant variants. *Nucleic Acids Research* **44**:D862–D868.

[236] Langmead B. and Salzberg S. L. (2012) Fast gapped-read alignment with Bowtie 2. *Nature Methods* **9**:357–359.

[237] Lappalainen I., et al. (2015) The European Genome-phenome Archive of human data consented for biomedical research. *Nature Genetics* **47**:692–695.

[238] Larson N. B. and Fridley B. L. (2013) PurBayes: Estimating tumor cellularity and subclonality in next-generation sequencing data. *Bioinformatics* **29**:1888–1889.

[239] Lawrence M. S., et al. (2013) Mutational heterogeneity in cancer and the search for new cancer-associated genes. *Nature* **499**:214–218.

[240] Le Scouarnec S. and Gribble S. M. (2012) Characterising chromosome rearrangements: Recent technical advances in molecular cytogenetics. *Heredity* **108**:75–85.

[241] Lee D., et al. (2015) A method to predict the impact of regulatory variants from DNA sequence. *Nature Genetics* **47**:955–961.

[242] Lee H., et al. (2014) Clinical exome sequencing for genetic identification of rare Mendelian disorders. *JAMA* **312**:1880–1887.

[243] Lee O.-H., et al. (2006) Expression of the receptor tyrosine kinase Tie2 in neoplastic glial cells is associated with integrin β1-dependent adhesion to the extracellular matrix. *Molecular Cancer Research* **4**:915–926.

[244] Lee S., et al. (2009) MoDIL: Detecting small indels from clone-end sequencing with mixtures of distributions. *Nature Methods* **6**:473–474.

[245] Leinonen R., Sugawara H., and Shumway M. (2011) The sequence read archive. *Nucleic Acids Research* **39**:D19–D21.

[246] Lek M., et al. (2016) Analysis of protein-coding genetic variation in 60,706 humans. *Nature* **536**:285–291.

[247] Lentaigne C., et al. (2016) Inherited platelet disorders: Toward DNA-based diagnosis. *Blood* **127**:2814–2823.

[248] Leong I. U. S., et al. (2015) Assessment of the predictive accuracy of five in silico prediction tools, alone or in combination, and two metaservers to classify long QT syndrome gene mutations. *BMC Medical Genetics* **16**:34.

[249] Ley T. J., et al. (2008) DNA sequencing of a cytogenetically normal acute myeloid leukaemia genome. *Nature* **456**:66–72.

[250] Li H. (2011) Tabix: Fast retrieval of sequence features from generic TAB-delimited files. *Bioinformatics* **27**:718–719.

[251] Li H. (2013) Aligning sequence reads, clone sequences and assembly contigs with BWA-MEM. **arXiv**:303.3997 [q–bio.GN].

[252] Li H. and Durbin R. (2009) Fast and accurate short read alignment with Burrows–Wheeler transform. *Bioinformatics* **25**:1754–1760.

[253] Li H. and Durbin R. (2010) Fast and accurate long-read alignment with Burrows–Wheeler transform. *Bioinformatics* **26**:589–595.

[254] Li H., et al. (2009) The Sequence Alignment/Map format and SAMtools. *Bioinformatics* **25**:2078–2079.

[255] Li H. and Homer N. (2010) A survey of sequence alignment algorithms for next-generation sequencing. *Briefings in Bioinformatics* **11**:473–483.

[256] Li J., et al. (2015) mirTrios: An integrated pipeline for detection of de novo and rare inherited mutations from trios-based next-generation sequencing. *Journal of Medical Genetics* **52**:275–281.

[257] Li J., et al. (2016) MicroRNA-200b inhibits the growth and metastasis of glioma cells via targeting ZEB2. *International Journal of Oncology* **48**:541–550.

[258] Li M.-X., et al. (2012) A comprehensive framework for prioritizing variants in exome sequencing studies of Mendelian diseases. *Nucleic Acids Research* **40**:e53.

[259] Li R., et al. (2009) SOAP2: An improved ultrafast tool for short read alignment. *Bioinformatics* **25**:1966–1967.

[260] Li Z., et al. (2014) Vindel: A simple pipeline for checking indel redundancy. *BMC Bioinformatics* **15**:359.

[261] Lipman D. J. and Pearson W. R. (1985) Rapid and sensitive protein similarity searches. *Science* **227**:1435–1441.

[262] Liu D., et al. (2010) Tie2/TEK modulates the interaction of glioma and brain tumor stem cells with endothelial cells and promotes an invasive phenotype. *Oncotarget* **1**:700–709.

[263] Liu Q., et al. (2012) Steps to ensure accuracy in genotype and SNP calling from Illumina sequencing data. *BMC Genomics* **13 Suppl 8**:S8.

[264] Liu X., et al. (2013) Variant callers for next-generation sequencing data: A comparison study. *PLoS One* **8**:e75619.

[265] Liu X., et al. (2016) dbNSFP v3.0: A one-stop database of functional predictions and annotations for human nonsynonymous and splice-site SNVs. *Human Mutation* **37**:235–241.

[266] Liu Y., et al. (2014) A gradient-boosting approach for filtering de novo mutations in parent-offspring trios. *Bioinformatics* **30**:1830–1836.

[267] Livingstone C. D. and Barton G. J. (1993) Protein sequence alignments: A strategy for the hierarchical analysis of residue conservation. *Computer Applications in the Biosciences* **9**:745–756.

[268] Lo K. C., et al. (2008) Comprehensive analysis of loss of heterozygosity events in glioblastoma using the 100K SNP mapping arrays and comparison with copy number abnormalities defined

by bac array comparative genomic hybridization. *Genes, Chromosomes and Cancer* **47**(3):221–237.

[269] Lu H., Giordano F., and Ning Z. (2016) Oxford Nanopore MinION sequencing and genome assembly. *Genomics, Proteomics & Bioinformatics* **14**:265–279.

[270] Lynch H. T., Snyder C., and Casey M. J. (2013) Hereditary ovarian and breast cancer: What have we learned? *Annals of Oncology* **24**:viii83–viii95.

[271] Mack S., et al. (2014) Epigenomic alterations define lethal cimp-positive ependymomas of infancy. *Nature* **506**:445–450.

[272] Macmanes M. D. (2014) On the optimal trimming of high-throughput mRNA sequence data. *Frontiers in Genetics* **5**:13.

[273] Magi A., et al. (2014) H3M2: Detection of runs of homozygosity from whole-exome sequencing data. *Bioinformatics* **30**:2852–2859.

[274] Magi A., et al. (2012) Read count approach for DNA copy number variants detection. *Bioinformatics* **28**:470–478.

[275] Mamanova L., et al. (2010) Target-enrichment strategies for next-generation sequencing. *Nature Methods* **7**:111–118.

[276] Mardis E. R. (2008) Next-generation DNA sequencing methods. *Annual Review of Genomics and Human Genetics* **9**:387–402.

[277] Mardis E. R. (2012) Genome sequencing and cancer. *Current Opinion in Genetics & Development* **22**:245–250.

[278] Margulies M., et al. (2005) Genome sequencing in microfabricated high-density picolitre reactors. *Nature* **437**:376–380.

[279] Marine R., et al. (2011) Evaluation of a transposase protocol for rapid generation of shotgun high-throughput sequencing libraries from nanogram quantities of DNA. *Applied and Environmental Microbiology* **77**:8071–8079.

[280] Martelotto L. G., et al. (2014) Benchmarking mutation effect prediction algorithms using functionally validated cancer-related missense mutations. *Genome Biology* **15**:484.

[281] Martins F. C., et al. (2012) Evolutionary pathways in BRCA1-associated breast tumors. *Cancer Discovery* **2**:503–511.

[282] Marusyk A., Almendro V., and Polyak K. (2012) Intra-tumour heterogeneity: A looking glass for cancer? *Nature Reviews Cancer* **12**:323–334.

[283] Marx V. (2015) The DNA of a nation. *Nature* **524**:503–505.

[284] Mathelier A., et al. (2016) JASPAR 2016: A major expansion and update of the open-access database of transcription factor binding profiles. *Nucleic Acids Research* **44**:D110–D115.

[285] McKenna A., et al. (2010) The Genome Analysis Toolkit: A MapReduce framework for analyzing next-generation DNA sequencing data. *Genome Research* **20**:1297–1303.

[286] McKusick V. A. (2001) The anatomy of the human genome: A neo-Vesalian basis for medicine in the 21st century. *JAMA* **286**:2289–2295.

[287] McLaren W., et al. (2016) The Ensembl variant effect predictor. *Genome Biology* **17**:122.

[288] McMurry J. A., et al. (2016) Navigating the phenotype frontier: The Monarch Initiative. *Genetics* **203**:1491–1495.

[289] Medvedev P., Stanciu M., and Brudno M. (2009) Computational methods for discovering structural variation with next-generation sequencing. *Nature Methods* **6**:S13–S20.

[290] Merico D., et al. (2015) Compound heterozygous mutations in the noncoding RNU4ATAC cause Roifman syndrome by disrupting minor intron splicing. *Nature Communications* **6**:8718.

[291] Michor F., Iwasa Y., and Nowak M. A. (2004) Dynamics of cancer progression. *Nature Reviews Cancer* **4**:197–205.

[292] Miller C. A., et al. (2011) ReadDepth: A parallel R package for detecting copy number alterations from short sequencing reads. *PLoS One* **6**:e16327.

[293] Miller C. A., et al. (2014) SciClone: Inferring clonal architecture and tracking the spatial and temporal patterns of tumor evolution. *PLoS Computational Biology* **10**:1–15.

[294] Minoche A. E., Dohm J. C., and Himmelbauer H. (2011) Evaluation of genomic high-throughput sequencing data generated on Illumina HiSeq and genome analyzer systems. *Genome Biology* **12**:R112.

[295] Modest D. P., et al. (2013) KRAS allel-specific activity of sunitinib in an isogenic disease model of colorectal cancer. *Journal of Cancer Research and Clinical Oncology* **139**:953–961.

[296] Molster C., et al. (2016) Survey of healthcare experiences of Australian adults living with rare diseases. *Orphanet Journal of Rare Diseases* **11**:30.

[297] Moreau Y. and Tranchevent L.-C. (2012) Computational tools for prioritizing candidate genes: Boosting disease gene discovery. *Nature Reviews Genetics* **13**:523–536.

[298] Morin S. J., et al. (2017) Translocations, inversions and other chromosome rearrangements. *Fertility and Sterility* **107**:19–26.

[299] Mostovoy Y., et al. (2016) A hybrid approach for de novo human genome sequence assembly and phasing. *Nature Methods* **13**:587–590.

[300] Mungall C. J., et al. (2017) The Monarch Initiative: An integrative data and analytic platform connecting phenotypes to genotypes across species. *Nucleic Acids Research* **45**:D712–D722.

[301] Nagahashi M., et al. (2016) Genomic landscape of colorectal cancer in Japan: Clinical implications of comprehensive genomic sequencing for precision medicine. *Genome Medicine* **8**:136.

[302] National Human Genome Research Institute. The Human Genome Project completion: Frequently Asked Questions. https://www.genome.gov/11006943. Accessed: 2017-06-13.

[303] Nature editorial staff. (2010) The human genome at ten. *Nature* **464**:649–650.

[304] Ng C. K., et al. (2015) Intra-tumor genetic heterogeneity and alternative driver genetic alterations in breast cancers with heterogeneous HER2 gene amplification. *Genome Biology* **16**:107.

[305] Ng P. C. and Henikoff S. (2002) Accounting for human polymorphisms predicted to affect protein function. *Genome Research* **12**:436–446.

[306] Ng S. B., et al. (2010) Exome sequencing identifies MLL2 mutations as a cause of Kabuki syndrome. *Nature Genetics* **42**:790–793.

[307] Ng S. B., et al. (2010) Exome sequencing identifies the cause of a Mendelian disorder. *Nature Genetics* **42**:30–35.

[308] Ngeow J. and Eng C. (2016) Precision medicine in heritable cancer: When somatic tumour testing and germline mutations meet. *NPJ Genomic Medicine* **1**:15006.

[309] NHLBI GO Exome Sequencing Project (ESP). Exome Variant Server. `http://evs.gs.washington.edu/EVS/`. Accessed: 2017-03-10.

[310] Ni S. and Stoneking M. (2016) Improvement in detection of minor alleles in next generation sequencing by base quality recalibration. *BMC Genomics* **17**:139.

[311] Nielsen R., et al. (2011) Genotype and snp calling from next-generation sequencing data. *Nature Reviews Genetics* **12**:443–451.

[312] Northcott P. A., et al. (2014) Enhancer hijacking activates GFI1 family oncogenes in medulloblastoma. *Nature* **511**:428–434.

[313] Nowell P. (1976) The clonal evolution of tumor cell populations. *Science* **194**:23–28.

[314] Nuzzo F., et al. (2015) Characterization of an apparently synonymous F5 mutation causing aberrant splicing and factor V deficiency. *Haemophilia* **21**:241–248.

[315] Oesper L., Satas G., and Raphael B. J. (2014) Quantifying tumor heterogeneity in whole-genome and whole-exome sequencing data. *Bioinformatics* **30**:3532–3540.

[316] Olivier M., Hollstein M., and Hainaut P. (2010) TP53 mutations in human cancers: Origins, consequences, and clinical use. *Cold Spring Harbor Perspectives in Biology* **2**:a001008.

[317] O'Rawe J., et al. (2013) Low concordance of multiple variant-calling pipelines: Practical implications for exome and genome sequencing. *Genome Medicine* **5**:28.

[318] Oren M. and Rotter V. (2010) Mutant p53 gain-of-function in cancer. *Cold Spring Harbor Perspectives in Biology* **2**:a001107.

[319] Oti M., et al. (2006) Predicting disease genes using protein-protein interactions. *Journal of Medical Genetics* **43**:691–698.

[320] Pabinger S., et al. (2014) A survey of tools for variant analysis of next-generation genome sequencing data. *Briefings in Bioinformatics* **15**:256–278.

[321] Palmirotta R., et al. (2015) DNA fingerprinting for sample authentication in biobanking: Recent perspectives. *Journal of Biorepository Science for Applied Medicine* **3**:35–45.

[322] Park C.-K., et al. (2015) Genomic dynamics associated with malignant transformation in IDH1 mutated gliomas. *Oncotarget* **6**:43653–43666.

[323] Patel R. K. and Jain M. (2012) NGS QC toolkit: A toolkit for quality control of next generation sequencing data. *PLoS One* **7**:e30619.

[324] Patil S. and Chamberlain R. S. (2012) Neoplasms associated with germline and somatic NF1 gene mutations. *The Oncologist* **17**:101–116.

[325] Pauling L. and Itano H. A. (1949) Sickle cell anemia a molecular disease. *Science* **110**:543–548.

[326] Peng G., et al. (2013) Rare variant detection using family-based sequencing analysis. *Proceedings of the National Academy of Sciences of the United States of America* **110**:3985–3990.

[327] Pengelly R. J., et al. (2013) A SNP profiling panel for sample tracking in whole-exome sequencing studies. *Genome Medicine* **5**:89.

[328] Petrovski S., et al. (2015) The intolerance of regulatory sequence to genetic variation predicts gene dosage sensitivity. *PLoS Genetics* **11**:e1005492.

[329] Petrovski S., et al. (2013) Genic intolerance to functional variation and the interpretation of personal genomes. *PLoS Genetics* **9**:e1003709.

[330] Pfeifer G. P. (2006) Mutagenesis at methylated CpG sequences. *Current Topics in Microbiology and Immunology* **301**:259–281.

[331] Philippakis A. A., et al. (2015) The Matchmaker Exchange: A platform for rare disease gene discovery. *Human Mutation* **36**:915–921.

[332] Pippucci T., et al. (2014) Detection of runs of homozygosity from whole exome sequencing data: State of the art and perspectives for clinical, population and epidemiological studies. *Human Heredity* **77**:63–72.

[333] Piro R. M. (2012) Network medicine: linking disorders. *Human Genetics* **131**(12):1811–1820.

[334] Piro R. M. and Di Cunto F. (2012) Computational approaches to disease-gene prediction: Rationale, classification and successes. *FEBS Journal* **279**:678–696.

[335] Piro R. M., et al. (2011) Evaluation of candidate genes from orphan FEB and GEFS+ loci by analysis of human brain gene expression atlases. *PLOS ONE* **6**(8):1–10.

[336] Piro R. M., et al. (2010) Candidate gene prioritization based on spatially mapped gene expression: An application to XLMR. *Bioinformatics* **26**(18):i618–i624.

[337] Piro R. M., et al. (2013) Disease-gene discovery by integration of 3D gene expression and transcription factor binding affinities. *Bioinformatics* **29**(4):468–475.

[338] Pirooznia M., Goes F. S., and Zandi P. P. (2015) Whole-genome CNV analysis: Advances in computational approaches. *Frontiers in Genetics* **6**:138.

[339] Pirooznia M., et al. (2014) Validation and assessment of variant calling pipelines for next-generation sequencing. *Human Genomics* **8**:14.

[340] Pollard K. S., et al. (2010) Detection of nonneutral substitution rates on mammalian phylogenies. *Genome Research* **20**:110–121.

[341] Pontikos N., et al. (2017) Phenopolis: An open platform for harmonization and analysis of genetic and phenotypic data. *Bioinformatics* **doi**:10.1093/bioinformatics/btx147. [Epub ahead of print].

[342] Posey J. E., et al. (2015) Molecular diagnostic experience of whole-exome sequencing in adult patients. *Genetics in Medicine* **18**:678–685.

[343] Prandi D., et al. (2014) Unraveling the clonal hierarchy of somatic genomic aberrations. *Genome Biology* **15**:439.

[344] Pruitt K. D., et al. (2014) RefSeq: An update on mammalian reference sequences. *Nucleic Acids Research* **42**:D756–D763.

[345] Pulst S. M. (1999) Genetic linkage analysis. *Archives of Neurology* **56**:667–672.

[346] Qiao Y., et al. (2014) SubcloneSeeker: A computational framework for reconstructing tumor clone structure for cancer variant interpretation and prioritization. *Genome Biology* **15**:443.

[347] Quinlan A. R. and Hall I. M. (2010) BEDTools: A flexible suite of utilities for comparing genomic features. *Bioinformatics* **26**:841–842.

[348] Radim J. (2014) The role of the TP73 gene and its transcripts in neuro-oncology. *British Journal of Neurosurgery* **28**:598–605.

[349] Rainer W. and Bodenreider O. (2014) Coverage of phenotypes in standard terminologies. In *Proceedings of the Joint BioOntologies and BioLINK ISMB2014 SIG session Phenotype Day* pages 41–44.

[350] Ramoni R. B., et al. (2017) The Undiagnosed Diseases Network: Accelerating discovery about health and disease. *American Journal of Human Genetics* **100**:185–192.

[351] Ramos A. H., et al. (2015) Oncotator: Cancer variant annotation tool. *Human Mutation* **36**:E2423–E2429.

[352] Ramu A., et al. (2013) DeNovoGear: De novo indel and point mutation discovery and phasing. *Nature Methods* **10**:985–987.

[353] Rands C. M., et al. (2014) 8.2% of the human genome is constrained: Variation in rates of turnover across functional element classes in the human lineage. *PLoS Genetics* **10**:e1004525.

[354] Ratan A., et al. (2010) Calling SNPs without a reference sequence. *BMC Bioinformatics* **11**:130.

[355] Rathod S. S., et al. (2014) Tumor suppressive miRNA-34a suppresses cell proliferation and tumor growth of glioma stem cells by targeting Akt and Wnt signaling pathways. *FEBS Open Biology* **4**:485–495.

[356] Rauch A., et al. (2012) Range of genetic mutations associated with severe non-syndromic sporadic intellectual disability: An exome sequencing study. *Lancet* **380**:1674–1682.

[357] Rausch T., et al. (2012) Genome sequencing of pediatric medulloblastoma links catastrophic DNA rearrangements with TP53 mutations. *Cell* **148**:59–71.

[358] Rausch T., et al. (2012) DELLY: Structural variant discovery by integrated paired-end and split-read analysis. *Bioinformatics* **28**:i333–i339.

[359] Reese M. G., et al. (1997) Improved splice site detection in Genie. *Journal of Computational Biology* **4**:311–323.

[360] Resnik P. (1995) Using information content to evaluate semantic similarity in a taxonomy. In *Proceedings of the 14th International Joint Conference on Artificial Intelligence* pages 448–453.

[361] Resnik P. (1999) Semantic similarity in a taxonomy: An information-based measure and its application to problems of ambiguity in natural language. *Journal of Artificial Intelligence Research* **11**:95–130.

[362] Retterer K., et al. (2015) Assessing copy number from exome sequencing and exome array CGH based on CNV spectrum in a large clinical cohort. *Genetics in Medicine* **17**:623–629.

[363] Reuss D. E., et al. (2013) Secretory meningiomas are defined by combined KLF4 K409Q and TRAF7 mutations. *Acta Neuropathologica* **125**:351–358.

[364] Rhoads A. and Au K. F. (2015) PacBio sequencing and its applications. *Genomics, Proteomics & Bioinformatics* **13**:278–289.

[365] Rieber N., et al. (2013) Coverage bias and sensitivity of variant calling for four whole-genome sequencing technologies. *PLoS One* **8**:e66621.

[366] Ritchie G. R. S., et al. (2014) Functional annotation of noncoding sequence variants. *Nature Methods* **11**:294–296.

[367] Rivière J.-B., et al. (2012) De novo mutations in the actin genes ACTB and ACTG1 cause Baraitser–Winter syndrome. *Nature Genetics* **44**:440–444.

[368] Roberts N. D., et al. (2013) A comparative analysis of algorithms for somatic SNV detection in cancer. *Bioinformatics* **29**:2223–2230.

[369] Robertson G., et al. (2007) Genome-wide profiles of STAT1 DNA association using chromatin immunoprecipitation and massively parallel sequencing. *Nature Methods* **4**:651–657.

[370] Robinson D. R., et al. (2013) Identification of recurrent NAB2-STAT6 gene fusions in solitary fibrous tumor by integrative sequencing. *Nature Genetics* **45**:180–185.

[371] Robinson P. N. and Bauer S. (2011) *Introduction to Bio-Ontologies.* Chapman & Hall/CRC.

[372] Robinson P. N., et al. (2008) The Human Phenotype Ontology: A tool for annotating and analyzing human hereditary disease. *American Journal of Human Genetics* **83**:610–615.

[373] Robinson P. N., et al. (2014) Improved exome prioritization of disease genes through cross-species phenotype comparison. *Genome Research* **24**:340–348.

[374] Robinson P. N., Krawitz P., and Mundlos S. (2011) Strategies for exome and genome sequence data analysis in disease-gene discovery projects. *Clinical Genetics* **80**:127–132.

[375] Rode A., et al. (2016) Chromothripsis in cancer cells: An update. *International Journal of Cancer* **138**:2322–2333.

[376] Rödelsperger C., et al. (2011) Identity-by-descent filtering of exome sequence data for disease-gene identification in autosomal recessive disorders. *Bioinformatics* **27**:829–836.

[377] Roncarati R., et al. (2013) Doubly heterozygous LMNA and TTN mutations revealed by exome sequencing in a severe form of dilated cardiomyopathy. *European Journal of Human Genetics* **21**:1105–1111.

[378] Rope A. F., et al. (2011) Using VAAST to identify an X-linked disorder resulting in lethality in male infants due to N-terminal acetyltransferase deficiency. *American Journal of Human Genetics* **89**:28–43.

[379] Rosenbloom K. R., et al. (2015) The UCSC Genome Browser database: 2015 update. *Nucleic Acids Research* **43**:D670–D681.

[380] Royer-Pokora B., et al. (1986) Cloning the gene for an inherited human disorder–chronic granulomatous disease–on the basis of its chromosomal location. *Nature* **322**:32–38.

[381] Ruderfer D. M., et al. (2016) Patterns of genic intolerance of rare copy number variation in 59,898 human exomes. *Nature Genetics* **48**:1107–1111.

[382] Ryland G. L., et al. (2015) Loss of heterozygosity: What is it good for? *BMC Medical Genomics* **8**:45.

[383] Samuels D. C., et al. (2013) Finding the lost treasures in exome sequencing data. *Trends in Genetics* **29**:593–599.

[384] Sanger F., et al. (1977) Nucleotide sequence of bacteriophage phi X174 DNA. *Nature* **265**:687–695.

[385] Sattler M. and Griffin J. (2003) Molecular mechanisms of transformation by the BCR-ABL oncogene. *Seminars in Hematology* **40**:4–10.

[386] Schaefer M. H., et al. (2012) HIPPIE: Integrating protein interaction networks with experiment based quality scores. *PLoS One* **7**:e31826.

[387] Schloss J. A. (2008) How to get genomes at one ten-thousandth the cost. *Nature Biotechnology* **26**:1113–1115.

[388] Schraivogel D., et al. (2011) CAMTA1 is a novel tumour suppressor regulated by miR-9/9* in glioblastoma stem cells. *The EMBO Journal* **30**:4309–4322.

[389] Schubert M., Lindgreen S., and Orlando L. (2016) AdapterRemoval v2: Rapid adapter trimming, identification, and read merging. *BMC Research Notes* **9**:88.

[390] Schwarz J. M., et al. (2010) MutationTaster evaluates disease-causing potential of sequence alterations. *Nature Methods* **7**:575–576.

[391] Schweizer L., et al. (2013) Meningeal hemangiopericytoma and solitary fibrous tumors carry the NAB2-STAT6 fusion and can be diagnosed by nuclear expression of STAT6 protein. *Acta Neuropathologica* **125**:651–658.

[392] Setlow R. B. and Carrier W. L. (1966) Pyrimidine dimers in ultraviolet-irradiated DNA's. *Journal of Molecular Biology* **17**:237–254.

[393] Sherry S. T., et al. (2001) dbSNP: The NCBI database of genetic variation. *Nucleic Acids Research* **29**:308–311.

[394] Shi Y. and Majewski J. (2013) FishingCNV: A graphical software package for detecting rare copy number variations in exome-sequencing data. *Bioinformatics* **29**:1461–1462.

[395] Shigemizu D., et al. (2015) Performance comparison of four commercial human whole-exome capture platforms. *Scientific Reports* **5**:12742.

[396] Shiraishi Y., et al. (2015) A simple model-based approach to inferring and visualizing cancer mutation signatures. *PLOS Genetics* **11**(12):e1005657.

[397] Siepel A., et al. (2005) Evolutionarily conserved elements in vertebrate, insect, worm, and yeast genomes. *Genome Research* **15**:1034–1050.

[398] Simon M., et al. (1999) Functional evidence for a role of combined CDKN2A (p16-p14(ARF))/CDKN2B (p15) gene inactivation in malignant gliomas. *Acta Neuropathologica* **98**:444–452.

[399] Smedley D., et al. (2015) Next-generation diagnostics and disease-gene discovery with the Exomiser. *Nature Protocols* **10**:2004–2015.

[400] Smedley D., et al. (2014) Walking the interactome for candidate prioritization in exome sequencing studies of Mendelian diseases. *Bioinformatics* **30**:3215–3222.

[401] Smedley D., et al. (2013) PhenoDigm: Analyzing curated annotations to associate animal models with human diseases. *Database* **2013**:bat025.

[402] Smedley D. and Robinson P. N. (2015) Phenotype-driven strategies for exome prioritization of human Mendelian disease genes. *Genome Medicine* **7**:81.

[403] Smedley D., et al. (2016) A whole-genome analysis framework for effective identification of pathogenic regulatory variants in Mendelian disease. *American Journal of Human Genetics* **99**:595–606.

[404] Smith C. L. and Eppig J. T. (2012) The mammalian phenotype ontology as a unifying standard for experimental and high-throughput phenotyping data. *Mammalian Genome* **23**:653–668.

[405] Smith K. R., et al. (2011) Reducing the exome search space for Mendelian diseases using genetic linkage analysis of exome genotypes. *Genome Biology* **12**:R85.

[406] Smith S. D., Kawash J. K., and Grigoriev A. (2015) GROM-RD: Resolving genomic biases to improve read depth detection of copy number variants. *PeerJ* **3**:e836.

[407] Soehn A. S., et al. (2016) Uniparental disomy of chromosome 16 unmasks recessive mutations of FA2H/SPG35 in 4 families. *Neurology* **87**:186–191.

[408] Song S., et al. (2012) qpure: A tool to estimate tumor cellularity from genome-wide single-nucleotide polymorphism profiles. *PLoS One* **7**:e45835.

[409] Stenson P. D., et al. (2017) The Human Gene Mutation Database: Towards a comprehensive repository of inherited mutation data for medical research, genetic diagnosis and next-generation sequencing studies. *Human Genetics* **136**:665–677.

[410] Stephens P. J., et al. (2011) Massive genomic rearrangement acquired in a single catastrophic event during cancer development. *Cell* **144**:27–40.

[411] Stratton M., Campbell P., and Futreal P. (2009) The cancer genome. *Nature* **458**:719–724.

[412] Strino F., et al. (2013) TrAp: A tree approach for fingerprinting subclonal tumor composition. *Nucleic Acids Research* **41**:e165.

[413] Stritt S., et al. (2016) A gain-of-function variant in DIAPH1 causes dominant macrothrombocytopenia and hearing loss. *Blood* **127**:2903–2914.

[414] Su X., et al. (2012) PurityEst: Estimating purity of human tumor samples using next-generation sequencing data. *Bioinformatics* **28**:2265–2266.

[415] Szklarczyk D., et al. (2017) The STRING database in 2017: Quality-controlled protein-protein association networks, made broadly accessible. *Nucleic Acids Research* **45**:D362–D368.

[416] Tabaska J. E. and Zhang M. Q. (1999) Detection of polyadenylation signals in human DNA sequences. *Gene* **231**:77–86.

[417] Talwalkar A., et al. (2014) SMaSH: A benchmarking toolkit for human genome variant calling. *Bioinformatics* **30**:2787–2795.

[418] Tammimies K., et al. (2015) Molecular diagnostic yield of chromosomal microarray analysis and whole-exome sequencing in children with autism spectrum disorder. *JAMA* **314**:895–903.

[419] Tan A., Abecasis G. R., and Kang H. M. (2015) Unified representation of genetic variants. *Bioinformatics* **31**:2202–2204.

[420] Tan R., et al. (2014) An evaluation of copy number variation detection tools from whole-exome sequencing data. *Human Mutation* **35**:899–907.

[421] Tarasov A., et al. (2015) Sambamba: Fast processing of NGS alignment formats. *Bioinformatics* **31**:2032–2034.

[422] Tattini L., D'Aurizio R., and Magi A. (2015) Detection of genomic structural variants from next-generation sequencing data. *Frontiers in Bioengineering and Biotechnology* **3**:92.

[423] Tatusova T. A. and Madden T. L. (1999) BLAST 2 sequences, a new tool for comparing protein and nucleotide sequences. *FEMS Microbiology Letters* **174**:247–250.

[424] Tavtigian S. V., et al. (2008) In silico analysis of missense substitutions using sequence-alignment based methods. *Human Mutation* **29**:1327–1336.

[425] Taylor A. D., et al. (2016) The path(way) less traveled: A pathway-oriented approach to providing information about precision cancer medicine on My Cancer Genome. *Translational Oncology* **9**:163–165.

[426] Taylor J. C., et al. (2015) Factors influencing success of clinical genome sequencing across a broad spectrum of disorders. *Nature Genetics* **47**:717–726.

[427] Thériault B. L., et al. (2014) The genomic landscape of retinoblastoma: A review. *Clinical & Experimental Ophthalmology* **42**:33–52.

[428] Thompson R., et al. (2014) RD-Connect: An integrated platform connecting databases, registries, biobanks and clinical bioinformatics for rare disease research. *Journal of General Internal Medicine* **29 Suppl 3**:S780–S787.

[429] Thorvaldsdóttir H., Robinson J. T., and Mesirov J. P. (2013) Integrative Genomics Viewer (IGV): High-performance genomics data visualization and exploration. *Briefings in Bioinformatics* **14**:178–192.

[430] Thusberg J., Olatubosun A., and Vihinen M. (2011) Performance of mutation pathogenicity prediction methods on missense variants. *Human Mutation* **32**:358–368.

[431] Tian S., et al. (2016) Impact of post-alignment processing in variant discovery from whole exome data. *BMC Bioinformatics* **17**:403.

[432] Tomasetti C., Vogelstein B., and Parmigiani G. (2013) Half or more of the somatic mutations in cancers of self-renewing tissues originate prior to tumor initiation. *Proceedings of the National Academy of Sciences* **110**:1999–2004.

[433] Türkmen S., et al. (2009) CA8 mutations cause a novel syndrome characterized by ataxia and mild mental retardation with predisposition to quadrupedal gait. *PLoS Genetics* **5**:e1000487.

[434] Turro E., et al. (2016) A dominant gain-of-function mutation in universal tyrosine kinase SRC causes thrombocytopenia, myelofibrosis, bleeding, and bone pathologies. *Science Translational Medicine* **8**:328ra30.

[435] Valentini G., et al. (2014) An extensive analysis of disease-gene associations using network integration and fast kernel-based gene prioritization methods. *Artificial Intelligence in Medicine* **61**:63–78.

[436] Valouev A., et al. (2008) A high-resolution, nucleosome position map of C. elegans reveals a lack of universal sequence-dictated positioning. *Genome Research* **18**:1051–1063.

[437] Van Allen E. M., Wagle N., and Levy M. A. (2013) Clinical analysis and interpretation of cancer genome data. *Journal of Clinical Oncology* **31**:1825–1833.

[438] Van der Auwera G. A., et al. (2013) From FastQ data to high confidence variant calls: The Genome Analysis Toolkit best practices pipeline. *Current Protocols in Bioinformatics* **43**:11.10.1–11.10.33.

[439] van Dijk E. L., et al. (2014) Ten years of next-generation sequencing technology. *Trends in Genetics* **30**:418–426.

[440] Varga E. A. and Moll S. (2004) Cardiology patient pages. Prothrombin 20210 mutation (factor II mutation). *Circulation* **110**:e15–e18.

[441] Veltman J. A. and Brunner H. G. (2012) De novo mutations in human genetic disease. *Nature Reviews Genetics* **13**:565–575.

[442] Venter J. C., et al. (2001) The sequence of the human genome. *Science* **291**:1304–1351.

[443] Vissers L. E. L. M., et al. (2010) A de novo paradigm for mental retardation. *Nature Genetics* **42**:1109–1112.

[444] Vogels A. and Fryns J.-P. (2006) Pfeiffer syndrome. *Orphanet journal of rare diseases* **1**:19.

[445] Vulto-van Silfhout A. T., et al. (2013) An update on ECARUCA, the European Cytogeneticists Association Register of Unbalanced Chromosome Aberrations. *European Journal of Medical Genetics* **56**:471–474.

[446] Wagner A. H., et al. (2016) DGIdb 2.0: Mining clinically relevant drug-gene interactions. *Nucleic Acids Research* **44**:D1036–D1044.

[447] Walters-Sen L. C., et al. (2015) Variability in pathogenicity prediction programs: Impact on clinical diagnostics. *Molecular Genetics & Genomic Medicine* **3**:99–110.

[448] Wang J., et al. (2011) CREST maps somatic structural variation in cancer genomes with base-pair resolution. *Nature Methods* **8**:652–654.

[449] Wang J., et al. (2015) Genome measures used for quality control are dependent on gene function and ancestry. *Bioinformatics* **31**:318–323.

[450] Wang K., Li M., and Hakonarson H. (2010) ANNOVAR: Functional annotation of genetic variants from high-throughput sequencing data. *Nucleic Acids Research* **38**:e164.

[451] Wang Q., et al. (2013) Detecting somatic point mutations in cancer genome sequencing data: A comparison of mutation callers. *Genome Medicine* **5**:91.

[452] Ward A. J. and Cooper T. A. (2010) The pathobiology of splicing. *The Journal of Pathology* **220**:152–163.

[453] Ware J. S., et al. (2015) Interpreting de novo variation in human disease using denovolyzer. *Current Protocols in Human Genetics* **87**:7.25.1–7.2515.

[454] Watson J. D. and Crick F. H. (1953) Molecular structure of nucleic acids; a structure for deoxyribose nucleic acid. *Nature* **171**:737–738.

[455] Wei Q., et al. (2015) A bayesian framework for de novo mutation calling in parents-offspring trios. *Bioinformatics* **31**:1375–1381.

[456] Wei Z., et al. (2011) SNVer: A statistical tool for variant calling in analysis of pooled or individual next-generation sequencing data. *Nucleic Acids Research* **39**:e132.

[457] Wendl M. C. and Wilson R. K. (2008) Aspects of coverage in medical DNA sequencing. *BMC Bioinformatics* **9**:239.

[458] Wiestler B., et al. (2013) ATRX loss refines the classification of anaplastic gliomas and identifies a subgroup of IDH mutant astrocytic tumors with better prognosis. *Acta Neuropathologica* **126**:443–451.

[459] Wildeman M., et al. (2008) Improving sequence variant descriptions in mutation databases and literature using the Mutalyzer sequence variation nomenclature checker. *Human Mutation* **29**:6–13.

[460] Wilks C., et al. (2014) The Cancer Genomics Hub (CGHub): Overcoming cancer through the power of torrential data. *Database* **2014**:bau093.

[461] Wolfe D., et al. (2013) Visualizing genomic information across chromosomes with PhenoGram. *BioData mining* **6**:18.

[462] Wong K. H., Jin Y., and Moqtaderi Z. (2013) Multiplex Illumina sequencing using DNA barcoding. *Current Protocols in Molecular Biology* **101**:7.11.1–7.11.11.

[463] Worthey E. A., et al. (2011) Making a definitive diagnosis: Successful clinical application of whole exome sequencing in a child with intractable inflammatory bowel disease. *Genetics in Medicine* **13**:255–262.

[464] Wright C. F., et al. (2015) Genetic diagnosis of developmental disorders in the DDD study: A scalable analysis of genome-wide research data. *Lancet* **385**:1305–1314.

[465] Wu J., Li Y., and Jiang R. (2014) Integrating multiple genomic data to predict disease-causing nonsynonymous single nucleotide variants in exome sequencing studies. *PLoS Genetics* **10**:e1004237.

[466] Xiong H. Y., et al. (2015) The human splicing code reveals new insights into the genetic determinants of disease. *Science* **347**:1254806.

[467] Xu H., et al. (2014) Comparison of somatic mutation calling methods in amplicon and whole exome sequence data. *BMC Genomics* **15**:244.

[468] Yadav V. K. and De S. (2015) An assessment of computational methods for estimating purity and clonality using genomic data derived from heterogeneous tumor tissue samples. *Briefings in Bioinformatics* **16**:232–241.

[469] Yandell M., et al. (2011) A probabilistic disease-gene finder for personal genomes. *Genome Research* **21**:1529–1542.

[470] Yang H., et al. (2012) IDH1 and IDH2 mutations in tumorigenesis: Mechanistic insights and clinical perspectives. *Clinical Cancer Research* **18**:5562–5571.

[471] Yang X., et al. (2013) HTQC: A fast quality control toolkit for Illumina sequencing data. *BMC Bioinformatics* **14**:33.

[472] Yang Y., et al. (2015) Databases and web tools for cancer genomics study. *Genomics, Proteomics & Bioinformatics* **13**:46–50.

[473] Yang Y., et al. (2013) Clinical whole-exome sequencing for the diagnosis of Mendelian disorders. *New England Journal of Medicine* **369**:1502–1511.

[474] Yang Y., et al. (2014) Molecular findings among patients referred for clinical whole-exome sequencing. *JAMA* **312**:1870–1879.

[475] Ye K., et al. (2009) Pindel: A pattern growth approach to detect break points of large deletions and medium sized insertions from paired-end short reads. *Bioinformatics* **25**:2865–2871.

[476] Yen J., et al. (2017) A variant by any name: Quantifying annotation discordance across tools and clinical databases. *Genome Medicine* **9**:7.

[477] Yoon S., et al. (2009) Sensitive and accurate detection of copy number variants using read depth of coverage. *Genome Research* **19**:1586–1592.

[478] Yu X. and Sun S. (2013) Comparing a few SNP calling algorithms using low-coverage sequencing data. *BMC Bioinformatics* **14**:274.

[479] Zahreddine H. and Borden K. (2013) Mechanisms and insights into drug resistance in cancer. *Frontiers in Pharmacology* **4**:28.

[480] Zankl A., et al. (2004) Novel mutation in the tyrosine kinase domain of FGFR2 in a patient with Pfeiffer syndrome. *American Journal of Medical Genetics. Part A* **131**:299–300.

[481] Zemojtel T., et al. (2014) Effective diagnosis of genetic disease by computational phenotype analysis of the disease-associated genome. *Science Translational Medicine* **6**:252ra123.

[482] Zhang F., et al. (2009) Copy number variation in human health, disease, and evolution. *Annual Review of Genomics and Human Genetics* **10**:451–481.

[483] Zhang J., et al. (2011) International Cancer Genome Consortium Data Portal–a one-stop shop for cancer genomics data. *Database* **2011**:bar026.

[484] Zhao M., et al. (2013) Computational tools for copy number variation (CNV) detection using next-generation sequencing data: Features and perspectives. *BMC Bioinformatics* **14 Suppl 11**:S1.

[485] Zhao X.-N. and Usdin K. (2015) The repeat expansion diseases: The dark side of DNA repair. *DNA Repair* **32**:96–105.

[486] Zhi D. and Chen R. (2012) Statistical guidance for experimental design and data analysis of mutation detection in rare monogenic Mendelian diseases by exome sequencing. *PLoS One* **7**:e31358.

[487] Zhou J. and Troyanskaya O. G. (2015) Predicting effects of non-coding variants with deep learning-based sequence model. *Nature Methods* **12**:931–934.

[488] Zhou W., et al. (2014) Bias from removing read duplication in ultra-deep sequencing experiments. *Bioinformatics* **30**:1073–1080.

[489] Zhou X. and Rokas A. (2014) Prevention, diagnosis and treatment of high-throughput sequencing data pathologies. *Molecular Ecology* **23**:1679–1700.

[490] Zhu N., et al. (2015) Strategies to improve the performance of rare variant association studies by optimizing the selection of controls. *Bioinformatics* **31**:3577–3583.

[491] Zhu X., et al. (2015) Whole-exome sequencing in undiagnosed genetic diseases: Interpreting 119 trios. *Genetics in Medicine* **17**:774–781.

[492] Zook J. M., et al. (2014) Integrating human sequence data sets provides a resource of benchmark SNP and indel genotype calls. *Nature Biotechnology* **32**:246–251.

Index

1000 Genomes Project, 172, 175
10x Genomics, 291
3' end adenylation, 23

A-overhang, 23
acrocentric chromosome, 263
adapter clipping, 77, 82
adapter contamination, 37, 75, 77, 81–83
adapter ligation, 23, 27
adapter read-through, 83
adapter trimming, 77, 82
adjacency matrix, 357
alignable scaffold-discrepant position, 108
alignment match, 115
allele bias, 29
allele fraction, 8
alternate contig, 98
alternate locus, 107
amplification, 261
annotation propagation rule, 373
array CGH, 264, 384, 413
ASCII, 60, 62
autosomal dominant, 301, 303, 308, 396
autosomal recessive, 299, 306, 405
awk, 147, 167, 288, 442, 500

BAM, 111, 125, 146, 153
BAM index, 134
bam-readcount, 442
barcoding, 24, 25, 36, 39, 41, 57, 63

base calling, 34, 57
base quality, 34, 58, 69, 70, 141
base quality score recalibration, 141
BaseQRankSum, 189
BaseSpace platform, 57
BCF file, 165
BCFtools, 165, 167, 171, 187, 290
BCL file, 57
bcl2fastq, 58, 61
BED file, 154
BEDtools, 155
bgzip, 173
BiocParallel, 453
bitflag, 114, 118, 132, 436
BLOSUM matrices, 333
bridge amplification, 30, 31
Burrows–Wheeler transform, 101
BWA, 102, 105, 115, 153, 178
bwakit, 97

cancer, 8, 419
Cancer Gene Census, 426
cancer genes, 422
cancer predisposition, 423
capture efficiency, 151
CASAVA, 57, 62
cavitation, 22
cBioPortal, 428
chastity filter, 34, 63
chemotherapy, 8, 460
chromothripsis, 425
CIGAR, 112, 114
Circos plot, 425, 427

ClinVar, 330
clipping, 115
ClippingRankSum, 189
clonal evolution of tumors, 420,
   429
CNVnator, 287
compound heterozygous, 300, 307
CoNIFER, 276
coordinate conversion, 100
copy number variant (CNV), 261,
   276, 451
CopywriteR, 451
Corpasome, 47, 70, 90
cosegregation, 297
COSMIC, 426
coverage, 149, 150, 152
coverage analysis, 152
coverage plot, 157
CRAM, 53, 125
CrossMap, 101
cryptic splicing, 220

dbSNP, 142, 175, 322, 325, 428
de novo, 396
de novo mutation, 247, 248, 254
decoy sequence, 98
deep phenotyping, 367
deletion, 195, 216, 217, 237, 272,
   274, 275
DELLY2, 290
Denovo Genotype Quality
   (DGQ), 255
Denovo Quality (DQ), 255
depth, 149
dictionary, 137
dideoxynucleotide, 12, 13
DNA fingerprinting, 430
DP4, 473
DrGaP, 481
driver mutation, 420, 422, 477

duplicate reads, 73, 131
duplicate removal, 75
duplication, 272

end repair, 22
enrichment PCR, 28
Epstein-Barr virus (EBV), 98
European Genome-phenome
   Archive, 50, 428
ExAC, 231, 322, 323
excess heterozygosity, 190
exome capture, 19, 151
ExomeWalker, 364, 398
Exomiser, 387
exon skipping, 220, 335

FASTA, 58
FASTA index, 137
FASTQ, 58, 61, 70
fastq-dump, 51
FastQC, 68
FastQC k-mer content, 76, 77
FastQC per base sequence
   content, 73
FastQC per base sequence
   quality, 71
FastQC per tile sequence quality,
   78
FastQC sequence duplication
   levels, 75, 76
FisherStrand (FS), 167
flow cell, 30
FM-index, 102, 103
fragment size, 271
fragmentation, 22, 71
free lunch, 87
FreeBayes, 181, 194
frequency-based variant filtering,
   321
fusion gene, 426

GATK, 129, 135, 163, 178, 185
GATK AnalyzeCovariates, 143
GATK ApplyRecalibration, 177
GATK BaseRecalibrator, 142,
    143
GATK CombineVariants, 181
GATK CountReads, 135
GATK GenotypeGVCFs, 201
GATK HaplotypeCaller, 163,
    199, 250
GATK IndelRealigner, 136, 137
GATK LeftAlignAndTrimVari-
    ants,
    198
GATK PrintReads, 143
GATK ReadBackedPhasing, 194
GATK RealignerTargetCreator,
    139
GATK VariantFiltration, 171
GATK VariantRecalibrator, 173
GC content, 73, 266
GeneWanderer, 363
genic intolerance, 351
Genome in a Bottle Consortium,
    48
Genome Reference Consortium,
    96, 99, 107
genome reference sequence, 95
genomic instability, 420, 422, 425
Genomiser, 387, 400
genotype, 149, 152, 191, 192
genotype quality, 192, 200
golden path, 106
Grantham score, 332
GRCh37 (hg19) reference
    genome, 96, 98
GRCh38 (hg38) reference
    genome, 96, 106
gVCF, 199, 249

hallmarks of cancer, 421
Hamming distance, 40, 45
haploinsufficiency, 262
haplotype, 164, 193, 194
haplotype information (HP), 194
HapMap, 174
hard filtering, 164, 170, 188
Hardy–Weinberg equilibrium, 327
hemizygous, 311
hereditary cancer, 423
het-hom ratio, 232
HGVS, 198, 210, 226
homozygous, 300
hs37d5.fa, 99
hs38DH.fa, 97
Human Gene Variation Society,
    198
human genome, 5, 95
human leukocyte antigen (HLA),
    98
Human Phenotype Ontology
    (HPO), 367, 368, 379,
    388, 405
hybridization, 264

Illumina, 21
Illumina FASTQ file naming
    scheme, 61
Illumina FASTQ read naming
    scheme, 63
indel calling, 196
indel realignment, 136, 138
indexing VCF files, 236
information content, 372, 373
insert, 36
insert size, 27, 37, 38, 43, 114,
    270, 271
insertion, 195, 216, 217, 237, 274,
    275
Integrative Genomics Viewer

(IGV), 116, 146, 233, 251, 416, 445, 451, 460
International Cancer Genome Consortium, 426
intersection filtering, 317
intron retention, 220
inversion, 263, 273

Jannovar, 203, 211, 212, 214, 304, 323, 388

k-mer content, 75

laser capture microdissection, 420
library preparation, 21, 73
library size, 37
liftOver, 100
linkage analysis, 7, 312, 362
long-read sequencing, 18, 291
loss of heterozygosity, 414, 440, 469

Mammalian Phenotype Ontology (MPO), 388
mappability, 266
mapping quality (MAPQ), 112, 117, 167, 190
MappingQualityRankSumTest (MQRankSum), 169, 190
mate-pair sequencing, 43, 270
Maven, 204
Mendelian disease, 6
Mendeliome, 400
metastasis, 422
minor allele frequency, 178, 327
missense variant, 215, 237
mitochondrial DNA (mtDNA), 412
mitochondrial inheritance, 304
most informative common ancestor (MICA), 374

multiplexing, 25, 36, 40
MuSiC, 485
Mutalyzer, 226
mutation nomenclature, 210
mutational process, 486
mutational signature, 486
MutationTaster, 336
MutSig, 485

named FIFO pipe, 439
nebulization, 22
next-generation sequencing, 16
non-allelic homologous recombination, 260
non-homologous end joining, 260
nonsense-mediated decay, 335
nucleotide Distribution, 70

Omni, 174
on-target ratio, 151
OncodriveFM, 485
oncogene, 422, 478
oncology, 8, 460
Oncotator, 445
Online Mendelian Inheritance in Man (OMIM), 356
optical duplicates, 74, 133
overclustering, 69

PacBio, 18, 291
paired-end sequencing, 35, 38, 63, 70, 113, 270
paralog, 411
passenger mutation, 422, 429, 477
pathogenicity, 329
patterned flow cell, 30, 69
PCR duplicates, 29, 74, 75, 132
PED file, 254, 298, 303, 310, 311, 398
pedigree, 297
phase error, 34, 69

phasing, 193, 194
phasing quality (PQ), 194
PhenIX, 400
PhenogramViz, 383
Phenomizer, 379, 383
phenotype, 367
phenotypic similarity, 379
Phred score, 35, 59, 60, 62, 69,
        70, 82
Picard, 129
Picard AddOrReplaceRead-
        Groups,
        105
Picard BuildBamIndex, 134
Picard CreateSequenceDictionary,
        137, 173
Picard LiftoverVcf, 101
Picard MarkDuplicates, 104, 131
Picard SortSam, 131
Picard ValidateSamFile, 130
pileup, 179, 431, 437
Poisson distribution, 150, 268
PolyPhen, 340
positional cloning, 7
post-alignment processing, 129
precision medicine, 9, 460
prioritization, 349
protein–protein interaction, 355
pseudodominant inheritance, 299
pseudogene, 411
PurBayes, 470
PurityEst, 469

QualByDepth (QD), 166, 190
quality trimming, 82

random contig, 97
random walk with restart, 355,
        358, 390
rare variant association testing,
        319

rare-variant intolerance score, 352
read depth, 190, 252, 265, 267
read group, 103, 121, 125
ReadPosRankSumTest, 170, 191
realignment, 136, 138
recurrent mutation, 478
recurrently mutated gene, 478
recurrently mutated pathway, 481
reference genome, 95
repeat expansion, 412
RMSMappingQuality (MQ), 167,
        190
RPKM, 278

SAM, 111
SAM tags, 122
Sambamba, 145
SAMBLASTER, 146
SAMtools, 121, 137, 178, 179
samtools faidx, 137, 173, 180
samtools flagstat, 121
samtools index, 134
samtools mpileup, 180, 437, 439
samtools tview, 180
samtools view, 105, 121, 133, 179
Sanger sequencing, 12
scree plot, 280
secondary alignment, 120
semantic similarity, 372, 373, 375,
        380
sequence dictionary, 137
sequence match, 115
Sequence Ontology, 212
Sequence Read Archive, 50, 428
sequencing by synthesis, 32
sequencing error, 141
single-end sequencing, 35, 111
single-nucleotide polymorphism
        (SNP), 178

single-nucleotide variant (SNV), 178
singular value decomposition, 277
soft clipping, 87, 116, 136
somatic mutation, 8, 247, 421, 426, 429
SomaticSignatures, 489
sonication, 22
splicing, 218, 389, 448
split read, 273
sporadic tumor, 423
SRA Toolkit, 50
strand bias, 167, 190, 191
structural variation, 259, 413, 425, 451, 456
supplementary alignment, 121
Symmetric Odds Ratio (SOR), 191

tabix, 142, 173
tandem duplication, 273
TARGET, 428
target enrichment, 19, 151
target region, 19, 151, 152
targeted therapy, 9, 460
The Cancer Genome Atlas, 426
Ti/Tv ratio, 229
transition, 229
translocation, 263, 274, 276, 425, 426
transversion, 229
trimming, 81
Trimmomatic, 81, 83
TrioDeNovo, 254
true path rule, 373
tumor evolution, 8
tumor heterogeneity, 419, 467
tumor purity, 419, 467
tumor subclones, 420
tumor suppressor, 422, 478
tumorigenesis, 420

uniparental disomy, 413

variant annotation, 209, 445
variant calling, 163, 177, 248, 436
variant filtering, 440
variant normalization, 195, 198, 224, 323
variant quality, 165, 166
variant quality score log-odds (VQSLOD), 171
variant quality score recalibration, 164, 171, 175
variant types, 212
Variant Validator, 226
VarScan2, 436, 451
VCF file, 164, 165, 183, 211, 236, 248, 249

X–chromosomal dominant, 303
X–chromosomal recessive, 310
X–linked inheritance, 302

Y–linked inheritance, 303
YAML, 396

z-score, 279

Printed and bound by CPI Group (UK) Ltd, Croydon, CR0 4YY

24/10/2024

01778308-0016